Palgrave Studies in Media and Environmental Communication

Series Editors

Anders Hansen
Department of Media and Communication
University of Leicester
Leicester, UK

Steve Depoe
McMicken College of Arts & Sciences
University of Cincinnati
Cincinnati, OH, USA

Drawing on both leading and emerging scholars of environmental communication, the Palgrave Studies in Media and Environmental Communication Series features books on the key roles of media and communication processes in relation to a broad range of global as well as national/local environmental issues, crises and disasters. Characteristic of the cross-disciplinary nature of environmental communication, the books showcase a broad variety of theories, methods and perspectives for the study of media and communication processes regarding the environment. Common to these is the endeavour to describe, analyse, understand and explain the centrality of media and communication processes to public and political action on the environment.

More information about this series at
http://www.springer.com/series/14612

Bridie McGreavy · Justine Wells
George F. McHendry, Jr.
Samantha Senda-Cook
Editors

Tracing Rhetoric and Material Life

Ecological Approaches

palgrave
macmillan

Editors
Bridie McGreavy
Department of Communication and Journalism
University of Maine
Orono, ME, USA

George F. McHendry, Jr.
Department of Communication Studies
Creighton University
Omaha, NE, USA

Justine Wells
Department of English
New Mexico State University
Las Cruces, NM, USA

Samantha Senda-Cook
Department of Communication Studies
Creighton University
Omaha, NE, USA

Palgrave Studies in Media and Environmental Communication
ISBN 978-3-319-65710-3 ISBN 978-3-319-65711-0 (eBook)
https://doi.org/10.1007/978-3-319-65711-0

Library of Congress Control Number: 2017950721

Cover credit: Frans Lanting Studio/Alamy Stock Photo

Printed on acid-free paper

This Palgrave Macmillan imprint is published by Springer Nature
The registered company is Springer International Publishing AG
The registered company address is: Gewerbestrasse 11, 6330 Cham, Switzerland

The original version of the book was revised: Editor affiliation has been updated. The erratum to the book is available at https://doi.org/10.1007/978-3-319-65711-0_14

Contents

Editors and Contributors

About the Editors

Bridie McGreavy is an Assistant Professor of Environmental Communication in the Department of Communication and Journalism and an affiliate faculty with the Senator George J. Mitchell Center for Sustainability Solutions at the University of Maine. She is broadly interested in how communication shapes and sustains relationships between humans and ecosystems and she explores this interest in diverse contexts, including: large-scale sustainability science collaborations; marine resource and shellfish management; and inland freshwater systems associated with lakes, rivers, and wetlands. McGreavy's engaged research and teaching draws from qualitative, quantitative, and creative methods to link science and critical practice with decision making.

Justine Wells is an Assistant Professor of English at New Mexico State University, where she teaches courses in rhetorical theory, the history of rhetoric, cultural studies, and sensory rhetoric. Her scholarship in language perception and cognitive technology has appeared in *Cognitive Psychology*, *Behavior Research Methods*, and *AI & Society*. Her current book manuscript, *Sensory Rhetoric Beyond the Human*, re-engages the Western rhetorical tradition to explore how human and nonhuman agents of sensation entwine to build material realities, as evident in contemporary food activism. Other work in progress considers Karl Marx's theory of sensory emancipation, W.E.B. Du Bois's ecological approach to race conservation, and material rhetorics of race in the Postbellum period.

George F. McHendry, Jr. is an Assistant Professor in the Department of Communication Studies at Creighton University. Working at the intersection of rhetorical theory, critical/cultural studies, and performance studies, he investigates the affective and/or material dimensions of rhetoric. His research examines affects in airport security checkpoints, attending to the ways security spaces and performances limit the potential for resistance to the Transportation Security Administration. McHendry is also very active with a local animal rescue group and often can be found working to rehabilitate and rehome neglected Boston Terriers.

Samantha Senda-Cook is an Associate Professor in the Department of Communication Studies and an affiliated faculty member with the Environmental Science and Sustainability programs at Creighton University. She studies rhetorical theory and analyzes environmental communication and materiality in the contexts of social movements, outdoor recreation, and urban spaces/places. Her current research focuses on constructions of space/place in urban environments, specifically Omaha, Nebraska, and tactical acts of resistance. When she is not researching or teaching, she is probably reading a mystery novel, cooking, or riding the hills of Omaha on her bike.

Contributors

John M. Ackerman is jointly appointed in the Program for Writing and Rhetoric and the Department of Communication at the University of Colorado at Boulder. He teaches graduate courses on materiality and public space, agentive technologies and participatory design. He currently writes about cultural and economic change in late-industrial neighborhoods premised on theories of rhetoric as everyday life. He brings qualitative and critical methods to bear on how economic performance, collective memory, and material circulation help to constitute resilient communities, along the front range in Colorado and in the greater Cleveland area.

Brian Cozen is an Assistant Professor of environmental communication at California State University, Fresno. Cozen's research examines the rhetorical strategies involved when people talk about energy consumption and social practices related to mobility. This research focus includes interviewing nuclear energy professionals, theorizing commuter and computer buses, and analyzing media case studies, including the energy-focused collaboration between the United Nations and Revolution (TV). His work can be found in *Argumentation and Advocacy*, *Critical Studies in Media Communication*, *Communication Theory*, and *Environmental Communication*, among others. When not researching or teaching, Cozen enjoys cooking from his seasonal vegan cookbook.

Danielle Endres is a Professor and Chair in the Department of Communication at the University of Utah. Her research focuses on the rhetoric of science/environmental controversies and social movements, with a particular focus on environmental justice, American Indian activism, nuclear technologies, decolonization, place, and energy. Endres is the co-author of *Participatory Critical Rhetoric: Theoretical and Methodological Foundations of Studying Rhetoric In Situ* (Lexington, 2015). The National Science Foundation and Department of Energy have funded her research. Outside of work, Endres enjoys exploring local mountains and venturing to national parks with her partner and two kids.

Joshua P. Ewalt is an Assistant Professor in the Department of Communication at the University of Utah. His scholarship concerns rhetorics of space and place, including regional rhetoric, places of public memory, the way place factors into organizational rhetoric, identity and identification, cartographic rhetorics, and environmental rhetoric, and employs a combination of textual and field-based rhetorical criticism. His secondary research area is in critical social theory, with particular attention devoted to the work of Jacques Rancière and Doreen Massey. Combining these two areas, his latest line of work combines spatial and new materialist theory to theorize and study more-than-human rhetorics.

Caroline Gottschalk Druschke is an Assistant Professor of Composition and Rhetoric in the Department of English and faculty affiliate in the Nelson Institute for Environmental Studies at the University of Wisconsin-Madison, where she collaborates with a variety of humans and fish. Funded by the National Science Foundation, National Park Service, and US Environmental Protection Agency, Druschke's peer-reviewed research is published across disciplines including in *Frontiers in Ecology and the Environment*, *Technical Communication Quarterly*, *Environmental Communication*, and *Ecology and Society*.

Diane M. Keeling is an Assistant Professor in the Department of Communication Studies at the University of San Diego. Her recent projects consider the ontological qualities of rhetorical tropes across different scientific disciplines. Keeling has collaborated with ecologists and biologists on interdisciplinary projects to amplify disciplinary interactions and opportunities. She is currently working on a speculative orientation to rhetorical problem-solving. Her publications have engaged rhetoric's intellectual history, posthumanist trends in rhetorical theory and criticism, and the rhetoric of media and everyday life.

Jennifer C. Prairie is an Assistant Professor in the Department of Environmental and Ocean Sciences at the University of San Diego. Her research combines laboratory experiments, field work, and mathematical modeling to gain insight into how planktonic organisms interact with their environment at the smallest scales and the implications of these interactions for ocean processes and the global carbon cycle. Prairie's research and teaching both draw on her interdisciplinary background in mathematics and biology, and she has recently been funded to develop educational initiatives integrating these two disciplines as part of her CAREER Award from the National Science Foundation.

Candice Rai is an Associate Professor of English and the Director of the Expository Writing Program at the University of Washington. She is the author of *Democracy's Lot: Rhetoric, Publics, and the Places of Invention*, an urban ethnography that explores the dynamics, pitfalls, and possibilities of everyday democracies in a diverse and gentrifying Chicago neighborhood. Her work engages in space and place-based inquiry to study public rhetoric and writing, political discourse and action, and argumentation.

Thomas Rickert is a Professor of English at Purdue University. His areas of interest include histories and theories of rhetoric, critical theory, network culture, and ecology. He published his first book, *Acts of Enjoyment: Rhetoric, Zizek, and the Return of the Subject*, with the University of Pittsburgh Press in 2007. His second book, *Ambient Rhetoric: The Attunements of Rhetorical Being*, was published by the University of Pittsburgh Press in 2013. He has produced numerous essays on rhetorical theory and history. His current book project investigates how we might conceive a prehistory of rhetoric before its consolidation in fifth-century Greece.

Nathaniel A. Rivers is an Associate Professor in the Department of English at Saint Louis University. His current research addresses the posthuman's impact on public rhetorics such as environmentalism and locative media. He is at work on a book project currently titled *Geocomposition*, which describes and reflects upon a framework for and pedagogy of rhetoric and composition designed to explore how writing and rhetoric move and how this movement shapes both rhetorical activity and the locations it inhabits. Together with Paul Lynch, he edited *Thinking with Bruno Latour in Rhetoric and Composition* (SIUP 2015), which explores the impact of Bruno Latour on rhetoric and composition. His work has appeared in journals such as *Rhetoric Society Quarterly, College Composition and Communication, Technical Communication Quarterly, enculturation*, and *Kairos.*

Anthony Stagliano is an Assistant Professor of English at New Mexico State University, where he researches the relationship between technology, rhetoric, and activist politics, especially media art interventions into the world's problems, and what those can teach us about rhetoric's material and technological nature. He also maintains an active film and media making practice with films and videos screened in festivals and galleries around the world, and a feature film released in the US.

Nathan Stormer is a Professor of Communication and Journalism at the University of Maine. His primary line of research is medical rhetoric about abortion in the USA and his secondary line of research is rhetorical theory, principally about studying rhetoric's material diversity. He teaches undergraduate courses in rhetorical theory, rhetorical criticism, argument, and persuasion, and graduate courses in rhetorical theory, rhetorical criticism, and historiography of rhetoric.

List of Figures

CHAPTER 1

Introduction: Rhetoric's Ecologies

Justine Wells, Bridie McGreavy, Samantha Senda-Cook and George F. McHendry, Jr.

A material fact about human experience on Earth is that we are, mostly, confined to a terrestrial existence. Unless we make a living at sea as merchants, fishers, or voyagers, our experiences with aquatic territories tend to be far more limited than those on land. Yet, with water covering three quarters of the Earth's surface and with our own bodies mostly

The original version of this book was revised: Editor affiliation has been updated. The erratum to this chapter is available at https://doi.org/10.1007/978-3-319-65711-0_14

J. Wells (✉)
Department of English, New Mexico State University, Las Cruces, NM, USA
e-mail: jbwells@nmsu.edu

B. McGreavy
Department of Communication and Journalism, University of Maine, Orono, ME, USA
e-mail: bridie.mcgreavy@maine.edu

S. Senda-Cook · G.F. McHendry, Jr.
Department of Communication Studies, Creighton University, Omaha, NE, USA
e-mail: SamanthaSenda-Cook@creighton.edu

G.F. McHendry, Jr.
e-mail: GMcHendry@creighton.edu

B. McGreavy et al. (eds.), *Tracing Rhetoric and Material Life*, Palgrave Studies in Media and Environmental Communication, https://doi.org/10.1007/978-3-319-65711-0_1

composed of this material, our very existence is much more fluid than we might, at first, perceive. As environmentalist Rachel Carson once remarked, this fluidity connects us to ecologies across time and space:

> To stand at the edge of the sea, to sense the ebb and flow of the tides, to feel the breath of a mist moving over a great salt marsh, to watch the flight of shore birds that have swept up and down the surf lines of the continents for untold thousands of years, to see the running of the old eels and the young shad to the sea, is to have knowledge of things that are as nearly eternal as any earthly life can be.[1]

Of course, while Carson's ecological attunement was novel and persuasive for her particular audiences, indigenous inhabitants of what is now known as North America have long cultivated sensibilities to the mutual permeability of ecosystems and bodies.[2] The Nez Perce, for instance, practice *weyekin*, or embodied listening, by which they ground themselves in their lived environments, sensing deeply, searching for resonance among bodily and natural rhythms, and finding ways to mimic those rhythms through bodily performance.[3] For Blackfeet people listening is a communicative act that can open "deeply significant consultations with non-human spirits and powers that are active in the world."[4] And for Penobscot natives, narratives of the river reposition the order of time—as Penobscot elder Butch Phillips explains, "Most stories usually began with simply 'a long time ago' and events that happened hundreds of years ago might be interpreted by the listener to have taken place during the elder's lifetime."[5] Here, embodied listening and storytelling attune to an interactional ecology spanning time and space, in practices that indigenous communities have maintained in the face of centuries of colonial effort to marginalize them.

Following these and countless other examples, this book explores how ecological care and ecological attunement can be cultivated in tandem. Like Carson's attunement, first to the sea and then to the devastating effects of biocides, the chapters in this volume have been, in different ways, shaped by the perceived fragility of earth's living and non-living systems at a critical moment in the planet's material history. On a daily basis the intensification of contemporary capitalism is eroding our ability to depend on the very substances with which we subsist. Our entrapment in the material logic of fossil fuel economies and transportation systems has us trenching the earth to pipe oil through watersheds and ecosystems. The pesticide use Carson protested advances on a vast scale, in conjunction with disproportionate cancer rates and the decimation of bee colonies. Mass deforestation continues to satiate land development and meat consumption; the circulation of plastics outlasts the year of their production more than ten-fold; and oceanic vitality is dampened by thick sheens of oil and waves of plastic debris.[6] Through all this resounds our undeniable, but much denied, mass alteration of the world's climate.[7] Today, our material, ecological interdependencies are felt across scales, with scientists sounding notes still more apocalyptic than Carson's.[8]

Meanwhile, modes of addressivity have been evolving. While recent years have seen a rise in mass actions, from Occupy, to Standing Rock, to Black Lives Matter, the effects are often more energeic than institutional: these efforts have resulted in complex networks of coalitional investments and sparked new approaches to resistance, with participants shifting from rhetorics of protest to rhetorics of occupation, mattering, and protection. In these diverse venues, water is championed as part of native land rights, men don pink ribbons alongside women marchers, and hashtags decorate bodies, protest signs, and sidewalks. Movements for change, in short, are becoming more ecological—more mobile, micro, coalitional, and hybrid. And with this, a space is opening for greater ecological care.

Scholarship in rhetorical studies and environmental communication offers rich potential for affirming this opening, yet this potential has not been fully realized. In seeking entryways for reckoning with today's myriad crises, environmental communication scholars have valued the ecological perspectives of scientists, indigenous elders, activists, and many others. From environmental communication scholarship we have learned how ecological attunement enables deep engagement with, and care for, the natural and human worlds. Meanwhile, in rhetorical theory, richly ecological perspectives have been forwarded to explore connections between interdependency and addressivity. In the process, rhetoric as traditionally conceived has become more networked, material, hybrid, immersive, and consequently dynamic. However, those developments have not always made connections to ecological care, and they have often been isolated from one another via rhetoric's institutional divides across speech and writing.

The purpose of the current volume is to entangle these often independent lines of ecological thinking in order to advance an ecological approach to care. In Maria Puig de la Bellacasa's terms, care is "a living terrain" and "a thick, non-innocent requisite of collective thinking in interdependent worlds" wherein vibrant attachments among people and things can take root.[9] Seeking to enrich the "living terrain" of ecological care in rhetorical studies, we gather diverse contributors from environmental communication, environmental rhetoric, rhetoric, communication, composition, and environmental studies. Our chapters unfold ecological approaches in contact with a range of contemporary exigencies, including climate change, nuclear testing and waste, oceanic pollution and acidification, immigration politics, community activism, local foodways, river dams, and fossil fuel dependency.[10] Together these chapters reveal how ecological orientations to our tactics of intervention, modalities of investment, and approaches to human-nonhuman well-being enrich our capacities to care for and with the world. Although there is surely room for more inter- and intradisciplinary diversification, especially connecting with indigenous and non-Western approaches, the current collection makes an important first step in situating the *ecology* as a productive trope for such connective, collective work.

In this, we are not seeking to produce a unified notion of the ecological. Indeed, the trope of the ecology is neither singular nor static—rather it is itself ecological or tropic, "continuously reconfigured through repetition and difference, expressing rhetoric's creative potential."[11] The diversity across the chapters as contributors variously engage with rhetoric becoming ecological and ecology becoming rhetorical, as Nathan Stormer describes in the Afterword, is both intentional and tensional. What we mean by this is that the shifting, and sometimes conflicting, notions of rhetoric and ecology across the chapters is an enactment of the stochasticity that drives creative change. Our hope is that this volume will catalyze further ecological turnings in ways that engage rhetoric's human and nonhuman participants in a richer milieu of care.

To set the stage for this project, this introduction takes a moment to look back at the recent history of rhetorical studies and environmental communication, tracing the emergence of several complementary but fairly disparate ecological "turns." We trace these turns genealogically and tropically—not only to describe or even re-describe them, but also to participate in their ongoing turning. As Diane Keeling reminds us, the "'same' bodies of work" produce different turns and tropes, and here we encounter key works with a heightened attunement to their *ecological* tendencies.[12] In the spirit of genealogy, we follow these tendencies not as linear paths forged by intending authors, but rather as overdetermined orientations emergent from diverse concepts, practices, and objects of study. We also turn these paths in new directions, angling them to unfamiliar audiences and prompting their entanglement. Individually and together, our genealogies show the possibility of being more ecological than we are and also reveal how ecological we have always been.[13]

The ecological turns we identify coincide with a larger turning in recent interdisciplinary work that embraces *ecology* as distinct from *environment*. In the next section ("Ecological Matters"), we first explore this distinctive sense of the ecological via recent "new materialisms." Then, in "Rhetoric's Ecological Turns," we trace the emergence of resonant ecological approaches to rhetoric that largely predate discussions of new materialism. We do not (and could not) comprehensively cover rhetoric's many ecological trajectories, but rather articulate three genealogical paths as initial provocations:

1. The development of models of "constitutive rhetoric" in the past several decades of communication scholarship, especially as they have sparked ecological attunements through discussions of articulation and transhumanism.
2. The emergence of ecological models of composition and rhetorical invention among composition and rhetoric scholars reaching back to the 1970s, particularly as they have produced complex views of process, systems, and composition itself.
3. The evolution of "*in situ*" methodologies of rhetorical field work in the past three decades at the nexus of the objects, places, and bodies of engaged rhetorical practice.

In some ways it is quite unecological to trace out these separate lines of inquiry, and doing so underrepresents the crossover among them. However, until recent years, that crossover has been relatively limited, with institutional and theoretical divisions resulting in rather isolated approaches to rhetorical inquiry. We describe separate trajectories, then, to reveal both the multiple origins of today's ecological leanings and the potential for further multiplication. We thus note moments where separate lines of thought have produced similar shifts, particularly in understandings of rhetoric's more-than-human constitutive force. As noted in our final section ("Toward Ecological Care"), this volume intensifies those shifts, ultimately cultivating ecological care in three arenas of rhetorical being—those of change, ethics, and justice. Though themselves overlapping, these three conceptual-practical sites loosely organize the sections of the book, and we preview the chapters to be found in each.

Ecological Matters

In humanistic thinking over the past several decades, previously marginalized concerns such as race, gender, and sexuality have taken center stage; rather than specialized or secondary areas, these are increasingly recognized as starting points for inquiry and action in general. In like fashion, today's so-called "new" materialisms and posthumanisms introduce a centering of previously marginalized environmentalist concerns. Yet, in the process, those concerns have been reframed as "ecological."

This "ecological" turn accompanied a two-part recognition of the importance and the limits of the social turn. As Stacy Alaimo puts it, social construction has been crucial for "critiquing the naturalized and oppressive categories of race, class, gender, sexuality, and ability." Yet these critiques have their limits from the standpoint of ecological care—for instance, while the concept of gender did important work to show how social categories are constructed rather than natural, it often positioned the "sexed" body as a fixed, unchanging reality, dismissing the liveliness of the biological world.[14] "From an environmentalist perspective," Alaimo notes, an emphasis on social construction "may bracket or minimize the significance, substance, and power of the material world."[15] For Jane Bennett, a more vibrant, lively, and material sensibility demands the revising of centuries-old understandings of matter as an inert substance that only comes to be animated by an outside vital force.[16] If we can come to see matter as something not animated by a god, infused with human meaning, or socially constructed, but rather as intrinsically generative and self-animated—as "vibrant, vital, energetic, lively, quivering, vibratory, evanescent, and effluescent"—this would mark a new articulation of environmentalism better termed "vital materialism."[17]

Evoking the rhythmic attunement of the Nez Perce peoples and Carson's attunement to the edge of the sea, recent ecological approaches encounter matter as process, rather than a fully articulated physical substance. Understood as process, the "material" evokes not so much discrete, perceivable physical entities—whether trees, cars, rocks, or subatomic particles—but rather the energies at work in the ongoing constitution of such things. Such energies are not assumed to originate in an external, "social" realm. Rather, the "material" in and of itself evokes the "excess, force, vitality, relationality, or difference" that "renders matter active, self-creative, productive, unpredictable."[18] To speak of "material life," then, emphasizes not what enlivens matter, but rather the liveliness of the material—a liveliness which, though long familiar in many venues, is newly gaining prominence in Western thought. In embracing such vitality, many have retreated from the term "environmentalism," which can tend to position "the environment" as a fixed and passive container separate from human activity. Instead, environmentalist commitments have been rearticulated with terms like "vital materialism," "postenvironmentalism," "ecological politics," "cosmopolitics," and "ecology without nature."[19]

Just as ecological approaches have reframed classical environmental concerns, so also have they recast "social" issues. Thus, Karen Barad takes up the concept of performativity but broadens it, inviting us to see all matter as performative.[20] Stacy Alaimo extends cultural studies' embrace of the body to highlight "trans-corporeality," pointing out that the body does not stop at the individual human.[21] Meanwhile, black studies scholars such as Kodwo Eshun and Alexander Weheliye intensify humanistic conceptions of soul toward "postsoul" and "hypersoul" affirmations of human–machine entwinement.[22] Generalizing such intensifications, Latour urges that it is time to cultivate a view that reality, rather than "socially" constructed, is simply constructed—and that we should take care with these constructions.[23] From such a vantage point, traditionally "social" arenas of care are reformulated. Sara Ahmed, for instance, argues that race does not belong to individual bodies, but emerges from how bodies are able to take up space and what spaces press in on them: race is "a question of what is within reach, what is available to perceive and to do 'things' with."[24] Such human–nonhuman entanglement has long been apparent to those who experience and study disability as something which exceeds the individual human body, involving material surroundings and discourses.[25]

Attuning to such human-nonhuman ecologies is crucial in a moment of proliferating nature-culture "hybrids," including ozone depletion, deforestation, hurricanes, and climate change.[26] As Alaimo asserts in her study of environmental toxins, "elaborate, colossal human practices, extractions, transformations, productions, and emissions have provoked heretofore unthinkable intra-actions at all levels."[27] In what has been termed the Anthropocene, "humans now wield a geological force," and with this "the distinction between human and natural histories…has begun to collapse."[28] Of course, the geological force of the human is regularly overlooked and denied, especially as it manifests in somewhat elusive forms, such as environmental pollutants and climate change. While there is an urge to assert the reality of that force by appealing to the authority of science, to do so risks ultimately becoming alienated from it. As Latour helps us see, the most realistic approach to ecological care would recognize today's ecological crises not as "matters of fact"—objects at a distance—but rather as "matters of concern"—enveloping problems of which we are a part.[29] An ecological orientation, in short, is needed for our profoundly ecological times.

Rhetoric's Ecological Turns

Long before the term "new materialism" came into use, rhetorical studies was proving well-equipped to host an ecological orientation to evolving "matters of concern" in distinction to fixed "matters of fact." Of course, rhetoric may not initially seem a fitting site to think of human–nonhuman entanglement, given the centuries-old Western positioning of rhetorical capacity as precisely what distinguishes humans from nonhumans. However, beginning with the sophistic revival in the 1980s, the rhetorical tradition has been re-encountered as a long-running alternative to essentialist, logocentric, and patriarchal habits of thinking that have dominated Western thought. As new work is increasingly showing, ancient rhetorical concepts evoke a vital, lively sense of matter. In modern rhetorical theory, Kenneth Burke's explicit treatments of ecology reach back to the 1930s.[30] And in contemporary work, a wealth of distinctly "ecological" approaches to rhetoric can be identified. In the brief genealogies to follow, we trace how ecological orientations have emerged in three key contemporary areas: discussions of constitutive rhetorics among communication and environmental communication scholars; ecological models of composition and invention among scholars of composition and rhetoric; and practices of *in situ* methods among rhetorical scholar-practitioners and their objects.

Constitutive Rhetorics: Practices of Articulation and Transhumansim

Beginning in the 1970s and 1980s, rhetorical scholars in communication studies began to draw from Burke to forward "materialist" accounts of rhetoric. Though Burke's explicit engagement with ecology was not typically foregrounded, these accounts nonetheless had ecological tendencies. In 1982, for instance, Michael McGee urged scholars to refocus on describing the material reality of rhetorical practice, rather than prescribing an ephemeral ideal of rhetoric. McGee's eminently ecological approach reconnected with rhetoric not as an ideal end *product*, but a real, everyday *process*—a lived, bodily experience enacted in actual, everyday speech.[31] As scholars analyzed the ideological forces at play in that everyday process, an Althusserian strain in the field began to complicate deterministic understandings of ideology. It was in this vein that Maurice Charland forwarded a constitutive model of rhetoric. As he noted, "attempts to elucidate ideological or identity-forming discourses as persuasive" face the challenge that audiences are "already constituted with an identity and within an ideology."[32] For instance, rather than simply persuading audiences to support Quebec sovereignty, rhetorical efforts actually constituted audiences as *peuple Quebecois*. Widely taken up, this theory highlighted rhetoric's ecological workings at the fundamentally generative level of subject formation: in a constitutive model, rhetoric works not by persuading fixed subjects, but in the potent flux of ideological forces that prompt the constitution of subjects in the first place.

To help track this generativity, Charland had drawn on the term "articulation" that was gaining currency in cultural studies.[33] Resisting economic and ideological determinisms alike, this term enabled cultural theorists to speak of elements articulating, or linking together in creative ways, without a single origin or a fixed endpoint. Thus in 1989 Barbara Biesecker noted not only how rhetorical texts articulate subjects but also how texts, audiences, and even contexts articulate along with subjects.[34] Poststructuralist elaborations of articulation such as Biesecker's stimulated an ecological mindset in the great success they had in retraining rhetoricians to think outside of fixed identities and environments. As DeLuca put it, articulation's two aspects—speaking forth elements and linking elements—help delineate moments of discursive stability from a postmodern (and, we would add, ecological) perspective: "In a world without foundations, without a transcendental signified, without given meanings," the notion of articulation helped rhetoricians trace "the struggle to fix meaning and define reality temporarily."[35]

As it troubled deterministic understandings of rhetoric, the concept of articulation prompted rethinkings of power. For example, DeLuca called environmentalist groups to "disavow an essentialist identity" and articulate with other movements, thereby "link[ing] the different antagonisms that give rise to environmental struggles, workers' struggles, feminist struggles, and anti-racist struggles." Articulation, then, opened the way for understanding rhetorical intervention on ecological terms, as contingent coalitions of human and nonhuman commitments. Such intervention could enable not an overthrow, but rather a "disarticulation of the hegemonic discourse that constructs these various groups in relations of oppression."[36] Articulation models duly retreated from oppressor-oppressed views of power. As Ronald Walter Greene argued in 1998, influence models of rhetoric suggest that if Gulf War protests had been given proper media voice, the public would have realized the problems with the war and resisted. An articulation model, in contrast, recognizes entire apparatuses of power that generated consent for the war in the first place—it recognizes, in short, the immanence of rhetoric, its operation as a generative ecology of forces that precede and shape formal discourse.[37] In such a scenario, revolution and critique have limited rhetorical power—instead, rhetorical action might seek opportunities to affirm and intensify those forces in new directions.

As rhetoric's immanence was elaborated, the trope of articulation carried views of constitutive rhetoric beyond the arena of speech. In 2004, Stormer explored how realities articulate not only as "enunciation or clear speech," but also physically—as, for instance, "the linkage between parts of a body or a train."[38] With articulation's multiplicity in mind, Stormer proposed that rhetoric scholars might adopt a "posthumanist historiographic perspective," using a genealogical approach to our objects of study to ask how they are articulated and what they articulate. Just as performativity, soul, and construction have come to exceed purely human realms, then, so articulation expanded from the manifestation of human identities to the articulation of worlds at large, greatly enlarging rhetoric's constitutive function.

Overall, the trope of articulation enabled ecological thinking in framing rhetoric's constitutive work as emergent, generative, coalitional, immanent, and more-than-human. Recently, attention to articulation has also opened ecological approaches within environmental communication, a field that demonstrates rhizomatic connections to rhetoric. This was, of course, evident in DeLuca's approach to environmental movements. Bridie McGreavy has also adopted an ecological orientation to articulation in demonstrating how concepts such as resilience are shaped and sustained in an interconnected set of discursive practices.[39] Karen Hutchins and Stormer's discussion of how forest-based identities are articulated provides yet another exemplar, as does William Kinsella's analysis of how nuclear power is produced through a "larger network of material and communicative elements."[40]

At the same time that articulation was expanding models of constitutive rhetoric, the related concept of transhumanism was provoking similar expansions in environmental communication. In broader circles both academic and popular, transhumanism typically refers to active efforts to transcend human limitations of intelligence, mobility, and mortality through the embrace of technology. But in environmental communication, the concept followed a different path, evoking not the human's potential to be more-than-human, but rather its already more-than-human status—and in this sense the concept invited what we might today call posthuman approaches.[41] Such treatments of transhumanism trace, in part, to Richard Rogers' seminal 1998 paper, which prompted environmental communication scholars to explore how constitutive models of rhetoric afford an overwhelming power to language.[42] Constitutive theories, Julie Kalil Schutten and Rogers argued, can tend to "posit that nature is made meaningful only through human communication, positioning it as a passive product of human activity."[43] A Rogerian transhuman approach challenged such dominant perspectives, broadening constitutive theories to examine the participatory quality of the world. By asking us to think across human and more-than-human rhetorical forces, the notion of transhumanism in environmental communication provided a way to "avoid conceiving of 'nature' as a benevolent, caring entity"—or indeed, "as a singular entity of any sort."[44] Echoing Alaimo, Bennett, and Barad, such transhuman approaches emphasized how constructivism (re)inscribes power relations that simultaneously privilege masculinity and reassert human dominance over nature.

Following this opening onto ecological thinking, environmental communication scholars turned to considering the participation of diverse forms of human–nature–animal relationality in the production of meaning and identity. Engaging Aldo Leopold's call that we, collectively, embrace a land community ethic, M. Nils Peterson, Markus Peterson, and Tarla Rai Peterson built from Rogers' critique of constructivism to forward a participatory approach to communication and democracy, inclusive of nonhuman voices in deliberative forums. This argument appeared in the inaugural edition of *Environmental Communication: A Journal of Nature and Culture*, helping set the expectation that the field's already nascent discussions of transhumanism be recognized and sustained. Advancing this, Tema Milstein's influential work advocated for ecological approaches to practices of identification, proposing that rhetors "explicitly take up individual animal's stories in interwoven ecological contexts."[45] Moving subtly from transhumanism to transcorporeality, Milstein advanced ecological thinking and care by locating meaning in the transitory spaces between sensate bodies, formed in contexts such as whale watching tours, zoos, and hospitals.[46]

The concepts of transhumanism and transcorporeality also stimulated ecological understandings of constitutive rhetorical forces in ways that are coming to fruition in recent work. Framing rhetoric across species, Emily Plec, in her 2013 edited volume addressing human–animal relationality, urged that communication analysis recognize that "[a]nimals, including humans, speak not only via vocalization but also in scent, posture, eye gaze, even vibration."[47] Indeed, investigating animals' rhetorical powers has recast understandings of social-environmental movements. For example, in a recent essay Caitlyn Burford and Schutten analyzed the participation of nonhuman actors in activist movements. They looked to the profound impact that the film *Blackfish* had in demonstrating how captive animals—in this case Seaworld's prisoner Orca "Tilikum"—participate in activist goals.[48]

Instead of distinct lines of argument, we can conceive of these developments in the concepts of articulation, transhumanism, and transcorporeality as entangled invitations to more ecological modes of thought. For example, in emphasizing how environmental rhetoric works in conjunction with the eventfulness of images, DeLuca gestured toward how images participate in rearticulating linkages across bodies and species.[49] Likewise, in her critique of a photograph of a taxidermied wild bear, Natasha Seegert highlighted how both bear and photograph are lively participants in the articulation of memory, pushing on simple distinctions between animate and inanimate matter at the threshold of mortality.[50] In a similar vein, Joshua Trey Barnett explored a "rhetoric of carnality" that allows a radical surrender of the individual to instead articulate with other entities as their literal "sustenance, especially to such 'lowly' others as fungi." Barnett's rhetoric of carnality is a paragon of ecological care—or, as he puts it, "a radical gesture of respect for oneself as a fleshy being, for others for whom one's flesh is sustaining, and to the ecologies that make these relations possible and necessary."[51]

In sum, transhumanist and articulation models of constitutive communication do more than simply shift the locus for meaning making wherein we temper a sense of control in the world and acknowledge the agency of nonhuman entities. In addition, these shifts provoke ecological transformations in what it means (or does not mean) to be human, what it is to exert power with/in the world, and, also, how fundamentally flawed the conceit that humans make the world through communication turns out to be.

Ecological Models of Composition: Toward Complex Processes and Systems

While speech scholars were engaging the tropes of articulation and transhumanism to generate often implicitly ecological approaches to constitutive rhetorics, writing theorists were directly mobilizing the trope of the ecology, along with ecological concepts of compositional "processes" and "systems," to describe rhetorical invention.[52] Such ecological models of composition emerged in part from responses to the "process turn" that had overtaken composition theory and instruction beginning in the 1960s. Evocative of McGee's later call for a materialist approach to speech, compositionists' sweeping concern for the writing process had been poised against overly prescriptivist, product-oriented treatments of writing. Process views indeed introduced implicit ecological tendencies insofar as they emphasized writing as inherently dynamic; however, as writing scholars explicitly engaged the trope of the ecology, they came to more complex views of compositional "processes" and "systems" that came to be called "post-process." Thus in his 1975 "Eco-Logic for the Composition Classroom," Richard Coe critiqued the analytic style of composition textbooks that "divide wholes into smaller parts to be discussed individually or serially."[53] For Coe, an "eco-logic" was needed to supplement an "analytic logic," and to avoid the fallacy of "discuss[ing] a subsystem without considering the whole system."[54] Elaborating Coe's critiques, in her classic 1986 piece, "The Ecology of Writing," Marilyn Cooper argued that the writing "process" was conceived in primarily mental and individualistic terms, wherein ideas were understood to appear "Athena-like, full grown, and complete, out of [the writer's] brow," to be "directed at an unknown and largely hostile other."[55] Ecological models avoided such an isolated and combative picture of writing by emphasizing not only that writing was part of a rich set of ecosystems, but that it was a participant in the ongoing *composition* of those systems: "An ecologist," Cooper noted, "explores how writers interact to form systems" that are "inherently dynamic," "constantly changing," and "concrete."[56] As Margaret Syverson later put it, the concept of "ecology" offered a retreat from "atomistic" treatments of composition "focusing on individual writers, individual texts, isolated processes, or artifacts."[57]

While Cooper's and Coe's "post-process" turn to ecologies largely featured the social systems emergent in writing, the trope of the ecology also came to inspire attention to the nonhuman elements of writing systems. For instance, in her 1999 *The Wealth of Reality*, Syverson explored writing as a "larger system" that encompasses "environmental structures," including "pens, paper, computers, books, telephones, fax machines, photocopiers, printing presses, and other natural or human constructed features."[58] Understood as a larger system of things and processes, the material and environmental reach of writing came to the fore. As Laura Micciche recently noted, for example, paper alone has been substantially explored as an active, more-than-human participant in the systems of writing, with compositionists variously examining its physical and visual affordances, availability, toxicity, and circulation as a commodity.[59] The seemingly ephemeral participants in writing systems also began to gain concreteness with ecological thinking: in recent work, images too have come to be seen as material, more-than-human "things" participating in compositional processes.[60]

Such more-than-human views of writing systems and processes can make ready connection to the classic concerns of environmentalism, as was apparent in the embrace of "ecocomposition" in the early 2000s.[61] As Sid Dobrin and Christian Weisser framed it, ecocomposition "continues the post-process move to understand relationships between writers and larger systems" by considering "the role of environment, place, nature, and location" and "examining the relationships between discourse and place."[62] In so doing, ecocomposition approached the systems of writing both as sites for environmental engagement and participants in environmentalist agendas. Environmentally-attuned work has since come to embrace the broader term of "ecology," and Dobrin's 2012 edited collection, *Ecology, Writing Theory, and New Media* "looks to establish ecological writing studies not just as a legitimate or important form of research, but as preeminent to the future of writing studies and writing theory."[63] Just as in larger circles, then, the concept of the ecology has come to highlight the often marginalized concerns of environmentalism, even in the potentially anthropocentric realms of writing and composition.

Crucial in decentering the human from compositional processes and systems were writing scholars' engagements of classic concepts in the

Western rhetorical tradition, which prompted an ecological rethinking of rhetorical invention. Reaching back to the 1950s, for example, the notion of *kairos* invited ways of resisting linear understandings of invention grounded in binaries of passive and active action, calling attention from a rhetor seizing the occasion, to occasion and rhetor mingling. According to Debra Hawhee, as in the kairotic movements of athletic competition, rhetoric's movements cannot be mapped to the purely active productions or passive receptions of discrete bodies; rather, invention emerges "in the middle," where subjects and objects are still coming to be.[64] Likewise helping position rhetorical invention in terms of the multidirectionality and multiplicity of response, Hawhee explored the octopus-like figure of *metis*, which has since prompted ecological understandings of rhetorical bodies and disability.[65] Meanwhile, reaching back to Gregory Ulmer's 1995 *Heuretics*, the notion of *chora* acquainted composition scholars with a realm of rhetorical invention that preceded the fixed "places" of the *topoi*.[66] And recently Aristotle's concept of *hyle*, which evokes matter not in itself but in its potent relational capacity, has also been engaged toward an ecological and material understanding of rhetorical invention.[67] Together, these concepts helped composition and rhetoric scholars develop a view of change that parts with the linear, causal, and elemental paradigms of Newtonian physics.

As ecological approaches broke with such paradigms, they moved toward a complex understanding of both composition and rhetorical invention. This was particularly notable in Jenny Edbauer Rice's landmark 2005 piece, which inaugurated an abiding interest in the "rhetorical ecologies" of public rhetoric. Building on Biesecker's critiques of the rhetorical situation and composition studies' engagement of ecologies, Edbauer Rice argued that the *situ* of situation evokes "the originary position of objects" and "a bordered, fixed space-location," and yet there is no fixed, original location that we can trace rhetorics back to.[68] The rhetorical milieu, she noted, is not a "relatively closed system," as "situation" suggests, but rather more like a weather system, encompassing "an agglomeration of processes."[69] The notion of ecologies helped evoke those never-complete processes, as Edbauer Rice demonstrated in analyzing the affective ecology of the "Keep Austin Weird" trope.[70] An ecological approach engaging the circulation of affect, she showed, invites us to recognize the fluidity of public rhetoric: "Situation bleeds into the concatenation of public interaction. Public interactions bleed into wider social processes. *The elements of the rhetorical situation simply bleed.*"[71]

As Rivers and Weber expanded this biological emphasis, "Rhetorical exchange is a bloody mess, a living thing, or, more accurately, a confluence of many living things: an ecology."[72]

While the complex, ecological character of composition and invention was being thought in bloody, affective, and biological paradigms, it was also being approached in digital and networked imaginaries. These approaches ultimately came to mingle, as exemplified in Byron Hawk's 2007 *A Counter-History of Composition*. As Hawk showed, the modern era marks a shift from simple to complex views of how vital, self-organizing systems are composed. In this development, systems of all sorts came to be characterized by emergent movements of composition—those seen, for instance, among flocks of birds or schools of fish, whose overall compositions can neither be predicted in advance nor reduced to component parts in retrospect. But complex understandings are not limited to classically "ecological" realms, Hawk made clear; rather, complexity emerges from relations of all sorts—discursive, machinic, biological, cellular, and so on: "Just as the bumblebee is part of the reproductive system of the clover, humans are part of the reproductive system of machines."[73] As such, Hawk offered a fully "posthuman" articulation of ecological systems, where the social, biological, and mechanical mingle in complex co-composition.

Understood in complex terms, the hybrid systems of composition were increasingly recognized as having unpredictable, unknowable, and withdrawn aspects. Using the example of ambient music, Rickert explored how Brian Eno's work makes audible the "moment of complexity" when "disparate elements combine to create a new level of order."[74] Here Eno is not fully in charge of the composition process, nor does he know just what will come of it.[75] Rather, in the course of auditory experimentation, complex systems can shift, manifesting acoustic affordances that were previously withdrawn. As such, the materiality of composition does not rest in the physicality or presence or concreteness of "nonhuman" entities like walls and echoes, but rather in processual, provisional, and never fully articulated realities. Turns to "object-oriented rhetoric" similarly encouraged a recognition that our sensory and epistemological practices can never fully capture the potentiality of ecological systems. Rhetoric, however, positions us to remain receptive to it. Indeed, as Scot Barnett emphasized, "rhetoric is distinctive precisely because of its capacity to attune human beings to the veiled backgrounds and subterranean worlds constituting being and relations in everyday life."[76] Seeking such attunement, Rivers recently called for "deep ambivalence" as a means for environmental rhetorics to reckon with "the irreducible strangeness of wild objects"—the fact that we can never fully know, sense, or control the ecologies of which we are a part.[77]

While Charland's constitutive model highlighted how rhetoric calls subjects into being, then, ecological models of composition and rhetorical invention prompted scholars to attune to how all sorts of things are called into being via complex compositional processes and systems. Indeed, if articulation was a productively ecological concept for speech scholars, for writing scholars the concept of composition itself has invited ecological approaches to the world's ongoing generativity.

Rhetoric in Situ

As rhetoricians have increasingly attended to rhetoric's materiality, they have encountered diverse entities that have inevitably and sometimes in unexpected ways shaped their work. Thus, while our first two genealogical journeys have primarily traced the emergence of ecological *concepts*—especially as they relate to trans-species, hybrid, emergent, and processual views of rhetoric's constitutive powers—this final section turns to the emergence of notably ecological *practices* of rhetorical inquiry. Of course, we could by no means comprehensively cover the myriad ways that ecological notes sound in rhetorical methodologies, ranging from queer archival research, to Mystory, to multimodal methods and much more.[78] We limit our focus to an abbreviated genealogy of "*in situ*" practices that have developed from scholar-activists' close contact with things and places, especially as they reach toward ecological care. In this contact, objects themselves have participated in the unfolding of distinctly ecological modes of thinking and doing, and this section foregrounds that participation.

Over the past several decades, memory objects have been some of the most prominent participants in the emergence of *in situ* methods, with monuments, memorials, museums, quilts, and archives inspiring rhetorical scholars to step away from their desks for on-site research.[79] These objects have invited an ecological understanding of public rhetoric by showing how physical things participate in constituting narratives of past and future, affective practices such as mourning and celebration, and individual and communal identities. If memory objects have largely grounded foci on human rhetorics, they have nonetheless inspired scholars to intimate *in situ* explorations of nonhuman forces. Thus, a germinal moment in rhetorical criticism's *in situ* turn was Carole Blair's insistence that "being there" matters.[80] In her landmark essay with Marsha S. Jeppesen and Enrico Pucci, Jr., Blair demonstrated how material structures such as memorials can be rhetorical.[81] A decade later, she traced the differences between examining a speech when standing with other audience members, when listening to it, and when reading a transcript of it after the fact.[82]

While memory objects were shaping *in situ* approaches to rhetorical practices that were often official and monumental, the lived realms of the vernacular, mundane, and everyday were also shaping participatory methods.[83] This development was motivated by a scholarly interest in transforming inequitable and unsustainable power relations that reaches back to the 1980s.[84] Everyday objects, places, and practices were important for this shift—as Dickinson notes, "it is in the interstices of the everyday, it is in the littlest actions of our daily lives, that we most thoroughly materialize ourselves and our bodies."[85] Coffee shops, for instance, help us explore how the ordinary spaces of commerce entangle us with human-nonhuman agricultural ecologies. Attention to the realm of the mundane also invited *in situ* methods that promoted ecological care. For example, McGreavy and her partners in coastal conservation efforts explored how a host of entities, including parking lots, septic systems, bacteria, and computer technologies participated in creating the process referred to as the Conservation Action Plan. They advocated for a mundane aesthetic approach to rhetorical field methods, which acknowledged "how a host of living and non-living entities shape [coastal conservation], how we might actively work with the world to encourage patterns to emerge in particular ways, and how, so attuned, we can learn to trust what we become from what we do."[86] Such an approach calls scholars' attention to *mondanus*, or worldly material belonging, as well as the material consequences of choices rhetors make when they commit to participation "*inside* invention and advocacy."[87] Having been important realms of inquiry in both composition studies and communication, the mundane and everyday are among many promising sites for interdisciplinary mingling to the ends of ecological care.[88]

Mundane objects also manifested as key rhetorical participants in the making of place, prompting situated analyses devoted to ecological care. In one such instance, Donovan Conley and Lawrence Mullen stood atop Turtlehead Peak in Red Rock Canyon to examine writings from a notebook kept at the summit. This clear example of vernacular text elucidated the utility, persistence, and ultimately the paradoxical consequences of the nature-culture binary that environmental communication scholars have long sought to undo. The notebooks provided a space for articulating a new form of relationality, a "becoming common," while at the same time failing to become [more] ecological, and thereby different.[89] Similarly, for Samantha Senda-Cook, maps and trails materialized as key rhetorical participants in shaping perceptions and expectations about what counts as nature, access, and risk.[90] As rhetorical sites of meaning-making, such objects prompted scholars to begin encountering places not as fixed sites or environments, but as enveloping, dynamic ecologies.

As the above suggests, although often encountered as symbolic vehicles for human meaning, appreciated as active sites of rhetoric and encountered via all the senses, objects and places began to demonstrate a rhetoricity that exceeded the symbolic or interpretive work of human agents. Intuiting this, Richard Marback called upon rhetoric scholars to "give objects their due"—and they increasingly did so to the ends of ecological care.[91] For example, Caroline Gottschalk Druschke argued that environmental agents such as watersheds prompt identification from human community members beyond their symbolicity. Indeed, for Druschke the watershed's very "ambiguity as symbol and material" gives it broad transformative power, "mark[ing] it with the capacity to change rhetorics, selves, and, ultimately, landscapes."[92] Staying with water, Stormer and McGreavy used an extended example of working the tides to illustrate how ecological thinking requires emphasizing capacity relative to agency, vulnerability to violence, and resilience to resistance.[93] Through articulatory practices, they argued, tides, mud, clams, seeds, boots, nets, and more objects still, swirl around each other to form patterns that, in this case, incarnate a distinct rhetorical style known as clam farming. Attending to mundane, localized patterns of addressivity such as these, they urged, may help rhetors adopt an ecological orientation to historicize the myriad ways in which rhetoric articulates matter.

As physical objects and places increasingly drew scholars' attention, bodies themselves emerged as a lively nexus of rhetorical forces. As Michael K. Middleton et al. noted, "rhetorical scholars could learn much from our performance colleagues by more rigorously considering our own bodies and how they interact with the interpretive frameworks and situations we enter when critiquing lived rhetorical experience."[94]

Such attention to the body complicated the assumption "that the products of the mind are the primary modes by which rhetorical criticism is done."[95] The critic's body, in short, was embraced as part of the generative connections and flows present in any ecology. Perhaps no one has demonstrated this embrace more vividly than Phaedra Pezzullo in her studies of toxic tours. Through her situated listening and activism with/in Louisiana's Cancer Alley, she helped rhetorical critics understand how to respond in their bodies in the field. Rhetorical field work, she noted, allows the place to, as Toni Morrison expressed, "'flood' our nerves and skin" in ways that help us, collectively, remember our connections within communities and with forgotten histories.[96] As in broader attention to transcorporeality, the resultant attention to bodies has invited concerns of justice and ethics into rhetoric's *in situ* turn.

This engagement with objects, bodies, and places as active sites of rhetorical becoming accompanied a shift from localizing rhetoric atomistically, in "rhetorical artifacts," to attending ecologically to "rhetorical exchanges."[97] In this vein, and echoing McGee, Senda-Cook found limitations in research that studies "the *products* of practices without attending to the practices in particular."[98] Likewise, George F. McHendry et al. forwarded the method of "immanent participation," in which embodied critics enter an ecology of interest and allow themselves to be taken up by its flows—physically, affectively, and ecologically. "To embrace vulnerability as one enters the field," they noted, "is to make our bodies open to the inscriptions and lashings encountered when we are sensitive to the immanent flows of desire present in the field." In this vulnerable attitude, ecological sensibility is exercised reflexively, turned to a "space between" "the critical commitments *in situ* rhetorical critics bring to the field and the 'live rhetorics' they select to take out of the field."[99]

As noted in the previous section, recent ecological perspectives urge us to attend to rhetoric's materiality as it exceeds the realm of the situated, the physical, and the perceptible. Perhaps ironically, as this section suggests, such attunement can be at least partly cultivated through intimate, sensorial involvement with objects, realms, places, and bodies. Indeed, given a space to unfold as part of rhetorical thinking and doing, the objects of participatory research have shown us how to gather, taught us how to engage, and entered our activist efforts as lively participants. In this, the "material" has come to extend beyond concrete physical objects and fixed places, to how we immerse ourselves with(in) the world and articulate with it.

Toward Ecological Care

In an oft-cited essay, historian of rhetoric George Kennedy proposed that we approach rhetoric not as a substance, but as an energy by which things communicate and affect one another. As Kennedy speculated, such energy could be recognized among not only humans and animals, but even plants, whose colors may attract pollinators or odors repel predators.[100] As the above genealogies suggest, scholars working before and after Kennedy have also been drawn to such ecological imaginaries, rethinking rhetoric as an affective, generative, and more-than-human force. In so doing, they have suggested how a rhetorical perspective is uniquely situated to heighten ecological care in both the academy and broader publics.

As we see it, these ecological approaches to rhetoric have begun to rework the terrain of ecological care in three significant domains, each of which this volume seeks to cultivate. First, ecological approaches have invited a fundamental rethinking of the operations and agents of "environmental change." As many of the above lines of thought conclude, change is no longer best understood in binary, oppositional frames, but rather more on a model of "surfing," or "cultivating a way of being in the world that affords…the capacity to respond kairotically to what comes next."[101] At an even more basic level, an elemental and causal paradigm does not adequately connect us to the complexities and emergent dynamics of rhetorical change. Indeed, as much of the above suggests, we cannot understand change if we examine the causal impacts of pre-constituted bodies, subjects, and environments—rather we must attune to the circulation of affective, relational, or ambient forces that constantly articulate and rearticulate these provisional entities.[102]

When change is no longer assumed to emanate from pre-constituted agents, it can no longer be thought of on separate registers of social change and environmental change. Instead, ecological approaches suggest that change emerges from enveloping ecologies of human and nonhuman forces. This becomes apparent when we consider more fully how images create and shape events, how hashtags participate in protest movements, and how watersheds matter for coordinated action.[103] As such, rhetorical interventions might seek changes not in people's attitudes, beliefs, or values, but in human–nonhuman partnerships. For Rickert, for instance, rather than simply embracing sustainable values, we must compose worlds—in partnership, for example, with cars and road systems.[104] If change is to be embraced in ontological, affirmative, and hybrid terms, we face new questions. Perhaps rather than inquire what will be most effective, we might ask what will be most potent. Maybe instead of staging protest, we will learn more from experimenting with how to protect, and occupy, and matter. And above all, we might attune not just to what "we" do as scholars and tacticians, but also what openings emerge from our ongoing co-composition with the more-than-human world.

In exploring such possibilities, we must remember that we cannot fully predict their outcomes, and this insight has been crucial to a second arena that ecological approaches have reformulated—that of environmental ethics. This concern with ethics and with ethical contingency was evident in the inaugural edition of *Environmental Communication* referenced earlier. In classifying environmental communication as an emergent crisis discipline, Robert Cox influentially called the field to adopt normative commitments that recognize the "failures, distortions, and/or corruption in human communication about environmental concerns." Ethical attention to contingency and precarity, he argued, would enable responses "to signals of environmental stress in ways that are appropriate to human and biological well-being."[105] In the same issue, Plec and Steve Schwarze helped grapple with the contingency in any kind of ethical claim by creating critical rhetorical processes that effectively allow participants to find ways to evaluate and weigh tradeoffs in difficult choices. And as Peterson et al. advocate, this fundamentally involves a broader land community of participants.[106]

As scholars have increasingly emphasized the entwinement of human and nonhuman realms, they have also questioned the celebration of pristine natural places free of human influence.[107] Adding to calls for attitudes of sublimity, reverence, and respect, comportments such as enchantment, wonder, curiosity, and puzzlement have been urged as ways into richer ethical investments in human-nonhuman ecologies.[108] Meanwhile, as ecological understandings of *ethos* have come to highlight, ethics itself no longer rests with the individual human agent. For Nedra Reynolds, for instance, *ethos* emerges in the "betweens" of "the space of the body, her geographical location, her shifting intellectual positions, her distance or closeness to others, to texts, to events."[109] As such, an environmental or ecological ethic becomes larger than a question of individual attitude, encompassing the worldly moods and modes conducive to human-nonhuman thriving.[110] In seeking such modes of being, we must ask how habits of sensing, thinking, and dwelling emerge not only from human sources but also in the "betweens," in our emergent positioning with(in) environments. This is a fundamentally experimental question, where ethics involves trying out new comportments and new modes of dwelling.

Of course, such modes of dwelling have larger political significance, and yet ecological approaches significantly rework a third arena involving questions of justice and "environmental justice." Whether in academia or in broader publics, environmental justice is often articulated as the equitable protection from the effects of environmental degradation, a commitment which requires a more inclusive notion of community and environment. On such a paradigm, calls for environmental justice can slip into an overriding concern for social justice that positions the environment as a resource.[111] Responding by reasserting the primacy of environmentalism over environmental justice, however, risks overlooking the pervasive and often institutionalized prejudices of race, class, gender, and more that have long worked in tandem with environmental degradation. An ecological approach asks whether, instead of oscillating between social or environmental well-being, we might seek a more substantial recognition of the entwinement of these two—an entwinement which, after all, has long been asserted in calls for environmentalism and environmental justice alike.

For example, more ecological approaches to environmental justice advocate a community orientation to "making linkages among the multiple aspects of the ecosystem, including the biophysical environment, the built environment, and the social environment. For [environmental justice] activists it is incomprehensible and inaccurate, as well as immoral, to separate them."[112] Of course, recognizing such linkages is no easy task. What would such a recognition look like, when trade-offs constantly present themselves that appear to pit human interests against nonhuman ones? And how might publics come to such recognitions? Further, what becomes of ideas of justice, rights, power, colonialism, and politics when humans are no longer their sole domain? Indeed, these ecological puzzles of justice, as well as of ethics and change, are imposing ones that will occupy us for some time. The current volume makes modest but valuable headway in engaging and extending them. Though each chapter undeniably contributes in more than a single of these overlapping areas, the three *topoi* of change, ethics, and justice provide a loose organizational structure for the chapters, as a way to highlight their contributions to capacities for ecological care.

Our first two sections, then, foreground how we might complicate existing understandings of change to the ends of ecological care. Section I makes immanent returns to the ancient terrains of rhetoric and ecology to do so. In Chap. 2, which arguably goes the furthest in inter- and intra-disciplinary collaboration, rhetorician Diane Keeling and marine ecologist Jennifer Prairie bring us deep into an ocean vent to attune to the "tropical" character of change at both classically rhetorical and classical ecological levels. Suggesting that those levels are not so distinct after all, they explore how tropic and trophic dynamics serve as forms of attachment within environments, helping cultivate mutualistic relationships between humans and worlds, and our shared abilities to address complexity. If Keeling and Prairie entangle rhetoric and ecology, Thomas Rickert pulls cosmology into the mix of ecological care in Chap. 3. Turning to Empedocles' cosmology of "roots" and Martin Heidegger's cosmology of "folds," he shows how both can guide our response to global change by entwining science and religion, matter and meaning. The resultant sense of "hale," Rickert argues, attunes us to how ecological change must emerge from the already existent folds and roots of the world, including technologies and power structures that we may ultimately hope to rework. Each of these contributions shows how an ecological approach changes how we conceive of capacities for becoming rhetorical, and who and what were involved in our earliest notions of the available means of persuasion.

Section II develops ecological understandings of change with intimate, *in situ* engagements of dynamism and relationality. In Chap. 4 Bridie McGreavy draws from an extended ethnography with clammers in intertidal ecosystems to show how affectability conditions attunement to change. In this case, biological metaphor shapes the ways we remember change by tuning to rhythms and moving into broader relationality within, among, and beyond bodies. Also highlighting rhythmic relations, in Chap. 5, John Ackerman explores how scholars, walkers, and community activists can trace the energies of change through attunements to rhythms of everyday life. Specifically, he offers the practice of *oscillation* as way to remain ready, within established structures of order and meaning, to "punch through" to new orderings and ecologically conditioned meanings. Like others in this collection, Ackerman sees the potential for change in things themselves, attuning to how bridges, too, oscillate between different articulations of meaning and doing. Chapter 6 also approaches change in the midst of movement, with Joshua

Ewalt navigating the roads and prairies of central Nebraska to examine the means by which various human and nonhuman matters articulate together in "regional arrangements." Attuning to the "elsewhere," or larger ecologies where such arrangements come and go, Ewalt examines how regions are less about geospatial boundaries and more about mobilities where diverse forms of matter traverse space. Together these chapters encourage participation with/in ecologies for the inventive possibilities therein, and the authors show how metaphors, narratives, and mimetics help "us"—ecologically conceived—move in complex, nonlinear, and unexpected ways.

Highlighting our second *topos* of ethics, Section III invites practical, embodied, and sensory practices of ethical attunement as pathways to ecological care. In Chap. 7, Nathaniel Rivers examines how encouragements to reduce our footprint have fostered a removed approach to nature, when what we actually need is richer ecological engagement. Rather than reject the footprint trope, however, Rivers asks how it can carry us to a more involved ecological ethic, where our footprints may become not lighter but better. Demonstrating just such a possibility, in Chap. 8 Caroline Gottschalk Druschke and Candice Rai offer a diadromous methodology, inviting us to wade into the river with the fish, dams, and myriad other participants in that assemblage and, in so doing, come to feel the cyborg character of world-making. Though they would gladly invite us into their river, this methodology is a challenge and invitation for rhetors to participate in our own assemblages, to articulate ourselves within home ecosystems and compose livable worlds, and to feel the hybrid forces of rhetoric as we do. In due spirit in Chap. 9, Samantha Senda-Cook and George F. McHendry, Jr. turn to their local assemblages of environmental and food activism, exploring how *habitus* can be engaged to support not only dominant social practices but also resistive ones. Their study shows how embodied rhetorical tactics at a micro-scale can promote eco-political transformations that move across sites and scales. Together, these chapters show how rhetoric provides modes of attunement for orienting to matters of concern, like river restoration, turtle conservation, and local food systems, and how an ecological approach requires diverse forms of participation to connect these concerns with practices of care.

The final Section IV considers how ecological perspectives shift our understanding of politics, social justice, and environmental justice. In Chap. 10, Danielle Endres visits the Nevada nuclear test site, exploring how its violence affects not only indigenous and other peoples, but also larger land communities. Drawing on indigenous perspectives, she expands the historically human-centered concept of nuclear colonialism to encompass such human-nonhuman ecologies. Such an expansion, she shows, encourages that more-than-human agents be recognized not only as victims of nuclear violence, but also as potential resisters. In Chap. 11 Anthony Stagliano turns to a more-than-human resister in the form of a contemporary artistic technology that reformulates a site of potent political engagement—the US–Mexico border. In the process, art and politics become entangled in a matrix of "geopoetics" that invites creative, tactical approaches to political and rhetorical intervention in just border relations. In Chap. 12, Brian Cozen suggests that our more-than-human approaches to politics include how we figure energy. Though energy is often imagined as a locus of positive transformation by rhetorical theorists and energy corporations alike, he argues, this image can rely on the naturalization of fossil fuel dependency. To better approach socially and environmentally just practices, he proposes, rhetorical scholars would do well to critically evaluate their paradigms of rhetorical energies. This section demonstrates the requisite contingencies that attend an ecological orientation to justice, showing how caring is always political and also, possibly, transformative.

All together, the chapters here form a transitional space, or ecotone, where diverse approaches to rhetorics and ecologies brush upon one another, as Nathan Stormer makes apparent in his Afterword. Reading these chapters not far from where Rachel Carson sought her own intertidal edge, Stormer attunes to the lively ebbs and flows of their ideas, their ongoing mattering, turning, and caring. Cast in this light, this collection reminds us that we are perhaps not so confined to terrestrial existence as we might think—indeed, the flows and tides of the sea can be felt even in the most distant mountaintops, which, as Carson once noted, "are reminders of those ancient seas that more than once lay over all this land."[113]

Notes

1. Rachel Carson, *Under the Sea-Wind: A Naturalist's Picture of Ocean Life* (New York: Simon and Schuster, 1941), foreword.
2. Broader treatments of those histories extend far beyond the scope of this introduction but are available in indigenous studies, postcolonial, and feminist scholarship.

3. Michael Salvador and Traceylee Clarke, "The Weyekin Principle: Toward an Embodied Critical Rhetoric," *Environmental Communication* 5, no. 3 (2011): 243–260.
4. Donal Carbaugh, "'Just Listen': 'Listening' and Landscape Among the Blackfeet," *Western Journal of Communication* 63, no. 3 (1999): 261.
5. Butch Phillips, "A River Runs through Us," *Penobscot River Restoration Trust* (2006). Retrieved from http://www.penobscotriver.org/assets/river_BP06_blue.pdf. Accessed Sept. 30, 2017.
6. Johan Rockström, *Big World, Small Planet: Abundance within Planetary Boundaries* (New Haven: Yale University Press, 2015).
7. For eloquent and accurate characterizations of these trends, see Naomi Klein, *This Changes Everything: Capitalism vs. the Climate* (New York: Simon and Schuster, 2015), and Elizabeth Kolbert, *The Sixth Extinction: An Unnatural History* (New York: Picador, 2015).
8. Johann Rockström et al., "A Safe Operating Space for Humanity," *Nature* 461, no. 7263 (2009): 472–475. For an analysis of apocalyptic narrative, including a focus on Carson's rhetorical appeals, see M. Jimmie Killingsworth and Jacqueline S. Palmer, "Millennial Ecology: The Apocalyptic Narrative from *Silent Spring* to Global Warming," in *Green Culture: Environmental Rhetoric in Contemporary America*, eds. Carl George Herndl and Stuart C. Brown (Madison: University of Wisconsin Press, 1996).
9. Maria Puig de la Bellacasa, *Matters of Care: Speculative Ethics in More than Human Worlds* (Minneapolis: University of Minnesota Press, 2017), 8, 19.
10. Alfred North Whitehead, *Modes of Thought* (New York: The Free Press, 1938), 34–36.
11. Diane Marie Keeling, "Of Turning and Tropes," *Review of Communication* 16, no. 4 (2016): 327.
12. Ibid., 324.
13. In this regard we reach toward the materialist style of genealogy Nathan Stormer discusses in "Rhetoric's Diverse Materiality." While our narratives often focus on human subjectivities in the field and thus do not offer the fully materialist style of genealogy Stormer lays out, we attune to the more-than-human forces that have participated in the field's ecological turning, from watersheds and paper to monuments and nature trails. This is particularly apparent in our section on rhetoric *in situ*. Nathan Stormer, "Rhetoric's Diverse Materiality: Polythetic Ontology and Genealogy," *Review of Communication* 16, no. 4 (2016): 299–316.
14. Donna Haraway, "Situated Knowledges: The Science Question in Feminism and the Privilege of Partial Perspective," *Feminist Studies*

14, no. 3 (1988): 575–599. See also Judith Butler, *Gender Trouble: Feminism and the Subversion of Identity* (New York: Routledge, 1989).
15. Stacy Alaimo, *Bodily Natures: Science, Environment, and the Material Self* (Bloomington: Indiana University Press, 2010), 8.
16. Jane Bennett, *Vibrant Matter: A Political Ecology of Things* (Durham: Duke University Press, 2010), ix.
17. Ibid., 112.
18. Diana Coole and Samantha Frost, *New Materialisms: Ontology, Agency, and Politics* (Durham: Duke University Press, 2010), 9.
19. Bruno Latour, "Love Your Monsters: Postenvironmentalism and the Anthropocene," *Breakthrough Journal* 2, Winter (2012): n.p.; Timothy Morton, *Ecology Without Nature: Rethinking Environmental Aesthetics* (Harvard: Harvard University Press, 2009); Isabelle Stengers, *Cosmopolitics* (Minneapolis: University of Minnesota Press, 2011); Bruno Latour, "Whose Cosmos, Which Cosmopolitics? Comments on the Peace Terms of Ulrich Beck" *Common Knowledge* 10 (2004): 450–462.
20. Karen Barad, "Posthumanist Performativity: Toward an Understanding of How Matter Comes to Matter," *Signs* 28, no. 3 (2003): 801.
21. Alaimo, *Bodily Natures*, 262.
22. Kodwo Eshun, *More Brilliant Than the Sun: Adventures in Sonic Fiction* (London: Quartet Books, 1998); Alexander G. Weheliye, "'Feenin': Posthuman Voices in Contemporary Black Popular Music," *Social Text* 20, no. 2 (2002): 21–47.
23. In fact, Latour thinks there is a useful way of broadening the "social" as a "movement of associations," but he concludes that the habit of thinking of the social as an independently existing quality of things is so strong that we are better off dispensing with the term altogether (see, e.g., *Reassembling the Social: An Introduction to Actor-Network-Theory* (Oxford: Oxford University Press, 2005); "The Promises of Constructivism," in *Chasing Technoscience: Matrix for Materiality*, eds. Don Ihde and Evan Selinger (Bloomington: Indiana University Press, 2003)). For a resonant discussion of how attention to the "agency of the spatial" can reorient communication studies' treatments of "social" issues, see Joshua P. Ewalt, "The Agency of the Spatial," *Women's Studies in Communication* 39, no. 2 (2016): 137–140.
24. Sara Ahmed, "A Phenomenology of Whiteness," *Feminist Theory* 8, no. 2 (2007): 154. See also Darrel Enck-Wanzer, "Tropicalizing East Harlem: Rhetorical Agency, Cultural Citizenship, and Nuyorican Cultural Production," *Communication Theory* 21, no. 4 (2011): 344–367.
25. To take just one example, in 1975 the Union of the Physically Impaired Against Segregation stated that "disability is something imposed on

top of our impairments, by the way we are unnecessarily isolated and excluded from full participation in society" (cited in Tom Shakespeare, "The Social Model of Disability," in *The Disability Studies Reader*, ed. Lennard Davis (New York: Routledge, 2006), 198); for a contemporary connection to actor-network theory, see Tobin Siebers' rendering of disability "not as a personal misfortune or individual defect but as the product of a disabling social and built environment" (Tobin Siebers, *Disability Theory* (Ann Arbor: University of Michigan Press, 2008), 3); for connections between disability and ecological care, see, for instance, Rhonda Breitkreuz et al. "Rethinking Resilience in Families of Children with Disabilities: a Socioecological Approach," *Community, Work and Family* 17, no. 3 (2014): 346–365.

26. Bruno Latour, *We Have Never Been Modern* (Cambridge: Harvard University Press, 2002); see also Nancy Tuana, "Viscous Porosity: Witnessing Katrina," in *Material Feminisms*, eds. Stacy Alaimo and Susan J. Hekman (Bloomington: Indiana University Press, 2008).
27. Alaimo, *Bodily Natures*, 21; See also Phaedra Pezzullo, *Toxic Tourism: Rhetorics of Pollution, Travel, and Environmental Justice* (Tuscaloosa: University of Alabama Press, 2009).
28. Dipesh Chakrabarty, "The Climate of History: Four Theses," *Critical Inquiry* 35, no. 2 (2009): 206, 207.
29. Bruno Latour, "Why Has Critique Run out of Steam?" *Critical Inquiry* 30, no. 2 (2004): 225–248. Puig de la Bellacasa builds from Latour to articulate matters of concern with matters of care, which requires a speculative commitment to affect, relationality, and emergent capacity (Puig de la Bellacasa, *Matters of Care*, 38–39, 57–67).
30. Marika A. Seigel "'One little fellow named Ecology': Ecological Rhetoric in Kenneth Burke's *Attitudes toward History*," *Rhetoric Review* 23, no. 4 (2004): 388.
31. Michael Calvin McGee, "A Materialist's Conception of Rhetoric," in *Explorations in Rhetoric: Studies in Honor of Douglas Ehninger*, ed. Raymie McKerrow (Dallas: Scott, Foresman, 1982), 23–48.
32. Maurice Charland, "Constitutive Rhetoric: The Case of the *Peuple Quebecois*," *Quarterly Journal of Speech* 73, no. 2 (1987): 134.
33. e.g., Stuart Hall; Ernesto Laclau and Chantal Mouffe.
34. Barbara Biesecker, "Rethinking the Rhetorical Situation from within the Thematic of 'Différance.'" *Philosophy and Rhetoric* 22, no. 2 (1989): 110–130.
35. Kevin DeLuca, "Articulation Theory: A Discursive Grounding for Rhetorical Practice," *Philosophy and Rhetoric* 32, no. 4 (1999): 334–335.

36. Kevin DeLuca, *Image Politics: The New Rhetoric of Environmental Activism* (New York: Guilford Press, 1999), 82.
37. Ronald Walter Greene, "Another Materialist Rhetoric," *Critical Studies in Mass Communication* 15, no. 1 (1998): 21–40.
38. Nathan Stormer, "Articulation: A Working Paper on Rhetoric and Taxis," *Quarterly Journal of Speech* 90, no. 3 (2004): 263.
39. Bridie McGreavy, "Resilience as Discourse," *Environmental Communication* 10, no. 1 (2016): 104–121.
40. Karen Hutchins and Nathan Stormer, "Articulating Identity in and through Maine's North Woods," *Environmental Communication* 7, no. 1 (2013): 24–41; William J. Kinsella, "Rearticulating Nuclear Power: Energy Activism and Contested Common Sense," *Environmental Communication* 9, no. 3 (2015): 347.
41. For a discussion of varied uses of transhumanism and posthumanism, see Cary Wolfe, *What Is Posthumanism?* (Minneapolis: University of Minnesota Press, 2011); for a recent rhetorical treatment of transhumanism, see Andrew Pilsch, *Transhumanism: Evolutionary Futurism and the Human Technologies of Utopia* (Minneapolis: University of Minnesota Press, forthcoming, 2017).
42. Richard A. Rogers, "Overcoming the Objectification of Nature in Constitutive Theories: Toward a Transhuman, Materialist Theory of Communication," *Western Journal of Communication* 62, no. 3 (1998): 264.
43. Julie Kalil Schutten and Richard A. Rogers, "Magick as an Alternative Symbolic: Enacting Transhuman Dialogs," *Environmental Communication* 5, no. 3 (2011): 264.
44. Rogers, "Overcoming the Objectification of Nature in Constitutive Theories," 264.
45. M. Nils Peterson, Markus J. Peterson, and Tarla Rai Peterson, "Environmental Communication: Why This Crisis Discipline Should Facilitate Environmental Democracy," *Environmental Communication* 1, no. 1 (2007): 74–86; Tema Milstein, "Nature Identification: The Power of Pointing and Naming," *Environmental Communication* 5, no. 1 (2011): 20, emphasis removed.
46. Tema Milstein, "When Whales 'Speak for Themselves': Communication as a Mediating Force in Wildlife Tourism," *Environmental Communication* 2, no. 2 (2008): 173–192; Tema Milstein, "'Somethin' Tells Me It's All Happening at the Zoo': Discourse, Power, and Conservationism," *Environmental Communication* 3, no. 1 (2009): 25–48; Tema Milstein and Charlotte Kroløkke, "Transcorporeal Tourism: Whales, Fetuses, and the Rupturing and Reinscribing of Cultural Constraints," *Environmental Communication* 6, no. 1 (2012): 82–100.

47. Emily Plec, "Perspectives on Human-Animal Communication: An Introduction," in *Perspectives on Human-Animal Communication*, ed. Emily Plec (New York: Routledge, 2013), 3.
48. Caitlyn Burford and Julie "Madrone" Kalil Schutten, "Internatural Activists and the 'Blackfish Effect': Contemplating Captive Orcas' Protest Rhetoric through a Coherence Frame" *Frontiers in Communication* 1, (Jan. 2017): 1–11.
49. DeLuca, *Image Politics.*
50. Natasha Seegert, "Queer Beasts: Ursine Punctures in Domesticity," *Environmental Communication* 8, no. 1 (2014): 81.
51. Joshua Trey Barnett. "Politics of Edibility: Reconceptualizing Ecological Relationality," *Environmental Communication*, Jan. 11, 2016: 11.
52. Burke can provide a starting point to ecological approaches to rhetoric. Appearing as he does across rhetoric's disciplinary divides, Burke invites treatment as an anchor point for diverse inquiries into the ecology of rhetoric, but such an inquiry is beyond the scope of this piece. For an overview of the ecological components of Burke's theory of rhetoric and how he developed them in the context of ecological science, see, e.g., Marika A. Seigel.
53. Richard M. Coe, "Eco-Logic for the Composition Classroom," *College Composition and Communication* 26, no. 3 (1975): 232; see also Richard M. Coe, "Rhetoric 2001," *Freshman English News* 3, no. 1 (1974): 1–13; Richard M. Coe, "'Rhetoric 2001' in 2001," *Composition Studies* 29, no. 2 (2001): 11–35.
54. Coe, "Eco-Logic," 233.
55. Marilyn M. Cooper, "The Ecology of Writing," *College English* 48, no. 4 (1986): 366.
56. Ibid., 368–369.
57. Margaret A. Syverson, *The Wealth of Reality: An Ecology of Composition* (Carbondale: Southern Illinois University Press, 1999), 8.
58. Ibid., 5.
59. Laura Micciche, "Writing Material," *College English* 76, no. 6 (2014): 488–505.
60. Lori Gries, *Still Life with Rhetoric: A New Materialist Approach for Visual Rhetorics* (Logan: Utah State University Press, 2015).
61. See, e.g., Sidney I. Dobrin and Christian R. Weisser. *Natural Discourse: Toward Ecocomposition* (Albany: State University of New York Press, 2002); Sidney I. Dobrin and Christian R. Weisser, *Ecocomposition: Theoretical and Pedagogical Approaches* (Albany: State University of New York Press, 2001); Sidney I. Dobrin, *Saving Place: An Ecocomposition Reader* (Boston: McGraw Hill, 2005); Sidney I. Dobrin and Christian R. Weisser, "Breaking Ground in Ecocomposition:

Exploring Relationships between Discourse and Environment," *College English* 64, no. 5 (2002): 566–589.

62. Dobrin and Weisser, "Breaking Ground," 575.
63. Sidney I. Dobrin, *Ecology, Writing Theory, and New Media: Writing Ecology* (London: Routledge, 2012): 2.
64. Debra Hawhee, *Bodily Arts: Rhetoric and Athletics in Ancient Greece* (Austin: University of Texas Press, 2004); see also Hawhee's treatment of the highly ecological concept of *metis*. For an overview and further ecological engagement of *kairos* among scholars of composition and rhetoric, see Chap. 2 of Thomas Rickert, *Ambient Rhetoric: The Attunements of Rhetorical Being* (Pittsburgh: University of Pittsburgh Press, 2013).
65. Jay T. Dolmage, *Disability Rhetoric* (Syracuse: Syracuse University Press, 2014); Jay Dolmage, "Metis, Metis, Mestiza, Medusa: Rhetorical Bodies Across Rhetorical Traditions," *Rhetoric Review* 28, no.1 (2009): 1–28.
66. For an overview and further ecological engagement of *chora* in composition and rhetoric, see Chap. 1 of Rickert's *Ambient Rhetoric*; see also Brian L. Ott and Diane Marie Keeling, "Cinema and Choric Connection: Lost in Translation As Sensual Experience," *Quarterly Journal of Speech* 97, no. 4 (2011): 363–386.
67. Ehren Helmut Pflugfelder, "Rhetoric's New Materialism: From Micro-Rhetoric to Microbrew," *Rhetoric Society Quarterly* 45, no. 5 (2015): 441–461.
68. Jenny Edbauer, "Unframing Models of Public Distribution: From Rhetorical Situation to Rhetorical Ecologies," *Rhetoric Society Quarterly* 35, no. 4 (2005): 9.
69. Ibid., 13.
70. Indeed, the concept of affect has informed theoretically diverse ecological orientations across rhetoric's disciplinary divisions. For exemplars of this diversity, see, e.g., D. Diane Davis, *Inessential Solidarity: Rhetoric and Foreigner Relations* (Pittsburgh: University of Pittsburgh Press, 2010); Catherine Chaput, "Rhetorical Circulation in Late Capitalism: Neoliberalism and the Overdetermination of Affective Energy," *Philosophy and Rhetoric* 43, no. 1 (2010): 1–25. See also Jenny Edbauer Rice, "The New 'New': Making a Case for Critical Affect Studies," *Quarterly Journal of Speech* 94, no. 2 (2008): 200–212.
71. Edbauer, "Unframing Models of Public Distribution," 9.
72. Nathaniel A. Rivers and Ryan P. Weber, "Ecological, Pedagogical, Public Rhetoric," *College Composition and Communication* 63, no. 2 (2011): 193.
73. Byron Hawk, *A Counter-History of Composition: Toward Methodologies of Complexity* (Pittsburgh: University of Pittsburgh Press, 2007), 163.

74. Rickert, *Ambient Rhetoric*, 110.
75. Rickert, *Ambient Rhetoric*, 109.
76. Scot Barnett, "Toward an Object-Oriented Rhetoric: A Review of Tool-Being: Heidegger and the Metaphysics of Objects and Guerrilla Metaphysics: Phenomenology and the Carpentry of Things by Graham Harman," *Enculturation* 7 (2010): n.p.
77. Nathaniel A. Rivers, "Deep Ambivalence and Wild Objects: Toward a Strange Environmental Rhetoric," *Rhetoric Society Quarterly* 45, no. 5 (2015): 432.
78. See for representative examples: E. Cram, "Archival Ambience and Sensory Memory: Generating Queer Intimacies in the Settler Colonial Archive," *Communication and Critical/Cultural Studies* 13, no. 2 (2016): 109–129; Charles E. Morris, "Archival Queer," *Rhetoric and Public Affairs* 9, no. 1 (2006): 145–151; K.J. Rawson and Charles E. Morris III, "Queer Archives/Archival Queers," in *Theorizing Histories of Rhetoric*, ed. Michelle Baliff (Carbondale: Southern Illinois University Press, 2013); Gregory Ulmer, "Mystory: The Law of Idiom in Applied Grammatology," *The Future of Literary Theory* (Routledge: 1989): 304–323; Jody Shipka, *Toward a Composition Made Whole* (Pittsburgh: University of Pittsburgh Press, 2011); Anne Wysocki et al., *Writing New Media: Theory and Applications for Expanding the Teaching of Composition* (Boulder: University Press of Colorado, 2007).
79. For an overview and examples, see: Greg Dickinson, Carole Blair, and Brian L. Ott, *Places of Public Memory: The Rhetoric of Museums and Memorials* (Tuscaloosa: University of Alabama Press, 2010); Richard Marback, "Unclenching the Fist: Embodying Rhetoric and Giving Objects Their Due," *Rhetoric Society Quarterly* 38, no. 1 (2008): 46–65.
80. Carole Blair, "Reflections on Criticism and Bodies: Parables from Public Places," *Western Journal of Communication* 63, no. 3 (2001): 271–294.
81. Carole Blair, Marsha S. Jeppesen. and Enrico Pucci, Jr., "Public Memorializing in Postmodernity: The Vietnam Veterans Memorial as Prototype," *Quarterly Journal of Speech* 77, no. 3 (1991): 263–288.
82. Blair, "Reflections on Criticism and Bodies."
83. See, e.g., Martin Nystrand and John Duffy, *Towards a Rhetoric of Everyday Life: New Directions in Research on Writing, Text, and Discourse* (Madison: University of Wisconsin Press, 2003); John M. Ackerman, "The Space for Rhetoric in Everyday Life," in Nystrand and Duffy; Enck-Wanzer, "Tropicalizing East Harlem"; Michael K. Middleton, Samantha Senda-Cook, and Danielle Endres, "Articulating Rhetorical Field Methods: Challenges and Tensions," *Western Journal of Communication* 75, no. 4 (2011): 386–406; Samantha Senda-Cook,

"Rugged Practices: Embodying Authenticity in Outdoor Recreation," *Quarterly Journal of Speech* 98, no. 2 (2012): 129–152; R.J. Topinka, "Resisting the Fixity of Suburban Space: The Walker as Rhetorician," *Rhetoric Society Quarterly* 42, no. 1 (2012): 65–84.

84. There are many who make this call, but for a distinctly materialist orientation to this commitment, see Raymie E. McKerrow "Critical Rhetoric: Theory and Praxis," *Communication Monographs* 56, no. 2 (1989): 91–111.
85. Greg Dickinson, "Joe's Rhetoric: Finding Authenticity at Starbucks," *Rhetoric Society Quarterly* 32, no. 4 (2002): 6; Robert Hariman, "Terrible Beauty and Mundane Detail: Aesthetic Knowledge in the Practice of Everyday Life," *Argumentation and Advocacy* 35, no. 1 (1998); Gerald Hauser, "Vernacular Dialogue and the Rhetoricality of Public Opinion," *Communication Monographs* 65, no. 2 (1998), 83–107; Kent Ono and John M. Sloop, "The Critique of Vernacular Discourse," *Communication Monographs* 62, no. 1 (1995), 19–46.
86. Bridie McGreavy et al., "Belonging to the World: Rhetorical Fieldwork as Mundane Aesthetic," in *Places of Persuasion: Studying Rhetoric in the Field*, eds. C. Rai and C. Gottschalk Druschke (Tuscaloosa, AL: University of Alabama Press, in press), 7–8.
87. Aaron Hess, "Critical-Rhetorical Ethnography: Rethinking the Place and Process of Rhetoric," *Communication Studies* 62, no. 2 (2011): 129. For a more extended discussion of the concept of the mundane and its etymological origins see Steve Woolgar and Daniel Neyland, *Mundane Governance: Ontology and Accountability* (Oxford: Oxford University Press, 2014).
88. In composition studies, for example, Rivers and Weber, in "Ecological, Pedagogical, Public Rhetoric," suggest an ecological model of rhetoric prompts us to recognize that significant change occurs not only through the "monumental" rhetorical exertions of heroic individuals, but also through mundane documents, objects, and proceedings. Also see *Rhetoric, Through Everyday Things*, eds. Scot Barnet and Casey Boyle (Tuscaloosa: University of Alabama Press, 2016).
89. Donovan S. Conley and Lawrence J. Mullen, "Righting the Commons in Red Rock Canyon," *Communication and Critical/Cultural Studies* 5, no. 2 (2008): 180–199.
90. Samantha Senda-Cook, "Materializing Tensions: How Maps and Trails Mediate Nature," *Environmental Communication* 7, no. 3 (2013): 355–371; Casey R. Schmitt, "Mounting Tensions: Materializing Strategies and Tactics on National Park 'Social Trails'," *Environmental Communication* 10, no. 4 (2015): 1–14.

91. Marback, "Unclenching the Fist," 46–65.
92. Caroline Gottschalk Druschke, "Watershed as Common-Place: Communicating for Conservation at the Watershed Scale," *Environmental Communication* 7, no. 1 (2013): 88; 89.
93. Nathan Stormer and Bridie McGreavy, "Thinking Ecologically about Rhetoric's Ontology: Capacity, Vulnerability, and Resilience," *Philosophy and Rhetoric* 50, no. 1 (2017): 1–25. See also Nathan Stormer, "Rhetoric's Diverse Materiality".
94. Middleton et al., "Articulating Rhetorical Field Methods," 395.
95. George F. McHendry, Jr. et al., "Rhetorical Critic(ism)'s Body: Affect and Fieldwork on a Plane of Immanence," *Southern Communication Journal* 79, no. 4 (2014): 302.
96. Phaedra C. Pezzullo, "Touring 'Cancer Alley,' Louisiana: Performances of Community and Memory for Environmental Justice," *Text and Performance Quarterly* 23, no. 3 (2003): 247.
97. McHendry et al., "Rhetorical Critic(ism)'s," 293, 294.
98. Senda-Cook, "Rugged Practices," 131.
99. McHendry et al., "Rhetorical Critic(ism)'s," 302, 296.
100. George A. Kennedy, "A Hoot in the Dark: the Evolution of General Rhetoric," *Philosophy and Rhetoric* 25, no. 1 (1992): 1–21.
101. Christine Harold, *Ourspace: Resisting the Corporate Control of Culture* (Minneapolis: University of Minnesota Press, 2007): 161, emphasis removed.
102. See, e.g., Brian Massumi, *Parables for the Virtual: Movement, Affect, Sensation* (Durham, NC: Duke University Press, 2007).
103. Danielle Endres and Samantha Senda-Cook, "Location Matters: The Rhetoric of Place in Protest," *Quarterly Journal of Speech* 97, no. 3 (2011), 257–282; DeLuca, *Image Politics*; Druschke, "Watershed."
104. Rickert, *Ambient Rhetoric*.
105. Robert Cox, "Nature's 'Crisis Disciplines': Does Environmental Communication Have an Ethical Duty?" *Environmental Communication* 1, no. 1 (2007): 18.
106. Steve Schwarze, "Environmental Communication as a Discipline *of* Crisis," *Environmental Communication* 1, no. 1: 96; Emily Plec, "Crisis, Coherence, and the Promise of Critical Rhetoric," *Environmental Communication* 1, no. 1 (2007): 49–57; Peterson et al., "Environmental Democracy."
107. These calls can be traced in venues ranging from the critique of the wilderness notion documented by J. Baird Callicott and Michael P. Nelson, to environmental communication and environmental rhetoric scholarship emphasizing the inability to distinguish nature and culture, to new

materialist work by Bennett, Latour, and others. Several of these calls are summarized and integrated in Rivers, "Deep Ambivalence."

108. See, e.g., Latour, Ingold, Bennett, Rivers.
109. Nedra Reynolds, "Ethos as Location: New Sites for Understanding Discursive Authority," *Rhetoric Review* 11, no. 2 (1993): 335–336. See also: Michael J. Hyde, *The Ethos of Rhetoric* (Columbia: University of South Carolina Press, 2004); Kathleen J, Ryan, Nancy Myers, and Rebecca Jones, *Rethinking Ethos: A Feminist Ecological Approach to Rhetoric* (Carbondale: Southern Illinois University Press, 2016).
110. Here, Rickert's discussion of *Stimmung* is instructive (e.g., *Ambient Rhetoric*, 144).
111. See, e.g., Ronald D. Sandler and Phaedra C. Pezzullo, *Environmental Justice and Environmentalism: The Social Justice Challenge to the Environmental Movement* (Cambridge, Mass: MIT Press, 2007); DeLuca, "A Wilderness Environmentalism Manifesto Contesting the Infinite Self-Absorption of Humans," in Sandler and Pezzullo, *Environmental Justice and Environmentalism.*
112. Giovanna Di Chiro, "Nature as Community: The Convergence of Environment and Social Justice," in *Uncommon Ground: Rethinking the Human Place in Nature*, ed. William Cronon (New York: W. W. Norton, 1996), 317.
113. Rachel Carson, "Road of the Hawks," in *Lost Woods: The Discovered Writing of Rachel Carson*, ed. Linda J. Lear (Boston: Beacon Press, 1998), 32.

Acknowledgments For their valuable feedback on versions of this chapter, the authors would like to thank Byron Hawk and the Rhetoric and Professional Communication Reading Group at NMSU (Lauren Rosenberg, Kellie Sharp-Hoskins, Anthony Stagliano, and Patti Wojahn).

PART I

Returning to Change

CHAPTER 2

Trophic and Tropic Dynamics: An Ecological Perspective of Tropes

Diane M. Keeling and Jennifer C. Prairie

North of the Galapagos Islands, at the tectonic boundaries of the seafloor where the Earth's plates are pulling apart, magma rises from our planet's scorching mantle, allowing for the formation of hydrothermal vents. Here, cold seawater travels through cracks in the Earth's crust, creating hot, mineral-rich plumes rising from the seafloor's spreading center. Despite harsh conditions from high pressure and the super-heated water, there are ecosystems teeming with mussels, crabs, and other large and fast-growing fauna living in utter darkness.[1] This phenomenon astounded scientists aboard the deep-sea submersible, *Alvin*, when they visited this

D.M. Keeling (✉)
Department of Communication Studies,
University of San Diego, San Diego, CA, USA
e-mail: keeling@sandiego.edu

J.C. Prairie
Department of Environmental and Ocean Sciences,
University of San Diego, San Diego, CA, USA
e-mail: jcprairie@sandiego.edu

B. McGreavy et al. (eds.), *Tracing Rhetoric and Material Life*,
Palgrave Studies in Media and Environmental Communication,
https://doi.org/10.1007/978-3-319-65711-0_2

previously unseen marine ecosystem for the first time in 1977. That visit reorganized scientific explanations about how energy and nutrition are transferred between organisms through feeding within ecosystems, otherwise known as "trophic dynamics."[2]

Prior to the late nineteenth century, it was believed that the deep sea was azoic—devoid of life. Even after initial samples of living organisms were brought up from great depths, it was still suggested that organisms in the deep sea were necessarily sparse and small in size due to the absence of light and thus a lack of food.[3] Life at these great distances was thought entirely to depend on the small fraction of organic matter that survived the long journey from the sun-lit surface waters. However, the flourishing ecosystem at the Galapagos Islands' hydrothermal vents aroused new questions concerning food webs in the deep sea: who are the primary producers in this ecosystem, and how do they acquire nutrition without the assistance of the sun? As ecologists eventually learned, they are chemosynthetic bacteria, and they use the energy from the mineral-rich water to convert inorganic matter into nutrition for themselves and other organisms.

These chemosynthetic bacteria, whose relatives are believed to be among the earliest forms of life on earth, are turning inorganic matter into organic life. The energy obtained by the bacteria from the sulfide-rich seawater is used to produce biomass, which can be transferred to larger organisms.[4] Giant tube worms, *Riftia*, with bright red plumes extending into the mineral-laden water, thrive at the hydrothermal vents; each can grow up to three meters in length. Lacking a mouth and a gut, they do not devour the chemosynthetic bacteria; rather, the bacteria live within them. The tube worm receives its nutrition through an endosymbiotic relationship—the living of one organism within another—and in this case the relationship produces a shared advantage. Bacteria benefit from a constant source of oxygen and sulfides that are carried through the plumes of the tube worm directly to the bacteria's dwelling. In turn, the tube worm is provided a constant source of nutrition from the chemosynthetic attribute of its bacterial inhabitants. This tight coupling between organisms greatly benefits marine ecosystems and allows for increased biological productivity. The volatile conditions of hydrothermal vents drive a diversity of adaptive morphologies and physiologies for organisms to thrive and survive in this unique environment.

That the ocean is filled with polymorphic life was no mystery to the archaic Greeks, whose mythology anthropomorphized ecological relationships. Their god of the ocean, Oceanus, and goddess of the sea, Tethys, produced a daughter known for her suppleness, malleability, and shape-shifting features: Metis. Her polymorphic powers made her more clever "than all mortals."[5] Metis' wily ways and cunning intelligence so impressed Zeus, king of the gods, that after marrying her he "craftily deceived her" and "put her away in his own belly."[6] Like the *Riftia* tube worm, rather than devour Metis, Zeus swallowed her and kept her in "the depths of his own stomach."[7] In being combined with the goddess, he could "take counsel" with her on all "things good and evil."[8] In other words, an endosymbiotic relationship developed between Zeus and Metis; Metis nourished him "with an attunement to contingencies" and "unexpected situations."[9] She endowed Zeus with the ability to turn in multiple ways, so as "not to be bound in [his] turn."[10] Turning offered the flexibility to escape capture by his adversaries and survive as king of the gods.

The concept of turning has a long relationship with its Greek root *trope*. The regularly rehearsed etymology of *trope* in rhetorical studies attends to Cicero's articulation of rhetorical style with the Ancient Greek figure "to turn" (τροπη)—as in direction, weather, and war, but also indicative of fashion, habit, and way of life. Considerations of tropes typically emphasize the ability to turn a phrase or pivot on an argument, as with metaphor, metonymy, or synecdoche. However, troping's decorative expression is not relegated to human speech. As we will argue, troping refers to the complex physical metamorphoses performed through interactions that create possibilities for reasoning and argumentation.

Following Metis and *Riftia*, in this chapter we seek to activate the life of tropes by demonstrating their resonance with ecological thinking.[11] An ecological approach is one that emphasizes the relationships between organisms and their environment. Tropes are some of the more ecologically oriented concepts in rhetoric's history since they are "fundamentally relationships" and "inherent to messy articulations of human-nonhuman engagements."[12] Cultivating ecological thinking through tropes can assist us as we think through the way human activities affect the natural world, both in how our expressions indicate entanglements with the world and how there are always non-human dimensions to our expression. Ecological thinking foregrounds the importance of complexity and nuance, which is particularly apropos of our contemporary political climate and its treatment of scientific research.

This chapter enacts an ecological perspective of tropes embedded in the mythological emergence of the rhetorical tradition. In particular, we attend to the way tropes nourish—trophe (τροφη)—the creation of socio-bio-physical environments. This ecological perspective contends that tropes—including figures of thought and diction—are dynamic modes of environmental expression. In our discussion we draw on the field of ecology, and in particular trophic dynamics, to exemplify the ecological qualities of tropes, while also emphasizing the ways tropes can inform an ecological perspective of rhetoric.

We proceed by first historicizing the relationship between the fields of rhetoric and ecology over the past two millennia, discussing the emergent dichotomous thinking embedded in conceptualizations of tropes to which our perspective responds. We then consider two ecological qualities of tropes in turn: their interactivity and polymorphism. We conclude by considering the benefit of an ecological perspective of rhetoric for conceptualizing how troping distills complex socio-bio-physical environments, how distillations can obstruct collaborative problem solving, as well as the importance of conceptualizing the social with the environmental in interdisciplinary research, inspired by matters that have emerged from this collaboration between a rhetorical scholar and a marine ecologist. This piece weaves together ecological understandings of trophic dynamics with classical rhetorical theory to build a contemporary ecological understanding of tropes (and rhetoric in general).

Ancient Troping and Disciplinary Entanglements

Our conceptualization of tropes as dynamic and polymorphic modes of expression emerges from a history of disciplinary entanglements. This section offers a history of the relationship between the fields of rhetoric and ecology with three interrelated purposes: (1) to note the disciplining and partitioning embedded in normalized understandings of both troping and the world; (2) to develop troping as a less disciplined and partitioned engagement with the world; and (3) to constitute troping as a *making-with* the world.

Histories are told in many ways, each version accenting different relationships as they evolve and inform the specious present. In this sense, histories are recursive; they bring forth a different past each time they are performed. Recursion, Nathan Stormer argues, depends on "the figurative action of tropes."[13] Tropes work recursively as they bring together particular events while folding over others. Folding time and space is a tropological event where the present is "a position between past and future held in suspension by 'innumerable secret folds'."[14] Any retrospection on history is therefore a recursive history, one where this time and place are troped or turned "into another virtually."[15] This historical component of tropes is a feature of early oral Greek cultures that did not appear to formalize, define, or separate rhetoric and ecology.

For early oral cultures, the world was filled with poetic expression. Tropes were the "necessary modes of expression of all the first poetic nations," performing the metamorphoses humans experienced in their everyday environments.[16] Spoken language did not mimic, copy, or stand-in for things; instead language "made use of physical substances endowed with life," most of which were "imagined to be divine."[17] Natural phenomena were attributed agencies as gods incarnate, where changing landscapes, weather, and water were infused with mystical forces.

Stories of the world's divine beginnings record human interactions with a shifting landscape through epic poetic form. Typically, epic poetry attracts humanities-oriented scholarship that treats the works as literary accomplishments rather than pre-disciplinary histories. Within the last century, however, scholars have delineated historical elements entangled in Greek mythical accounts, particularly as archeologists have revealed evidence of artifacts and buildings described in the *Odyssey* that date back to when it was written. These scholars have "demonstrated that it is rash to underestimate the historical value of folk memory" and that myth can "no longer be dismissed as mere poetic fancies."[18] Similarly, modern scientists are acknowledging the benefit of integrating traditional ecological knowledge from indigenous peoples with contemporary scientific understanding to address complex environmental problems.[19] Scientific historian Mott Greene agrees, arguing that it is possible to "recapture the natural-historical content" embedded in mythologies passed along through spoken poetry.[20] In this way, human experiences are cast through molds of natural events, and worlds are poetically created through impressions—lived connections.[21] The "semblance of lived history" is discernable only through the "active environment" of the person engaging in the poetry.[22] Allegory is one such poetic form that documents human making-with the world.

Allegorical form is a technique that extends and compounds memory, offering the longevity potential of other memory mediums used to create and pass on cultural knowledge. For oral cultures, memory depends on rhythmic sequences, mnemonic devices, concrete description, and anthropomorphism, while avoiding abstract conceptual vocabularies and subordinate clauses that are too complex for memory recall, such as this sentence. In studying myth for its natural-historical content, scholars "set aside the literary and structural theories" that celebrate an individual author's creative activity and instead attend to figural descriptions as indicative of relations between organisms and their environments.[23] In this sense, myths offer descriptive scientific accounts of phenomena in their states of becoming.

Hesiod's *Theogony* provides an exemplar. Greene argues that Zeus' multiple battles with the Cyclopes are historical accounts of the highly active Mediterranean volcanoes in the fifteenth century BC, accurately preserved "for more than seven hundred years" in the oral tradition.[24] Lacking today's scientific terms for these processes, early people likely anthropomorphized catastrophic eruptions through battle scenes. Employing contemporary volcanology, Greene pays close attention "to the sequence of events in the battles," including "their appearance, sound, and their effect on the physical world."[25] He argues that the Cyclopes are one-eyed monsters and their hundred arms are descriptions of a volcanic eruption. Zeus' clash (lightning) with the one-eyed monsters (volcanoes) and their hundred arms (lava flow) occurs amidst the boiling of the earth and sea, intense heat, thunderous cries, and grumbling grounds. Greene contends that the narrative's sequentially ordered events "are volcanic eruptions described so carefully and in such detail that the volcanoes in question can be identified and the particular eruptions of the volcanoes dated."[26] From this perspective, the *Theogony* is not just a genealogy of the Greek gods, which includes Metis, daughter of the ocean and endosymbiont of Zeus; it is a natural history of archaic people's relationship with volcanic eruptions.

Our records of this early expression document the active ordering of the cosmos and phenomenal world by combining familiar features in the environment with the self, typically the human body. Although anthropomorphism is less explicit in today's descriptions of the natural world, all such descriptions indicate relationships between the human and the natural world. Troping is indicative of our relationships to time, space, and sense, and directs the way we understand and study any given phenomenon. Early poetic expression transformed the relationship between the thing and the self.[27]

In Ancient Greece, then, troping was a making-use of the animate world, a world that did not *have* an order, but was in a constant process of ordering, changing, and taking shape. Tropes were not ordered into neat taxonomies[28] and they were not understood as separate from figures and schemes.[29] The Latin lexicon of "physical metamorphosis largely overlaps with the lexicon of linguistic change"[30]:

> Cicero uses *immutatio*, "a changing," as a label to cover the Greek terms *tropos*, "trope" (literally "a turning"), and *schema*, "figure" (*Brutus* 69). Quintilian defines a figure (*figura*, literally "shape") as "arte aliqua nouata forma dicendi," "a shape of speaking altered by some art" (*Institutio Oratoria* 9.1.14). Cicero's definition of allegory as "continuae tralationes," "continuous metaphors" (*Orator* 94) might equally be translated as "successive transformations," a precise description of the *Metamorphoses*.[31]

For classical Greeks, changes in language were physical transformations; language did not come before or after the physical, it was manifest in it. Today's language–matter dichotomy was not readily apparent in classical rhetorical practice; it had to be created.

By the seventeenth and eighteenth centuries, a bifurcation of materiality and language became prevalent in academic and lay thought, and this worked in conjunction with the perception that language was uniquely human.[32] At the height of the Enlightenment, there was a perceived boundary between humans and a world filled with discrete objects that could be ordered and classified. Latour refers to this as the development of Modern purification—that is, the creation of "two entirely distinct ontological zones: that of human beings on the one hand; that of nonhumans on the other."[33] Even philosophers who rejected "objectivity" accepted a bifurcation between materiality and human language. Myth and art were thought to be "ephemeral" products of a different "realm" or "domain."[34] Romantic poets "reinforced the dichotomy between truth and reason, on the one hand, and art and imagination, on the other" by embracing the subjectivism of which they were accused.[35]

Today, communication about the world is not commonly recognized as an expression of the world, as in Greek mythology. Rather, it is regularly conceived as a unique human capacity to transfer to others intellectual thoughts that are in some way spiritually separate from the world itself. This belief is manifest in the typical understanding of how communication functions: through a sender–receiver transmission model.

This is the assumption that ideas can accurately represent reality, be encoded into words, and be transferred to a receiver who can then decode them. It presumes that ideas and words are not immersed in biophysical relationships, but merely reflect or re-present them. Through representative theories, language became "the core of rhetoric" and has retained this position through either "the view of literary theorists that rhetoric is a quality of the use of language" or in theories of public discourse "in which cultural and political values find expression."[36]

The sciences and humanities, which were not so clear-cut in pre-Socratic Greece, became partitioned into specialized areas of study during the mid-seventeenth and eighteenth centuries. Science and its rational foundations were presented as most equipped to discover enduring knowledge. Rhetoric, previously a primary subject of inquiry and a mode of knowledge production, became relegated to linguistic ornamentation. This separation allowed scientists to assume that their formulation and practice of research could be free of rhetoric. Scientists used "literal language" that was presumed to accurately reflect objective reality, with the figural added after if artistic decoration was suitable.[37]

Divisions among disciplines further encourage the belief in a bifurcated world where the biological and physical are studied in one domain and the capacity of expression in another. The domains are perceived to have distinct methods, vocabularies, and specialties, perpetuating silos. While the scientific field of ecology primarily assumes a representative understanding of language—using definitions to create indexical relationships between words and phenomena—the way ecologists view the natural world is less prone to this bifurcation. Ecologists emphasize interactions between the biological and physical worlds that are not unidirectional, but systemic. Our human interaction and figuration in our own ecosystem is biological and physical, just as other organisms' biological and physical interactions are figural.

While ecological thinking is as old as Greek mythos, its institutionalization did not occur until Darwin's evolutionary theory of natural selection gained notoriety. The formal study of ecology dates back to the mid-nineteenth century, emerging as a subfield of biology. Ernst Haeckel coined "ecology" in his 1866 *General Morphology of Organisms*, a two-volume work that promoted Darwin's *Origin of Species*. Haeckel derived the name from the Greek *oikos* (the English prefix eco-, for house, dwelling place, habitat) and *logia* (for "the study of," a derivative of logos, one who speaks on or treats of the subject). As a scientific field, ecology is the study of the interconnectedness of organisms and their habitats. An ecological perspective of rhetoric thus disrupts the bifurcation of expression and materiality and attends to relational complexity; everything is affected by and affects the environment, thereby performing figural expression. Tropes in particular exemplify this, and next we explore two ways in which tropes enact an ecological perspective: through their interactivity and polymorphism.

Interactivity

Interaction is a fundamental concept in ecology, given that ecology is the study of how organisms interact with one another and their physical environment. All organisms dynamically respond to and shape their surroundings. Interactions, too, are distinctly rhetorical; tropes express interactive relationships between different figures. To treat tropes ecologically is to treat them as interactional forms that affect and are affected by their interactions, rhetorically. All living forms experience "an imperative to respond" resulting from an irreducible relationality and interactional dynamics.[38] The different modes of response, or response-abilities, are indicative of different tropes. Thus, the ways organisms interact with each other and their environments are tropic.

The modes of response that ecologists study can be understood rhetorically as tropic forms. Ecologists study, for example, how environmental stressors produce behavioral responses of organisms. An ecosystem with particularly intense environmental pressures is the rocky intertidal; here, organisms must cope not only with the extreme force of pounding waves but also with the twice daily receding of the tides. Since the organisms that inhabit this environment are marine, they are unable to feed when exposed to air and risk desiccation. When the tide goes out, the organisms enact certain behaviors—responses—to counter these hazards. Some shelled organisms, like limpets, cling tightly to the rocks, creating a watertight seal to keep moisture in. Other organisms, like sea anemones, retract their tentacles and adorn themselves with fragments of shells. And some organisms, like crabs, are mobile enough to simply seek out low-lying areas where they can remain underwater. From an ecological perspective of rhetoric, these recurring forms of response are tropes; dynamic relationships between organisms and environments. Tropic forms repeat, indicating a repetition of form, but simultaneously indicate a transformation, a relational change.[39]

The simultaneous repetition of form and relational change in an ecological perspective of troping is evident in mythopoeic writing, where communication is understood as a type of nourishment—trophe—to be consumed. Archaic communication in the Western tradition was not meaning-oriented or separate from the body, but a bio-physical transformation of bodies. The *Iliad* and *Odyssey* document this archaic communication process through two words that appear most frequently in the texts: *phrenes* and *thumos.* Communication took "place when one person breath[ed] their words [*thumos*] into the *phrenes* of another…the passage of words physically—bodily."[40] Not easily translatable, *phrenes* were likely to have been situated in the chest and were the locations where *thumos* were trapped. *Thumos* is deciphered as a "substance frequently 'poured' into the *phrenes*" and served as a trope for life generally, sharing similar features with breath and blood.[41]

As tropes of nourishment and life, *thumos* and *phrenes* regularly interact to instigate action and ignite solutions. In the *Iliad*, Ajax's *thumos* desires confrontation, "mine own *thumos* also within my breast is the more eager to war and do battle."[42] Hector trusts the advice of his *thumos*: "Listen to me, you Trojans and strong-grieved Acheans, while I speak what the *thumos* within my breast urges."[43] These forms of communication are corporeal interactions, where figurations are shared with and consumed by other bodies. The body is not an object distinct from its environment, but "an aggregate of organs and limbs," multiple interacting forms, which seem to possess autonomy, dynamic force, and in some cases an agency of their own.[44]

Troping thus understood relies on bodily interactions, and the complexity of these interactions makes it so multiple tropes are entangled in each action. While it is possible to fixate on a singular trope or turn, attending to its individuality in any sustained regard is like attending to a discrete Newtonian object: it can be done only by minimizing its interactional complexity. So, while Kenneth Burke gives the four master tropes—metaphor, metonymy, synecdoche, irony—a special "role in the discovery and description of 'the truth'," their promotion to "master" necessarily distills the various other interactions for how "truth" emerges.[45] This distillation was appropriate for Burke's particular rhetorical analyses. However, rhetorical scholars should not assume that these four tropes are "master" in every rhetorical analysis.[46]

An ecological perspective, then, treats tropes as interactive forms of expression where biological and physical relationships are tropic forms. The capacity to interact indicates the capacity to trope: to change in patterned ways creating aesthetic figuration, a taking and making of form.

Polymorphism

An ecological perspective of rhetoric regards tropes as both interactive and polymorphic. Ecologically, the interactions between organisms and their environment produce polymorphism, which is evident in the vast diversity of life forms on earth. The Greek *polymorphos*, from *poly*—many or multiple—and *morphe*—shape or form—has been translated as multiform, of many forms, and manifold. The latter translation of mani-*fold* is consistent with the function of tropes to fold together interactive forms, creating different figurations. Manifold indicates that folding is not a solitary activity, just as "folding is not a singular trope."[47] Rather, troping is manifold and continuously folds into forms.

The many turns, changes of shape, and transformations are indicative of Metis, who is not only a deity, but also a bodily mode of response and corporeal intelligence in Ancient Greece. Pedagogically, students of rhetoric enact Metis' polymorphism as they adapt to changing circumstances through figuration. Instruction books describe the fluidity and flexibility required to enact Metis' polymorphic capacities through examples of marine life.[48] In his *Treatise on Fishing* written in second century AD, Oppian writes about the behavioral adaptations of sea creatures that outwit entities "superior in size and strength" through tricks (*doiloi*).[49] He treats adaptations as *techne*, an evolved art and craft. For example, the mimic octopus, *Thaumoctopus mimicus*, displays postures and bodily patterns that mimic other animal movements, size, and color expressions. A mimic octopus off the coast of Indonesia was observed to produce the appearance of a poisonous sea snake by sticking six of its arms down a hole and undulating the other two in opposite directions. At other times it was seen with its arms positioned in a leaf-shaped wedge, presumably impersonating a swimming flatfish. Through manipulation of both shape and color patterns, it mimicked and turned into lion fish, damselfish, and sand anemones.[50] The cleverness, or *metis*, of the octopus inspired Oppian's praises. The octopus' *polutropos*, many turns, perform the adaptive capacity of survival in an organism's changing environment.

Due to the polymorphic quality of the natural world, scientists struggle over how to define and categorize life on earth; this is a profoundly rhetorical problem. Since the dynamics and processes that affect the natural world are extremely complex, scientists employ many ways of simplifying concepts and distilling information. Although the tendency to sort organisms and phenomena by particular properties or functions can provide a useful perspective, it also poses problems since organisms or other phenomena of study do not always fit neatly into the defined categories. For example, classifying organisms by trophic status—that is, by mode of nutrition—is one way to distinguish among groups of organisms. It was once assumed that organisms could be classified dichotomously as either *autotrophic* or *heterotrophic*. Autotrophs are organisms that can manufacture their own food from inorganic substances by using energy from the sun (e.g., plants) or other sources (e.g., chemosynthetic bacteria). Heterotrophs obtain their nutrition by consuming other organisms or the organic matter derived from them (e.g., animals). However, over time these dichotomous terms proved insufficient to fully describe the complex nature of all nutritional modes in organisms. Pfeffer coined the term *mixotrophy* in 1897 to describe plants that require the presence of organic molecules to photosynthesize—that is, to perform autotrophy using energy from the sun.[51] More recently, the term mixotrophy has evolved to indicate the simultaneous use of autotrophy and heterotrophy in organisms—organisms that utilize both "plant-like and animal-like nutritional modes."[52]

Examples like these emphasize the challenge of attempting to understand polymorphism in nature discretely, since natural phenomena occur along a kind of continuous spectrum. In fact, even the classification most fundamental to the field of biology—life—is difficult to categorize; while viruses possess many qualities of living organisms, they lack others and thus are considered entities that are both living and non-living.[53]

Ecologists cope with masses of information derived from environmental polymorphism through the development of models. Scientific models are, *by design*, synecdochal and do not account for all the complexities of the natural world. The concept of the trophic or biomass pyramid—which may be memorable from grade school textbooks—is a simplification that, while crude, is extremely useful in illustrating how energy is lost as it is passed up the food chain from one trophic level to another. The first trophic level, primary producers (e.g., plants) pass energy to the second trophic level (e.g., herbivores), which is then made available to higher consumers on the pyramid. However with each step in the energy transfer, energy is lost through respiration, waste, and other processes, which results in organisms at the lowest trophic level being in much greater abundance than those at higher trophic levels.[54]

The notion of a trophic level itself is a simplification, since food webs include a complex set of relationships between many different organisms, many of which occur on multiple trophic levels (e.g., omnivores). Although more complex models can be, and have been, constructed to incorporate these aspects, simple models distill an immense amount of information to accent a particular concept and its role in ecosystem dynamics.[55] While there are different stakes involved in each figuration, trophic dynamics become comprehensible and communicable through the troping of some polymorphic interactions and the folding over of others.

The tendency to simplify and reduce complexities inherent to the polymorphism in ecology is somewhat similar to the way rhetorical scholars catalog and arrange tropes according to their different features. Given the vast ways that forms can interact, there have been hundreds of tropes documented in the rhetorical tradition.[56] The various ways tropes can and do interact as figures of repetition and change creates great taxonomical diversity.[57] Classification systems emerged during the Enlightenment and into Modernity as a way to distill the diversity of tropes to ascertain essential features. Today, students of rhetoric tend to master a few tropes rather than several hundreds. For example, Burke's declaration of four exceptional tropes—metaphor, metonymy, synecdoche, and irony—instills a hierarchical order amidst an otherwise polymorphic assemblage.[58]

Rhetorical scholars who seek to classify, order, and distinguish tropes do so in relation to *figures*—pertaining to meaning—and *schemes*—pertaining to pattern. There is rarely agreement about whether tropes are figures or schemes, whether tropes are distinct from figures or schemes, or whether figures and schemes can be considered distinct, that is, whether meaning can be separated from pattern and form. These questions rest on false dichotomies. Complete taxonomical agreement is not possible since meaning "arises out of the matter/form configurations," explains Celeste Condit, "as they take on and move through specific relationships and relationship patterns that are specified by language."[59] Quintilian's attempt to separate tropes from figures by defining tropes as a change in meaning and figures as a change in form is inconsistent and unsustainable across books I, VIII, and IX of *Institutio Oratoria*.[60] Meaning and pattern are entangled; "changes in form and changes in meaning" as well as "the size or scope of the change" are difficult to distinguish.[61] The differences among tropes are subtle, and the subtle differences indicate their capacities to blend into each other, just as the boundaries between ocean and land, mountain and valley, river and bank are changing and fluid. There is no empty space between things, and so too is there no empty space between tropes. Everything is attached, a folding form of every sense.

Traditionally rhetorical scholars have privileged sound and sight to explain the structure of tropes, but tropes are polysensual figurations. Metis is a shape-shifter of all sense, responding to her ever-shifting terrain in tropes of every sensuous form. Though less discussed in rhetorical studies, this stimulus-driven quality of turning is evident in biologists' adaptations of the Greek *tropos.* Within biology, tropisms refer to the directed movement or growth of organisms incited by their environments: chemotropism (in response to chemicals); gravitropism (in response to gravity); hydrotropism (in response to water); phototropism (in response to light); thermotropism (in response to temperature); electrotropism (in response to electric fields); and thigmotropism (in response to touch). The biological sense of trope (τροπη) accents the responsive quality necessary to stimulate growth or trophe (τροφη). The relationship between these two terms is as entangled as their similar spelling.

Tropic and trophic dynamics emerge from the great complexity of the natural world that cannot be discretely categorized. In both rhetorical studies and ecology, there is a tendency to simplify this complexity in a way that allows it to be more useful and more easily understood. Although this distillation of polymorphism can be helpful and even necessary for community building and scientific inquiry, it is important to be aware of the world's polymorphism as interactive relationships fold form and create expressions that privilege some relationships over others.

Conclusion

With an attention to tropes, this chapter has considered the relationship between rhetoric and ecology. An ecological perspective of tropes accentuates their interactivity and polymorphism, emphasizing the world's dynamic figurations. This ecological perspective is one way to conceive of rhetoric's ontology.[62] Consistent with other rhetorical concepts that enact ecological thinking, tropes are forms of attachment; they cultivate an appreciation for diversity; and they indicate environmental affordances.[63] With humans as active participants in this interactivity and polymorphism, we have difficulty coping with all of its complexity.

To navigate these complex systems, we rely on familiar forms to distill the world's expression. Irony, or the tendency to think dialectically, for example, is one particularly prevalent form of distillation.[64] Distillation cannot and should not be prevented since it is our way of managing environmental complexity; however, we must be prepared to disrupt distillations and reconceive of complexity when distillations prevent collaborative orientations to problem solving.

We have recently witnessed distillations as obstructions to collaborative problem solving, both in the political realm when nuanced analyses are dismissed as unclear and complicated, and when scientists attempt to communicate uncertainty to public decision-makers. Uncertainty is not only unavoidable in scientific research but also an important foundational tenet of the scientific method. Typically, when non-experts circulate research for lay audiences, uncertainty is either completely disregarded, so that the research is presented as undisputable fact, or amplified, so that the research is presented as inconclusive. Many environmental scientists have struggled to communicate research on issues like climate change to the public in a way that is accurate but also appropriately conveys the risks and practical societal implications.[65]

A specific example of an environmental body that has struggled with conveying its research to lay audiences is the Intergovernmental Panel on Climate Change (IPCC), a collection of the world's leading experts on climate change stemming from many disciplines, who are tasked with producing regular assessments of the science of climate change, the impacts associated with it, and potential mitigation plans. Given that climate change is a complex global issue with scientific, economic, social, and political ramifications, the IPCC recognized the importance of carefully quantifying and communicating uncertainty, and a common approach was chosen.[66] However, despite these efforts, the IPCC's decision to describe uncertainty in their assessment reports using probabilistic statements such as very unlikely or virtually certain has been found to lead to systematic inaccuracies in the interpretations of the IPCC's findings by the public, which in turn can affect democratic decision-making.[67] Uncertainty is a necessary quality of how information is distilled. In order to understand uncertainty, audiences must be able to conceive of research as a distillation of complex systems. This is a shared goal of educators in both rhetoric and ecology.

Merely amplifying the interactive and polymorphic qualities of rhetoric and ecology's mythopoeic tradition does not itself help scholars address the wicked problems of the twenty-first century, such as climate change.[68] However, it does provide a shared orientation to begin collaborating on ways to address them. Tropes indicate relationships in the world and both rhetorical scholars and ecologists study these biological and physical relationships albeit with different theoretical and methodological approaches. Conceptualizing our communication processes as a making-with the world rather than as a representation of the world is one potential shared orientation for collaborating that would emphasize how humans affect and are affected by their environments.

We opened this chapter by sharing stories about two similar endosymbiotic forms, one between the chemosynthetic bacteria and the *Riftia* tube worm and the other between Metis and Zeus. Endosymbiosis is formally related to parembole, a traditional figure of speech. Similar to parenthesis, parembole is the inclusion of something else into the subject; however, unlike parenthesis, its inclusion indicates a direct connection to the subject. Figures of speech emerge from biological and physical interactions that are experienced and made useful. Thus, there is a "consistency between the visual and verbal," explains Fahnestock, that "helps to underscore the fundamental conceptual processes expressed by the figures."[69]

There are many other examples of parembole or symbiotic relationships in nature like the chemosynthetic bacteria that inhabit giant tube worms at hydrothermal vents. The relationship between the chemosynthetic bacteria and its tube worm host constitutes a *mutualistic* one since both organisms appear to benefit. However, it is less clear what kind of symbiotic relationship Zeus and Metis shared. It did not appear to be mutualistic, since Metis was provided no advantage from Zeus in the same way that she provided him with counsel and mutation; so Metis was either unaffected or harmed by the relationship. If given the chance to escape, we assume she would, since certainly in a non-mutualistic relationship we would make that choice.

Interdisciplinary interactions, such as those between rhetoric and ecology, can be viewed as symbiotic relationships. But the type of symbiotic relationship changes over time. In archaic Greece, there is evidence to suggest that there were no distinctions between disciplines. In contrast, today our disciplines appear more ironic than parembolic as rhetoric and ecology have borrowed vocabularies from one another without mutual awareness. Our goal here has been to develop a mutualistic relationship between the two such that both rhetorical studies and ecology are not only aware of how each is affected by the other (τροπη), but also thrive (τροφη) because of it.

Notes

1. James Childress and Charles Fisher, "The Biology of Hydrothermal Vent Animals: Physiology, Biochemistry, and Autotrophic Symbioses," *Oceanography Marine Biology Annual Review* 30 (1992).
2. Richard Lutz and Michael Kennish, "Ecology of Deep-sea Hydrothermal Vent Communities: A Review," *Reviews of Geophysics* 31, no. 3.
3. Eric Mills, "Problems of Deep-sea Biology: An Historical Perspective," in *Deep-sea Biology*, ed. Gilbert Rowe (Hoboken: John Wiley & Sons, Inc., 1983).
4. Colleen Cavanaugh, Zoe McKiness, Irene Newton, and Frank Stewart, "Marine Chemosynthetic Symbioses," *Prokaryotes* 1 (2006).
5. Hesiod, *Theogony*, lines 886–891.
6. Ibid.
7. Marcel Detienne and Jean-Pierre Vernant, *Cunning Intelligence in Greek Culture and Society*, trans. Janet Lloyd (Atlantic Highlands: Humanities Press, 1978).
8. Debra Hawhee, *Bodily Arts: Rhetoric and Athletics in Ancient Greece* (Austin: University of Texas Press, 2004), 49.
9. Ibid.
10. Detienne and Vernant, *Cunning Intelligence*, 5.
11. Thomas Rickert, *Ambient Rhetoric: The Attunements of Rhetorical Being* (Pittsburgh: University of Pittsburgh Press, 2013); Diane Davis, *Inessential Solidarity: Rhetoric and Foreigner Relations* (Pittsburgh: University of Pittsburgh Press, 2010); Nathan Stormer, "Recursivity: A Working Paper on Rhetoric and Mnesis," *Quarterly Journal of Speech* 99, no. 1 (2013): 27–50; Nathan Stormer and Bridie McGreavy, "Thinking Ecologically about Rhetoric's Ontology: Capacity, Vulnerability, and Resilience," *Philosophy & Rhetoric* 50, no. 1 (2017): 1–25.
12. Nathan Stormer, *Sign of Pathology: U.S. Medical Rhetoric on Abortion, 1800s–1960s* (University Park: Pennsylvania State University Press, 2015), 64.
13. Stormer, "Recursivity," 43.
14. Ibid.
15. Ibid., 42.
16. Giambattista Vico, *The New Science*, trans. Thomas Bergin and Max Fisch (Ithaca: Cornell University Press, 1968), 152.
17. Ibid., 128.
18. John Luce, *The End of Atlantis: New Light on an Old Legend* (Frogmore: Granada Publishing Ltd., 1970), 14.
19. Deborah McGregor, "Coming Full Circle: Indigenous Knowledge, Environment, and Our Future," *The American Indian Quarterly* 28, no. 3/4 (2004).

20. Mott Greene, *Natural Knowledge in Preclassical Antiquity* (Baltimore: The Johns Hopkins University Press, 1992), 49.
21. Susanne Langer, *Feeling and Form: A Theory of Art Developed from Philosophy in a New Key* (New York: Charles Schribner's Sons, 1953), 265.
22. Ibid., 272.
23. Greene, *Natural Knowledge*, 49.
24. Ibid., 54.
25. Ibid., 47.
26. Ibid.
27. Vico, *The New Science*, 130.
28. Jane Sutton and Mari Mifsud, *A Revolution of Tropes: Alloiostrophic Rhetoric* (London: Lexington Books, 2015).
29. Harry Caplan, *Ad C. Herennium de ratione dicendi (Rhetorica ad Herennium)*, (Cambridge: Harvard University Press, 1981).
30. Philip Hardie, "Metamorphosis, Metaphor, and Allegory in Latin Epic," in *Epic Traditions in the Contemporary World: The Poetics of Community*, eds. Margaret Beissinger, Jane Tylus, and Susanne Wofford (Berkeley: University of California Press, 1999), 90.
31. Ibid.
32. George Lakoff and Mark Johnson, *Metaphors We Live By* (Chicago: University of Chicago Press, 1980); Diana Coole and Samantha Frost, "Introducing the New Materialisms," *New Materialisms: Ontology, Agency, and Politics* (Durham: Duke University Press, 2010).
33. Bruno Latour, *We Have Never Been Modern*, trans. Catherine Porter (Cambridge: Harvard University Press, 1991), 10–11.
34. Friedrich Nietzsche, "On Truth and Lies in a Nonmoral Sense," in *The Rhetorical Tradition: Readings from Classical Times to the Present*, 2nd ed. (Boston: Bedford/St. Martin, 2001), 1177.
35. Lakoff and Johnson, *Metaphors*, 192.
36. George Kennedy, "A Hoot in the Dark: The Evolution of a General Rhetoric," *Philosophy & Rhetoric* 25, no. 1 (1992): 1.
37. Ernesto Grassi, "Rhetoric and Philosophy," *Philosophy & Rhetoric* 9, no. 4 (1976).
38. Diane Davis, "Creaturely Rhetorics," *Philosophy & Rhetoric* 44, no. 1 (2011): 89.
39. Diane Keeling, "Of Turning and Tropes," *Review of Communication* 16, no. 4 (2016).
40. Rob Wiseman, "Metaphors Concerning Speech in Homer," in *Theorizing Communication*, eds. Robert Craig and Heidi Muller (Thousand Oaks: Sage, 2007), 15.
41. Ibid., 8.
42. Homer, *Illiad* 13, lines 73–74.

43. Homer, *Illiad* 7, lines 67–68.
44. Wiseman, "Metaphors," 9.
45. Kenneth Burke, "Four Master Tropes," *Kenyon Review* 3, no. 4 (1941): 421.
46. For example, see Jeanne Fahnestock, *Rhetorical Figures in Science*, (Oxford: Oxford University Press, 1999); Stormer, *Sign of Pathology*.
47. Stormer, "Recursivity," 42.
48. Detienne and Vernant, *Cunning Intelligence*.
49. Ibid., 28.
50. Mark Norman, Julian Finn, and Tom Tregenza, "Dynamic Mimicry in an Indo-Malayan Octopus," *Proceedings of the Royal Society of London, Series B: Biological Sciences* 268, No. 1478 (2001).
51. S.C. Rittenberg, "The Obligate Autotroph—The Demise of a Concept," *Antonie van Leeuwenhoeck* 38, no. 1 (1972).
52. Robert Sanders, "Mixotrophic Protists in Marine and Freshwater Ecosystems," *Journal of Protozoological* 38, no. 1 (1991); see also Rittenberg, "The Obligate Autotroph."
53. David Moreira and Purificación López-García, "Ten Reasons to Exclude Viruses from the Tree of Life," *Nature Reviews Microbiology* 7, no. 4 (2009).
54. Raymond Lindeman, "The Trophic-Dynamic Aspect of Ecology," *Ecology* 23, no. 4 (1942).
55. Masahiko Higashi, Thomas Burns, and Bernard Patten, "Food Network Unfolding: An Extension of Trophic Dynamics for Application to Natural Ecosystems," *Journal of Theoretical Biology* 140, no. 2 (1989).
56. Bernardi Dupriez, *A Dictionary of Literary Devices, A-Z*, trans. A.W. Halsall (Toronto: University of Toronto Press, 1991); Richard Lanham, *A Handlist of Rhetorical Terms*, 2nd ed. (Berkeley: University of California Press, 1991).
57. Fahnestock, *Rhetorical Figures*; Sutton and Mifsud, *Revolution of Tropes*.
58. Burke, "Four Master Tropes."
59. Celeste Condit, "The Materiality of Coding: Rhetoric, Genetics, and the Matter of Life," in *Rhetorical Bodies*, eds. Jack Selzer and Sharon Crowley (Madison: University of Wisconsin Press, 1999), 332.
60. Fahnestock, *Rhetorical Figures*.
61. Lanham, *A Handlist of Rhetorical Terms*, 155.
62. Stormer and McGreavy, "Thinking Ecologically"; Stormer, "Recursivity."
63. Stormer and McGreavy, "Thinking Ecologically."
64. Georg Wilhelm Friedrich Hegel, *Encyclopedia of Philosophy*, trans. Gustav Mueller (New York: Philosophical Library, 1959).
65. Nick Pidgeon and Baruch Fischhoff, "The Role of Social and Decision Sciences in Communicating Uncertain Climate Risks," *Nature Climate Change* 1, no. 1 (2011).

66. Michael Mastrandrea, Christopher Field, Thomas Stocker, Ottmar Edenhofer, Kristie Ebi, David Frame, Hermann Held, Elman Kriegler, Katharine Mach, Patrick Matschoss, Gian-Kasper Plattner, Gary Yohe, and Francis Zwiers, "Guidance Note for Lead Authors of the IPCC Fifth Assessment Report on Consistent Treatment of Uncertainties," Intergovernmental Panel on Climate Change (2010).
67. David Budescu, Han-Hui Por, Stephen Broomell, and Michael Smithson, "The Interpretation of IPCC Probabilistic Statements Around the World," *National Climate Change* 4, no. 6 (2014).
68. Timothy Morton, *Dark Ecology: For a Logic of Future Coexistence* (New York: Columbia University Press, 2016).
69. Fahnestock, *Rhetorical Figures*, xi.

CHAPTER 3

Towards Ecosophy in a Participating World: Rhetoric and Cosmology in Heidegger's Fourfold and Empedocles' Four Roots

Thomas Rickert

The issue of climate change charges the contemporary public sphere, political debate, and international relations. While it is obviously a matter of science and policy, it is equally a matter of rhetoric. Persuasion galvanizes the true and actionable. One thing striking about the American climate change debate is the staunch conservative denial of climate science's conclusions that humans are causing substantial climate change that can and will generate various forms of poverty, devastation, and suffering. What, it is asked, would be persuasive to those who deny the human role in climate change? These questions are bundled with further considerations, since acknowledging climate change's challenges does not necessarily show the path forward. I will argue that while the West blames itself for its ecological destructions, there remain resources within it that we should cultivate and pursue. I look in particular to two figures, one ancient, the Greek proto-philosopher Empedocles, and one modern,

T. Rickert (✉)
Department of English, Purdue University, West Lafayette, IN, USA
e-mail: trickert@purdue.edu

B. McGreavy et al. (eds.), *Tracing Rhetoric and Material Life*,
Palgrave Studies in Media and Environmental Communication,
https://doi.org/10.1007/978-3-319-65711-0_3

the German thinker Martin Heidegger. Both of them are remarkably ecological in their thinking. They are as concerned with *where* we are as *who* we are. Their ontologies are marked by complex structures of entwinement, which take form not as singularities, binaries, or even trinities, but quadrates. They treat not just with descriptions of reality but with the simultaneously material and meaningful components of existence, including deep ecological enmeshments, transhuman values, and the divine. In this sense they are cosmological. I illuminate their conceptual and rhetorical resources with an eye towards what they can offer us today.

This is pressing because it is precisely on cosmological issues that so much climate science runs afoul. When conservative Christians, for instance, deny that humans have the power to transform the climate, we see that their persuasive ground lays nowhere near the scientific fieldhouse. In fact, it often appears reactionary against science's materialist, reductionist ontology. Such argumentative grounds are cosmological, since they revolve around questions concerning the divine, the world, and the human role within it. Rhetorics of climate change, what we might call forms of *ecosophy*, or ecological wisdom, need to grasp and enact fresh resources that also have more vital cosmological traction. I will conclude not with the usual exhortations to go green, revamp daily life, and curb technology, but rather with a Peter Sloterdijk-inspired consideration of the ontology of spacecraft. Contra environmentalist wariness concerning technology and progress, spacecraft, I will argue, implicitly generate ecosophy, and thereby supply different—we might say otherworldy—rhetorical appeals since they stem from within a technical legacy, in keeping with an affirmation at the heart of Empedocles' and Heidegger's thought.

Cosmological Schisms

On January 21, 2015 Washington's Republican controlled Senate, embroiled in debates about whether to approve the Keystone XL pipeline, voted 98 to 1 that climate change is real. But then, when a second bill was put forward to highlight the human role in climate change, the Senate voted 50 to 49 (largely on party lines) that it is *not* caused by humans.[1] In the face of overwhelming scientific evidence to the contrary, it was argued that humans are not powerful enough to change the climate.[2] As Senator Jim Inhofe (Rep., OK) stated, "Climate is changing and climate has always changed and always will...The hoax is that there are some people who are so arrogant to think they are so powerful they can change climate. Man can't change climate."[3] An otherwise unmemorable vote is striking for the clarity it brings to the climate issue. While scientists, academics, and a majority of other nations are largely in agreement that humans are significant contributors to climate change, in the United States the issue is divided sharply across the Democratic and Republican Parties.[4] Certainly, the evidence is powerful enough to have spurred the creation of a new name for our age, the Anthropocene.[5] Yet the scientific evidence remains unpersuasive for well over half of Americans. There is a curious irony here. The notion of the Anthropocene *highlights* the power of human beings in swerving the world's climate. However, the complex models that demonstrate this human role are built from a sensibility that *downplays* the power accorded the human in favor of systems logics, ecological entanglements, and other posthuman stances. Meanwhile, a powerful conservative population (many of them Christian) embraces the idea that humans cannot arrogate such power for themselves, and thus have come to their own version of a posthuman stance.

These stances on climate change and human agency illuminate more than the ironies of a term such as Anthropocene, demonstrating powerfully the role of cosmology in public affairs. Cosmology comes from the Greek; it puts *kosmos* (the universe and all that is in it in ordered arrangement) into *logos* (order, meaning, discourse). It is implicated in all forms of human dwelling.[6] Climate change spurs cosmological conflict because how one envisions the world, the very cosmos, and the human role within it all become the fundamental grounds for argument. Accordingly, conflicts re-flare between the cosmologies of science and religion, particularly evangelical Christianity. Taking a long view, if modernism can be loosely characterized as seeking rationalist grounds for knowledge *and* values, supplanting moral and religious narratives with technocratic ones operationalized through disciplinary practices, then we see more evidence that this modernist conflict continues to evolve. Historian Dipesh Chakrabarty points out that for much of recent intellectual history up to the mid-twentieth century, environmental change was acknowledged to happen, but so slowly that it need not be considered a historiographic matter.[7] Nor was it a matter to be integrated into political, urban, or economic planning. This view of the climate resonates well with a religious perspective that places humanity well down the great chain of being. Hence, Inhofe: "Man can't change climate." As a Christian evangelical, he means that only God can change the climate.[8]

It is tempting to align the Senate vote with Chakrabarty's argument about the role of the environment in historical and philosophical writing in the modern age. The environment was held to change too slowly to be a player on the stage of human history. Those who deny the role of the human in climate change seem to uphold this view, and in this sense could be accused of being conservative. On this read, what is needed is—ironically—enlightenment: the proper evidence that will finally convince the deniers. This is why, argues environmentalist Thomas Princen, even the most perspicacious of environmentalist thinkers and activists find themselves repeating the scientific evidence in order to substantiate mantras about going green, conserving, being less greedy, going activist, or, finally, appealing to resonant principles large numbers can rally around, such as sufficiency.[9] Once the facts are accepted, what remains is exhortation for changing social behavior.

The Senate vote indicates that an enlightenment-based, reason-and-facts approach has limited traction in the United States, as evidenced by the surveys cited above.[10] The facts approach ignores the deep appeal of cosmology; for the climate change deniers, this entails belief in a world of meaning and purpose that exceeds the human and centers on God. Science reaches its limit when it runs aground on divine order. Thus, the Senators represent a group perfectly willing to accept the consumer goods science produces but quick to reject science's atheological picture of humanity in the universe. They accept neither the reductively materialist, evolutionarily-driven sense of human beings' unimportant role in the universe, nor the meaningless, purposeless cosmic order science seems to offer. As philosopher Thomas Nagel argues, many stakeholders, and not just evangelicals, are uncomfortable with this "blind watchmaker" figuration of cosmology (the phrase comes from noted biologist Richard Dawkins).[11] Science's "naturalist *Weltanschauung*," in which the cosmos is a meaningless void sprung from random physical and chemical interaction, is anathema to many, and not only on religious grounds.[12] As Nagel notes, at the heart of these science versus religion conflicts is the age-old mind/body problem, rendered not as a local issue but a cosmological one.[13] Mind, contra science's materialist reductionism, is implicit to the cosmic order.[14] We will see this idea worked out in Empedocles and Heidegger. The larger point is that the Senate vote is as much a response to disenchantment as that of numerous thinkers and academics, including Nagel.

The conservative Senators are committed to unseating the human from its planetary throne, recognizing the divine in worldly affairs, and granting meaning to life and work. Insofar as a scientific consensus on global warming threatens an entire way of life (characterized by strong beliefs in capitalist markets and hard work alongside a Biblically inspired sense of divine order) one sees how science itself is open to strategic disregard. It should be unsurprising that reactionary stances are watered from rhetorically nourishing cosmological pools. Surveys demonstrate the link between the strength of religious belief and climate change denial, particularly for Protestants and Evangelicals in contrast with Catholic, Jewish, or other denominations.[15] Indeed, from the conservative religious perspective, scientists are often cast as atheists, furthering the sense that this is a cosmological clash.[16] The modernist techno-scientific impulse, which has disenchanted the world and elevated the human to mastery, has not, and perhaps cannot, foster the webs of meaning, ritual, and practice that would affirmatively enliven an ecological attunement. There is instead a constant mantra of doom and gloom combined with exhortations to GO GREEN, as Princen has argued, strategies that have demonstrated limited rhetorical power. Coming to grips with the cosmological stakes means that ecological progress has to be more than a matter of individual choice, or, perhaps even less effectual, an attempt at conceptual clarification, as if precision in meaning will of itself deliver socially relevant suasion.[17]

How, then, can we revitalize values and bearings in a cosmological key so as to better pursue ecologically balanced and sustainable ways of life? One representative example comes from ecological scholar David Abram, who argues for recovering ways of life built from "direct sensuous reality" and "regular contact with the tangible ground and sky," so as to provide "a set of powerful conceptual tools" for environmentalism.[18] Abram also seeks to diagnose what is wrong with Western culture. He argues that alphabetic writing gradually eroded the power of story to uphold adherence to the "dictates of the land" and maintain engagement with particular places.[19] For Abram, the story of philosophy is the movement from an ancient Greek *sophoi* such as Thales, who could still claim that "all things are full of gods," to Plato's and Aristotle's embracement of the alphabetic, the abstract, and the metaphysical, on down to today, where the modernist distance from the sensuous world runs wild and catastrophic.[20] There is a direct line from alphabetization and increasing abstraction to modern environmental exploitation and ruin. The exception, for Abram, is the advent of phenomenology, which puts us back in touch with "direct experience" and hence provides a ground for meeting environmentally friendly indigenous cosmologies.[21]

Abram is a generative thinker, but we might pause before accepting such critiques of the Western tradition. Consider, for instance, Pope Francis' 2015 Encyclical, *Laudato Si': On Care for Our Common Home*, which is one of the more striking recent statements on the relations between religion, cosmology, and climate change. Pope Francis argues that all things are related, that technologies and markets are not neutral but agentive, and that the modernist mindset is precisely what has led us to our predicament.[22] He summons us to the broad cosmological issuance that *all* beings share in Creation—they are not merely "potential resources to be exploited…they have values in themselves."[23] In so sharing, there is a moral dimension to being itself that marks and obligates the human to care for a "common home."[24] This moral dimension is opened up by a sense of the holy graced by God, and through the holy there is granted the possibility of the *hale*. The hale is a way of life that stems from a sensibility to how humans are part of the rich, mysterious tapestry of creation, and thus that certain values inhere in the world beyond human attribution. Finally, the Pope castigates human confidence in itself and its technologies, since they prove "incapable of seeing the mysterious network of relations between things."[25] Despite Abram's claim about the West, then, we see instead that Abram, the Pope, and Inhofe all call us to uphold in various ways the cosmological dimension in everyday life. However, they do so by elevating the religious dimension and blaming or downplaying science.

Empedocles and Heidegger remain important sources for cosmologically oriented thought because they sew ecological themes into the broadest possible weave, one that includes both religious and scientific groundings. While Heidegger can often be seen as a critic of technology, his critique was part of an affirmative project that included technoscience. The question becomes how to find a path more attuned to what is hale alongside science and beyond science's reductive *Weltanschauung*. Such a project requires rhetorical work. Heidegger and Empedocles create space for rhetorical work emergent with the lifeworld. They do so by granting a respect for materiality and place that dissolves the primacy accorded to the symbolic; acknowledging a sense of the ethical and the divine as being interior to life and rhetoric; and providing ecologically sophisticated understandings of the world and our place in it. I should also note that their *fourness* is not formal in the sense of providing a framework or grid. Rather, these quadrates are emergent and generative. As I will conclude, they can provide generative resources inclusive of technology for pursuing ecologically attuned ways of life, that is, for a sense of the hale.

Empedocles

Empedocles was a Sicilian sage who lived circa 495–430 BCE. He was a student of Parmenides, who resided in Elea, now Velia, on the western Italian coast—for this reason he is grouped with the Eleatics, which also included Zeno, Melissus, and perhaps Xenophanes, all of whom are predominately classified as pre-Socratic philosophers. Empedocles was the teacher of the sophist Gorgias, and Aristotle famously claimed that Empedocles discovered rhetoric.[26] Empedocles speaks of persuasion, magic, healing, and *mêtis*, rhetorical themes all. He also performed rhetorically intricate poetry with cosmological and ethical themes and a strongly protreptic intent. His role in rhetorical studies has been limited since he does not thematize rhetoric; and it is only recently, as we will see, that his focus on the "four roots" has attracted environmental interest.[27] But he is also of interest for his moral teaching, sense of civic service, and thoughts on the soul, all of which are bound up with his larger cosmological themes.

Empedocles' cosmology begins with the four roots (*rhizomata*) that he names as the sources for the entire universe: earth, aether (sometimes rendered incorrectly as air), water, and fire.[28] They are *archai*, governing powers.[29] The four elements intermix to create everything in the cosmos, but they are in turn motivated by two other great forces, Love (*philotes*) and Strife (*neikos*), about which more below. Empedocles makes a number of associative substitutes for this tetralogy, suggesting gods (Hera, Zeus, Nestis, and Aidoneus), physical phenomena (sun, stars, rain, and earth; but elsewhere sun, earth, heaven, and sea); and what the Greeks considered the four basic colors (white, black, yellow, and red).[30] Empedocles understands the four roots as partaking of both cosmos and logos, the material and the mindful. He states, "For know that all things have thought and a portion of understanding."[31] The roots, then, cannot be reduced to a contemporary scientific understanding of cosmology that emphasizes materiality and material forces as primary, intelligence and meaning as secondary. Empedocles' roots work simultaneously as symbols, personifications (or deifications), abstractions, and physical referents. They constitute the cosmos as being both ordering powers and as being parts of the arrangement. As powers, they come to physical presence in a variety of forms. For instance, Hera is affiliated with earth, establishing continuities between animate and inanimate, mundane and divine; aether stitches together sky, heaven, Zeus, and the soul.[32]

The elements also generate: "From them all things that were, that are, and that will be sprang."[33] This idea is implicit in the term *rhizomata*, for roots are themselves generative, sources for growth intimate already with all four elements. Roots dig into earth, drink water, send up shoots to the aether, and gather the fire of the sun. Roots are part of plants and, as Aristotle himself notes referring to a passage in Homer, plants are associated with the word *phusis*, nature, which soon after becomes a technical term for being.[34] And as one reads Empedocles' surviving fragments, it is striking how palpable and sensuous everything is. The sun is shining and hot; immortals are blazing, soaked in heat; rain lurks in everything, "dark and chilling"; the moon spins round like a chariot wheel; and so on throughout the fragments.[35]

This last point is worth underscoring. Part of what is remarkable about Empedocles is not simply that he posited a rational basis for the cosmos through underlying principles, but that this important step in logical abstraction remained phenomenological. Certainly we are given abstract principles, but these principles are always encountered. They are physically presenced in boundless variations, including figurations hidden and obscure to us, such as the divine. This conception of the cosmos embraces embodiment and emplacement. This emphasis on worldly encounter cultivates ecologically minded thought. This point returns us to the arguments of Abram, mentioned above, who decries Western abstraction and the ecological ruin it has brought with it. While he espies an ecological sensibility still surviving for a time in ancient Greece, in his view it fades once alphabetic-driven abstract thought takes hold. Thus, he reads figures such as Empedocles and Heraclitus as stretching between two orientations, one still indebted to the powers of sensuous nature and the other committed to exploring the possibilities of the new, literate reflection.[36]

Abram's point becomes particularly acute when we consider that when Plato and Aristotle read Empedocles, they translate *rhizomata*, roots, into *stoicheia*, or elements. In certain respects, this is a fruitful translation, since *stoicheia* also means letters. While Empedocles might be cast in a more painterly light, with roots as colors making a cosmic palette, Plato and Aristotle transliterate the roots into a writerly idiom, turning what was vibrantly encountered into the elemental book of life. In the *Timaeus* Plato likens the elements to letters, "the ABC of all things" (*stoichiea tou pantos*).[37] This can be a productive move, since abstract generalization grips the world differently and spurs new styles of thinking, particularly the theoretical mode. But ecological critics such as Abram also note the costs, such as a distant, disinterested engagement, and spatial abstraction.

The dynamic quality of Empedocles' cosmology also takes form as Love and Strife, the two opposing forces that power and scatter the four roots. In great sweeping movements, Love first brings everything together, then Strife drives them apart.[38] Empedocles says, "I shall speak a double tale: at one time they grew to become one alone from many, at another time it grew apart to be many from one."[39] When Love is ascendant, everything is brought tightly together, not just harmonized but bound—one image that comes to mind is the Universe as we understand it right before the Big Bang. While we tend to want to see the triumph of Love as a positive state, as Kingsley remarks, it is more akin to "a cosmic blood bath."[40] It is the movement from persuasion, to pressure, to coercion—no space, no separation, no difference. Our era, says Empedocles, is one of ascendant Strife. Things want to be different, to seek separation. This is better although not idyllic. Strife bespeaks Empedocles' refusal of mysticism or easy holism.

We see this point extended further since Empedocles' cosmology of conflicting forces buffeting four roots is interlocked with his other teachings. He maintains a rich sense of interconnection-in-tension throughout his textual and rhetorical performance, and it also informs his ideas about healing and ethics. Put differently, Empedocles' cosmology is not simply in service to curiosity or knowledge but to the arts of living. Thus, as Gencarella argues, achieving livable balances and harmonies within the world and with other people goes hand in hand with an attunement to the cosmic laws of Love and Strife as they scramble the four roots. Within this cosmic storm the challenge will be to find some bearing, some way to find a course through the always shifting elements. Since there is no escape, one must seek balance, direction, freedom. Such way-finding is practical and political, too, as seen in the recurring motif of

Empedocles' fights against tyranny and other forms of social injustice, which are portrayed as social pollutions in need of purification. There is a fittingness to be achieved in the complexities of interaction, one that is pulled away from injustice, impiety, tyranny. Such fittingness, a sense of the hale, extends to everyday life and beyond. Empedocles believed the soul was immortal, teaching (as he learned, likely from the Pythagoreans) that it could transmigrate from human to animals. Because of this metempsychosis Empedocles condemned animal sacrifice and meat eating. In a striking passage, Empedocles laments how those who eat meat are eating their relatives in newly incarnated form.[41] Since all share in being through the four elements, and the soul is the spark of the divine found in all life, it is easy to see how cosmology and ethics mutually condition each other. Interconnection calls for an ethical comportment towards other beings beyond an exclusively human valuation. Ethics is manifest both in the physical and the metaphysical: "what is lawful for all extends continuously both through wide-ruling aether and through vast sunlight."[42]

There is significant import for environmental thought here. As the roots mix and morph across the material and the mental, the mundane and the divine, the realm of fact and the realm of value commingle. This move grants the environment value beyond human attribution. I will come back to this point in the conclusion, but I raise it here: the tension Empedocles assigns between Love and Strife is mirrored in the performative tension between Empedocles' attempts to think the rationality of the cosmos, without, for all that, equating the cosmos with *human rationality*. Thus, we cannot align Empedocles with the modernist conception of a fragmented, meaningless cosmos generated exclusively through material interaction and biological evolution.[43] Rather, as Empedocles tells us, there is a "communion of all things," which begets and kills.[44] Meaning inheres in the cosmos beyond human reach, but humans can *share* in this meaning, because, as Simon Trépanier notes, "our ability to perceive and to think is a result of our material constitution being the same as that of the world."[45] This is no holistic panacea, however, as it is here that the real challenge emerges. The soul, embodied and embedded in the unceasing strife, finds itself confounded and unable to cope. Our plight is one of helplessness, not from a lack of direction as much as an overabundance of stimulation and suasion.[46] Put differently, being embedded in a "communion with all things" is only a partial answer to the existential challenges of human dwelling and ecological sensibility.

The problem of finding existential bearings is central to Empedocles' teaching. For instance, when he speaks to his student, Pausanias, Empedocles tells him to cultivate his teachings like seeds that need tending. For if he does not, then he will

> reach out instead after other kinds of things—after the ten thousand worthless things that exist among humans blunting their cares—then you can be sure they will only too gladly leave you with the circling of time, longing to return to their own dear kind. For you need to know that everything has intelligence and a share of awareness.[47]

There is much in this passage. It understands teaching as a kind of agriculture, in keeping with thematic of roots—but it further underscores a very particular kind of rhetoric. Persuasion here is a matter of care, of attending, and of attunement—to the fact that everything, being ensouled or in some sense aware in its place in the cosmos, takes part in everything else. Empedocles' rhetorical wisdom is thus rooted equally in words of power as in cosmic rationale. In a beguiling, confounding world of strife, the need to forge hale and attuned forms of life and community become paramount. Lacking these, we are at the mercy of all the worthless things that bedazzle, mislead, and fail us.[48] For Empedocles, rhetoric in the form he understands it is enmeshed within a complex quadrate, one that conjoins the material and the meaningful. Persuasion of every kind is knit into the cosmic churn. Nevertheless, the core of his teaching is that within this storm paths open up to more just, ecumenical, and harmonious ways of life. In this way, rhetoric springs from and responds to cosmology, worldly attunement, and ethical responsibility, bringing them all together, with and alongside a strife we can never escape. Thus, Empedocles provides an early cosmology that situates human being within an ecologically complex and meaningful world and challenges it to achieve some measure of the hale.

Heidegger

Heidegger worked within the phenomenological tradition initiated by Husserl, whose prime directive was "to the things themselves." This physicality establishes a connection between Empedocles and Heidegger beyond the serendipity of their fouring; they both attend to staying with things, to the sensuous world. They also share an awareness of the dynamic of *presencing*, the exploration of how things show forth as they do. And they both seek to find equitable, attuned ways of life. Much of this can be seen in Heidegger's famous example of the Black Forest farmhouse, where he shows the fourfold of earth, sky, mortals, and divinities entwined in everyday life. Heidegger writes:

> The house was built some two hundred years ago by the dwelling of peasants. Here the self-sufficiency of the power to let earth and heaven, divinities and mortals enter in simple oneness into things, ordered the house. It placed the farm on the wind-sheltered mountain slope looking south, among the meadows close to the spring. It gave it the wide overhanging shingle roof whose proper slope bears up under the burden of snow, and which, reaching deep down, shields the chambers against the storms of the long winter nights. It did not forget the altar corner behind the community table; it made room in its chamber for the hollowed places of childbed and the "tree of the dead"—for that is what they call a coffin there: the *Totenbaum*—and in this way it designed for the different generations under one roof the character of their journey through time. A craft which, itself sprung from dwelling, still uses its tools and frames as things, built the farmhouse.[49]

There is a distinctly ecological feel to this passage, although there are some oddities. More powerfully than in Empedocles, ecology unseats the traditional locus of agency, from human action to dwelling itself. It is not the peasants who built the farmhouse, but dwelling. Dwelling means the flourishing of life through attunement to the surrounding environment. The peasants are present, but so also the fourfold in all its physical manifestations: weather, soil, plants, animals, water, sun, tools, craft, and a sense of value or even the holy. This is a sustainable and mutually supporting ecology, one that Heidegger therefore characterizes as hale. However, just as in Empedocles, strife in the forms of harshness, pain, fear, and death are equally present and an inescapable part of dwelling.

The four do not appear solely on account of the peasants. There is no instrumental relation—the four are neither a *telos*, nor a means, nor an explanatory grid. They are emergent. We can draw a parallel from complexity theory with the notion of an *attractor*, in which a particular state or form finds its habituated role over time within a larger, dynamic system. That is, given evolving conditions, specific states or forms are more likely. These new forms of stability or habituation are akin to flowing into a "basin of attraction."[50] There is pull not push.[51] Action forms patterns and "grooves."[52] The fourfold are pulled into their forms, then, like a compass, pulled to polar north. The compass discloses but does not create north; rather, north emerges from a variety of worldly factors and forces, and, given the proper means, made present. The fourfold are woven into the tapestry of life and value—what constitutes *world*. World entails more than physicality: it includes growth, processes, pursuits, cares, and involvements. Accordingly, the fourfold are brought forth in the everyday: through death, birth, rituals, seasons, weather, plants and animals, farming, making, building, stories, and more. The human remains an important site for disclosure, but the world also discloses itself in its own ways—this too is part of the world's "worlding." This is why the fourfold transcends human reason. "Causes and grounds remain unsuitable for the world's worlding," Heidegger tells us, because explanation, or human cognition in general, "just does not reach" far enough.[53]

As in Empedocles, Heidegger's four intermix. Heidegger sees them as mirroring each other, or being engaged in a dance that rings their essential oneness. For instance, human beings are mortals, but as mortals they are given in the rest of the fourfold, since they are of the earth, under the sky, alive until death claims them, and cast before whatever glimmer of the divine is hinted. This same process can be followed for each of the fourfold, as each in turn opens into all the others. Sky as seasons, rhythms, forces turns back into matter and earth; earth plunges down to molten star stuff at the core, once again sky. Here we see that Heidegger's choice of poetic idiom as opposed to more precise terminology (technical terms, first principles, etc.) is deliberate. Poetic idiom reinforces that the fourfold are never simply revealed givens. They can take manifold forms.[54] For instance, the earth/sky vector can come to presence as forms of ordering/scattering, depth/surface, pattern/randomness, light/dark, known/unknown, packed/dispersed, encompassing/encompassed, perceptible/imperceptible, and—why not?—territorializing/deterritorializing. These patterns are not first principles but available disclosures brought forth through the motivations and selections ongoing in life. They are materialities, forces, processes, relations, and significances brought to presence through our everyday lifeways.

Because they are made manifest through being and living rather than given as grounding principles, there is an interpretative or disclosive leeway to the fourfold. This complicates the appeal of unmediated immersion in the environment, where transhuman values would speak directly to us. For instance, Abram invokes phenomenological thought to put us back in direct touch with the elemental world. While Abram is correct that we *are* placed in touch with the elemental world, it is still a leap to then claim that in some fashion the world will "dictate" what we are to value and do.[55] While we are meshed into the world in varying degrees of closeness, such proximity is never purely direct. The processes of disclosure cannot be circumvented.[56] Manifestation takes disclosive work, which generates space, or leeway, between manifestation and its coming to presence in such and such a way. This leeway generates difference and motivation. Thus, while Heidegger does not use the word "rhetoric," it nevertheless remains implicit in his fourfold, no less than for Empedocles. It emerges from a profound reflection on situationality—that we are, as Heidegger will say, be-thinged, or conditioned.[57] As with Empedocles, the world and the things in it make claims on us, shape us, call us, persuade us. They do not command. As Allen Scult perceptively remarks, even in terms of being situated, we are called to make existential choices about what to think, what to do, all of which betray what he calls "a very subtle rhetoric."[58]

This aspect of Heidegger may be made even clearer in his ruminations about what a thing is. Heidegger claims that things gather the fourfold, and by so gathering, they "thing" the world. When he considers what the jug gives, Heidegger notes the jug's giftly character, how it niches itself into the world and human life. By tracing this niching, we learn how the jug in turn reveals the world's fourfold character.[59] The jug holds things such as fluids, which discloses how holding things is already a worldly possibility—wells and lakes, for instance, are holding vessels. As a holding vessel created in a world where holding is a possibility and a necessity, the jug helps stitch water, milk, wine into the tapestry of human life. Meanwhile, water, milk, and wine are in turn granted by other things—rain, animals, plants. Yet rain, animals, and plants are themselves possible because of countless other forms of being: the spring in the rock; the earth as fertile for growing; the sun ripening the grapes; and so on.[60] These habituated forms and patterns take shape through their complex interdependencies. From our recognition of these interdependencies springs the first intimation of an ethic, posed as a question: why should the world grant itself in such way as to make dwelling possible? This is where we locate the giftly character of the jug; in a way so simple as to scarcely pass register, it bequeaths life, and in so bequeathing, illumines some small part of the richer, largely unknown tapestry that is needed for our world to be. We are thereby inclined towards seeing our debts to these gifts of the earth, which we only partially understand since they transcend our capacities. The jug thereby invokes something beyond us that beckons us with hints. The jug's giftly character is both material and meaningful—there are material dependencies complexly interlocked into each other; and such material dependencies in turn call for meaningful and existential responses, one of which is to *respond to this call as it calls*. This is why things, for all their mundanity, also open us to the world. In this gesture toward a sustaining world beyond comprehension, things show how they are woven into the tapestry of the sacred—from such thanks, from such recognition of the gift of being, stems further spiritual possibilities. And these, like all values, are not solely of human making, although, insofar as humans are one of the ways such values are disclosed and pursued, humans have their part.[61] And it is through these possibilities as opened through the fourfold that a sense of the hale becomes possible. The hale would be a way of life that responds to, even attends to, the giftly character of the world. But as we have seen in the Black Farmhouse example, this cannot be solely a matter of human doing—or Heidegger's for that matter.

If the fourfold are emergent then Heidegger's thoughts cannot be the sole entryway into their workings. We can note that he had several sources for the fourfold, including Empedocles but especially Hölderlin. However, I think it is worth considering the irrepressibility of the fourfold by seeing it emerge outside Heidegger's orbit. Consider the posthuman arguments of the Deleuzian philosopher Rosi Braidotti. Citing Lyotard, she points out that the ontological inhuman certainly springs from the human, but, she argues, it is unconvincing to render this as the sacred (or the divine); rather, she claims, we stretch out towards "endless cosmic energy."[62] Braidotti declines any religious affiliated language here, for personal and/or philosophical reasons, but the deeper connection to religion remains. There is an ontological point Braidotti skirts: we too are granted our existence in such endless cosmic energy—we are star stuff, as Carl Sagan has famously stated. Finding meaning and communion in what transcends humanity is a sense of the divine, since, as Heidegger has it, the godhead is withdrawn, not present—we are beckoned by their messengers, signs, and hints.[63] What divine glimmer is granted cannot and does not depend solely on human attribution. Braidotti, then, invokes what both transcends and permeates the human and the world. This is indeed how sacredness essences, and in this manner Braidotti discloses in her own idiom what the fourfold traces. Let us look to how she sketches a posthuman ethics across five features. First, human beings are "eco-sophical," marked by their interdependence with the environment; second, the human is plugged into a "variety of possible sources and forces"; third, the first two induce humans to "raise questions of ethical and political urgency"; fourth, human temporality needs to be "rethought beyond the metaphysics of mortality"; and fifth, there must be "considerations of power," especially as implicated in the current techno-capitalist moment.[64] Adding the fifth element of power is certainly helpful, but does little to mask what is otherwise a perspicaciously delivered fourfold: earth, sky, mortals, and gods respectively.

It is worth lingering on this moment. First, it is not a matter of trumping Braidotti with Heidegger, since Heidegger neither invented nor owns the fourfold. If Heidegger has any elevation here, it is the poetic simplicity with which he rendered it. The poetic idiom is especially powerful because it does not just evade the pretention of philosophical precision but calls it into question, and thereby leaves room for the claims of other forms of knowledge, such as arts, poetry, religion, all of which have their stake in the fourfold. As I have suggested, the fourfold are *attractors*. In complex, evolving situations including humans (and perhaps animals), the world habituates in such fourfold form. But, second, it is illuminating that it can appear in a thinker who has little use for Heidegger. This has the advantage of *decentering* Heidegger as author/ity about the fourfold, which opens it for stances different than Heidegger's. Braidotti is far more generous about technology, for instance, a point I have been emphasizing. Third, there is also the deeper lesson, given in both Braidotti and Heidegger, of an attentiveness to the vital and giving qualities of the cosmos that in turn spur new attunements. They foster a living reception to these vital and giving qualities and in this way open humans up to possibilities for more hale, ecologically attuned ways of being together.

Conclusion: Ecosophy, with Spacecraft

The fourfolds of Empedocles and Heidegger sketch a cosmology in which humans have a role but not an unconditioned one. They are a part of a complex weave that transcends human understanding and control, and that gestures towards a beyond to which we are called to respond. Finding the best course for human lifeways, they find, cannot finally be solely a matter of human knowledge and will. Consider Empedocles again: while metempsychosis is likely a difficult sell in the contemporary world, its greater import lies in the cosmology it bears up. All the world shares in meaning—there is a cosmic communion, even if buffeted by the great conflicts of Love and Strife. Empedocles' philosophy informs, guides, and grounds a total way of life, inclusive of morality, healing, and religious beliefs. Similarly, Heidegger's fourfold captures how all human doing is interwoven within the world, and made possible through it, not only materially but meaningfully, and not only meaningfully but sacredly.

From the standpoint of the contemporary drive for endless technical and economic growth, Empedocles and, in particular, Heidegger can appear overly rustic or nostalgic. Critics point to examples such as Heidegger's complaints about electric lighting, liquor at the bar, and his praise of artisan craftwork. Such criticisms miss Heidegger's larger point that interconnection written on a cosmological as opposed to human scale has its precarity. Take, for instance, Heidegger's complaint that we should not "turn night into day."[65] This seems reactionary until we consider that the rhythms of day and night define our situation as much as anything else. They include our biorhythms, rituals, daily schedules. Night and day also provide rich metaphoric resources. Artificial light changes these rhythms, rituals, and schedules. It also requires industry and infrastructure to build and sustain it, which entails still further consequences. We are given wonders, perils, and temptations as we step out into extended day and find new things to make and do with that time. As with the jug, artificial light is tightly stitched into the world. While artificial light is seemingly just one small thing, we see that it produces multiple effects beyond its immediate locus; some of these consequences we see but many we do not. Thus, Heidegger shows how our densely interconnected fourfold background will shift as we make changes and innovations. This corresponds to what, in the language of extended cognition, we might call "structural couplings," whereby "coupling with the environment constrains and enables world-making."[66] But shifts thereby change world-making. Such change often escapes notice, at least at first. Climate change exemplifies this sort of big swerve; it was never actively sought but emergent from advanced industrial life.[67] We might say, then, that Heidegger is one of the first thinkers of blowback.

Blowback can stem from any intervention, but here I am using it to mean unintended consequences resulting from changes to complex systems, such as the intricate webs of worldly, life-granting connection.

Both Empedocles and Heidegger attune us to these intricate webs, as a source of meaning and a necessary if mostly veiled background to life. Further, there is growing momentum for cultivating awareness about an interconnected, mutually implicating and sustaining world. Environmentalist scholar Leslie Paul Thiele, similarly to Braidotti above, advocates for an "ecosophic awareness" to convey this attuning to our larger environment, hinged on sensitivities to powerful interdependencies and unintended consequences.[68] Promoting her own version of ecosophy, Jane Bennett uses the example of a national power failure to illustrate how myriad entities can be assembled to form complex objects, such as an electrical grid, which in turn requires attentiveness to their emergent, unexpected behaviors.[69] More sweepingly, Jason Moore argues that capitalism itself must be understood as grounded in a participating world. It is not an economic system nor a social system, he argues, but rather "*a way of organizing nature.*"[70] This organization has operated on a dualism between nature and culture, with the result that capitalism has appropriated nature howsoever it pleases. However, Moore points out that if one looks again, with something approximating Thiele's ecosophic perspective, then forms of blowback quickly become apparent. The most obvious would be the generation of forms of negative capital, which sooner or later show up on the balance sheets.[71] Negative value, he argues, destabilizes surplus value, and this enables different paths forward, perhaps even stretching capitalism towards new formations.[72] Negative value and global warming are two forms of blowback to which we are becoming sensitized. They have at least demanded a response, even if for some that response so far is denial. One of our challenges, then, is to increase ecosophic awareness within the structures and precepts of contemporary capitalism and science.

Empedocles and Heidegger continually emphasized the need to forge a positive way forward, to work not just with who people are but with *where* they are. We cannot escape the infrastructures, industries, and lifeways people currently have but must work within and through them towards a more ecologically attuned dwelling. In this regard, science, technology, and capitalism will need to become part of that path, despite the fact that these are some of the most common targets for environmentalist critique. On these points I take inspiration from Peter Sloterdijk's digression on spaceships in his massive philosophical trilogy, *Spheres.* Sloterdijk argues that spaceships are more than technological objects—they are equally ontological interventions. The suturing of a thing to ontology harkens back to Heidegger, and we can see how Sloterdijk's spaceship updates Heidegger's jugs and bridges. But there is also an inversion: jugs and bridges evoke an interconnected whole to which we are all wedded and indebted; a spaceship, however, operates in a realm of extreme inhospitality. Space is ***nonconductive*** for nearly all life as we know it. Since no aspect of dwelling in space can be taken for granted, a spaceship "constitutes a world model," a kind of "immanence machine."[73] The level of explicatory detail necessary for surviving in such a harsh environment is vertiginous. To give one small example: astronauts must sleep with fans next to them, less their exhalations, which do not move in zero gravity, suffocate them in a bubble of nitrogen and carbon dioxide.[74] There is no resting on unexplicated assumptions, as is possible back on Earth where the ecological background can simply be assumed, as Moore exemplified through capitalism's cavalier appropriations of nature. Space lacks such "ontological comfort."[75]

The point, Sloterdijk argues, is that "space travel amounts to an ontological alphabetization: what follows from it is that the elements of being-in-the-world can, and must, be written."[76] Sloterdijk picks up the thread we saw connecting Empedocles and Aristotle in the movement from roots to elements, where increasing abstraction and alphabetization formalize as science and episteme. Sloterdijk also counters Abram's argument that abstraction goes hand in hand with the West's loss of intimacy with the environment. Rather, Sloterdijk taps a great affirmation resonant with Empedocles and Heidegger. It is the affirmation of the fourfold. The lesson is that abstraction and alphabetization also gather the fourfold and disclose the worldly depths of rich interconnection. Spaceships necessitate this insight. For life to maintain itself in space, bubbles of lifeworld must be constructed. To construct a bubble of lifeworld requires a profoundly rich explication of an ecological background we take for granted on earth, where our lifeworld is simply given. For spacecraft, these ABCs of lifeworlding must be spelled out and implemented through the technics of disclosure and building. As Heidegger has argued, to build is to dwell, and here we see how this manifests in even the most focused technological enterprise.[77]

The explicative disclosure necessary for space travel thus affirms some of environmentalism's biggest demons—abstraction, science, technology, capitalist progress. But this disclosive work pushes through their dangers and lights up their roles in the ecological lifeworld it is claimed they despoil. Still, it may be argued that space travel is a fleeting pursuit far from average life or environmental intervention, and hence of dubious resource. Yet technology, as Heidegger continually asserts, discloses. The pre-Socratic philosopher Heraclitus famously claimed that the cosmos was in tension like a back-turned bow, thereby pulling *logos* from *techne* and granting it cosmic sweep.[78] Sloterdijk makes a similar move in showing that a technology not only *does* something but also illuminates and *explains* the world. Such explanation is, he claims, modernism's most important undertaking because it demonstrates "what the coexistence of someone with someone and something in a shared whole means."[79] Here we see that spacecraft, just as much as, say, Heidegger's jug, makes explicit the rich ecological stitchwork supporting life. The advance on where Heidegger was able to bring his understanding of the fourfold is that technology, as dwelling through building, also returns us to what we euphemistically call "nature." But the twist is that the background we take for granted on earth is now more difficult to ignore; indeed, as Sloterdijk suggests, contemporary politics, insofar as it lazily and confidently relies on what it need not acknowledge or explain, is a form of "hermeneutical dreamtime."[80] We can be harsher; it is a willful ignorance gifted through the earth's irrepressible bounty. But such bounty is not infinite, and ignorance cannot forever ride high.

This, finally, is the idea that we are left with: ecosophy cannot be achieved only through exhortation, inducement, or training. These traditional understandings of rhetoric have their role to play, no doubt, but if they are to have greater sway in forging more ecologically attuned lifeways, they need to be grounded in cosmology. As a consequence, ecosophy cannot be solely a human pursuit, but rather, in keeping with the declination to elevate the human to master of the planet, the world must be a co-participant; its materially dense and evolving interconnections will have to be the counterthrow to the disclosive gesture we offer in turn. We do not *give* value to the world; we dwell, participating in the world, so that the world pulls us towards the way in which we will have been with it over time. In this sense, ecosophy, rendered sophistically across the lessons of the quadrate, is an attractor. Gratitude for a world that bears up life remains a source for attuning to what signs of the hale we are granted. The fourfolds of Empedocles and Heidegger teach us to cultivate awareness for these signs. The world is replete with them. As the water springs from the rock, giving life to human and vine, or as the event of spacecraft slowly peels back the ignorance we assume in crafting our earthly lifeworlds, we can find our own way to a new fittingness, one no longer conceived simply in terms of science versus religion, ancient versus modern, or human versus world. That fittingness, in a fourfold sense, will emerge not from contemplation, or some wrenching attempt at transvaluation, but rather through what drives life now, including technology, capitalist progress, and some sense of being part of a larger whole. It is here, on these cosmological grounds, that rhetorical resources gather, ones that have potential to sway even wayward constituencies. The power of such rhetorical resources rises precisely to the extent that they transcend the human and the plane of polemic, pulling us toward a larger weave we can no longer afford to ignorantly dream away, as if the world will always smile benignly.

Notes

1. Laura Barron-Lopez, "Senate Votes that Climate Change is Real," *The Hill*, January 21, 2015, accessed November 27, 2016, http://thehill.com/.
2. The NASA website provides a quick view of the extent of consensus, with links for further exploration. See "Scientific Consensus: Earth's Climate is Warming," NASA, accessed November 27, 2016, http://climate.nasa.gov/scientific-consensus/.
3. Suzanne Goldenberg, "US Senate Refuses to Accept Humanity's Role in Global Climate Change, Again," *The Guardian*, January 22, 2015, accessed November 27, 2016, https://www.theguardian.com/.
4. Richard Wike, "What the World Thinks about Climate Change in 7 Charts," Pew Research Center, April 18, 2016, accessed November 27, 2016, http://www.pewresearch.org/.
5. The term Anthropocene was first coined by Paul J. Crutzen in 2000. While a name can be proposed for an era, there is no guarantee that it will be accepted. The term Holocene, proposed by Sir Charles Lyell in 1933, took several decades to win acceptance. See Dipesh Chakrabarty, "The Climate of History: Four Theses," *Critical Inquiry* 25 (2009): 209. Usage of the term does not mean acceptance of it or its implicit claim. Those who oppose the idea that humans cause climate change thereby use the term, or its variations such as anthropogenic, negatively.
6. Cosmos is related to the Greek verb *kosmeo*, to set in order or arrange. For more on *kosmos*, see Charles H. Kahn, *Anaximander and the Origins of Greek Cosmology* (Indianapolis: Hackett, 1994), 219–222.
7. Chakrabarty, "The Climate of History," 198–222, 204. See also Bruno Latour, *We Have Never Been Modern*, trans. Catherine Porter (Cambridge: Harvard, 1993).

8. See Elizabeth Bruenig, "Why Do Evangelicals Like Jim Inhofe Believe That Only God Can Cause Climate Change?" *New Republic*, January 15, 2015, accessed November 28, 2016, https://newrepublic.com/. Note quotes from Inhofe and others on the evangelical position, which attributes major changes such as the climate to God, and thus even notions of End Times, rather than human activity—or human intervention.
9. Thomas Princen, *The Logic of Sufficiency* (Cambridge: MIT, 2005), vii–viii, 5–6.
10. Wike, "7 Charts."
11. Thomas Nagel, *Mind and Cosmos: Why the Materialist Neo-Darwinian Conception of Nature is Almost Certainly False* (New York: Oxford, 2012), 5, 10.
12. Ibid., 4.
13. Ibid., 3.
14. Ibid., 16.
15. Chris Mooney, "New Study Reaffirms the Link between Conservative Religious Faith and Climate Change Doubt," *The Washington Post*, May 29, 2015, accessed November 28, 2016, https://www.washingtonpost.com/.
16. Scott Rodin, "As a Conservative, Evangelical Republican, Why Climate Change Can't Be True (Even Though It Is)," *The Steward's Journey*, accessed November 28, 2016, http://thestewardsjourney.com/. Rodin gives a first-person account of evangelical perspectives, describing how environmental issues are cast as liberal and associated with laziness, alarmism, and godlessness.
17. See Thompson for an examination of how conceptual clarification proceeds; cf. David Abram, *The Spell of the Sensuous* (New York: Vintage, 1996).
18. Abram, *Spell of the Sensuous*, x.
19. Ibid., 185.
20. Ibid., 108.
21. Ibid., xi.
22. Cf. Latour.
23. Pope Francis, *Laudato Si': On Care for our Common Home* (New York: Melville House, 2015), 22.
24. Ibid., 13.
25. Ibid., 14, 15.
26. Diogenes Laertius, *Lives of Eminent Philosophers*, trans. R.D. Hicks (Cambridge: Harvard, 1925), 8.57.
27. For Empedocles and rhetoric, see Richard Leo Enos, "Aristotle, Empedocles, and the Notion of Rhetoric," *In Search of Justice: The Indiana Tradition in Speech Communication*, eds. R. Jensen and

J. Hammerjack (Amsterdam: Rodopi, 1987): 5–21; Stephen Olbrys Gencarella, "Purifying Rhetoric: Empedocles and the Myth of Rhetorical Theory," *Quarterly Journal of Speech* 96.3 (2010): 231–256. For Empedocles and environmentalism, see David Macauley, *Elemental Philosophy: Earth, Air, Fire, and Water as Environmental Ideas* (Albany: State University of New York Press, 2010).

28. Empedocles B6: "The four roots of all things hear first: shining Zeus, life-giving Hera, Aidoneus, and Nestis, who by her tears moistens the mortal spring." See Peter Kingsley, *Ancient Philosophy, Mystery, and Magic: Empedocles and the Pythagorean Tradition* (New York: Clarendon, 1995), 13, 47. While the question of which God equals which root remains contested, Kingsley argues compellingly for Zeus as aether, Hera as earth, Aidoneus (an alternate name for the god of the underworld, Hades) as fire, and Nestis as water.
29. Macauley, *Elemental Philosophy*, 107.
30. Ibid.
31. Empedocles B110.
32. Peter Kingsley, "Empedocles for the New Millennium," *Ancient Philosophy* 22 (2002): 282.
33. Empedocles B21.
34. Aristotle, *Metaphysics*, trans. Hugh Tredennick (Cambridge: Harvard University Press, 1933), 1014b–1015a.
35. Empedocles B21, B46.
36. Abram, *Spell of the Sensuous*, 108.
37. Plato, *Timaeus* 48B; Macauley, *Elemental Philosophy*, 107–108, 143.
38. Empedocles B17.16–20.
39. Empedocles B17.1–2.
40. Peter Kingsley, *Reality* (Point Reyes: Golden Sufi Center, 2003), 354.
41. Empedocles B136: "Will you not cease from ill-sounding bloodshed? Do you not see how you are devouring each other with heedless mind?" Eating animals, which contain souls, is tantamount, therefore, to eating each other. See also B137.
42. Empedocles B135.
43. Nagel, *Mind and Cosmos*, 4; Ursula Goodenough and Terrence W. Deacon "The Sacred Emergence of Nature," in *The Oxford Handbook of Religion and Science*, ed. Philip Clayton and Zachary Simpson (New York: Oxford, 2006), 853.
44. Empedocles B17.4.
45. Simon Trépanier, *Empedocles: An Interpretation* (New York: Routledge, 2004), 160.
46. See Kingsley, "Empedocles," 360. Kingsley provides a striking interpretation of a key passage in Empedocles' work, describing how narrow palms

(*steinopoi palamai*) have been poured over people's limbs, and "many worthless things keep crashing in, blunting their cares." The image of narrow palms metaphorically connotes how our senses are always open, and, without guidance, overwhelmed in the overflow.

47. Kingsley, *Reality*, 520.
48. Empedocles' sensitivity to communicative oversaturation suggests that he is the first thinker of the pitfalls of social media.
49. Martin Heidegger, *Poetry, Language, Thought* (New York: Harper & Row, 1971), 160.
50. Alicia Juarrero, *Dynamics in Action: Intentional Behavior as a Complex System*, (Cambridge: MIT, 1999), 153.
51. Ibid., 152. Juarrero succinctly defines attractors as "trajectories that converge on typical patterns" in the sense of being *pulled*.
52. Jane Bennett, "The Agency of Assemblages and the North American Blackout," *Public Culture* 17, no. 3 (2005): 457.
53. Heidegger, *Poetry Language, Thought*, 180.
54. Deacon, 157: "there are many possible ways that micro-details of structure and interaction can converge to produce the same higher-order properties."
55. Abram, *Spell of the Sensuous*, 185.
56. For a more thorough examination of the problems inherent to an unmediated environmentalism, see Iain Thomson, "Ontology and Ethics at the Intersection of Phenomenology and Environmental Philosophy," *Inquiry* 47, no. 4 (2004): 380–412.
57. Heidegger, *Poetry, Language, Thought*, 181.
58. Allen Scult, *Being Jewish/Reading Heidegger* (New York: Fordham, 2004), 58.
59. Heidegger, *Poetry, Language, Thought*, 172.
60. Ibid.
61. Ibid., 181.
62. Rosi Braidotti, *The Posthuman* (Malden: Polity Press, 2013), 185.
63. Heidegger, *Poetry, Language, Thought*, 178.
64. Braidotti, *The Posthuman*, 139–140.
65. Heidegger, *Poetry, Language, Thought*, 150.
66. John Protevi, "Beyond Autopoiesis," in *Emergence and Embodiment*, eds. Bruce Clarke and Mark B.N. Hansen (Durham: Duke, 2009), 108.
67. Chakrabarty, "The Climate of History," 210.
68. Leslie Paul Thiele, *Indra's Net and the Midas Touch* (Cambridge, MIT 2011), 2. Note that the emphasis on awareness underscores the point that while ecosophy is assumed to be based in philosophy, it is actually more attuned to sophism. It connotes not just wisdom but a receptiveness to environmental suasion.

69. Bennett, "The Agency of Assemblages," 446.
70. Jason Moore, *Capitalism in the Web of Life: Ecology and the Accumulation of Capital* (New York: Verso, 2015), 2.
71. Ibid., 270–276. Moore provides many examples of negative value. For instance, industrialized agriculture depletes the ground, requiring investment in fertilizer; produces superweeds, requiring greater chemical intervention; depletes water, requiring the importation of water from elsewhere; and so on. Fertilizers, pesticides, herbicides, and irrigation are finally unsustainable not only on their own demerits but on their growing expensiveness.
72. Ibid., 290.
73. Sloterdijk, *Spheres III: Foams*, trans. Wieland Hoban (South Pasadena: Semiotexte, 2016), 299.
74. Ibid., 304.
75. Ibid., 308.
76. Ibid., 309.
77. Heidegger, *Poetry, Language, Thought*, 146.
78. For a discussion of Heraclitus' B51 fragment and the explanatory power of technology see David Rothenberg, *Hand's End: Technology and the Limits of Nature* (Berkeley and Los Angeles: University of California Press, 1993), 2–3.
79. Sloterdijk, *Spheres III*, 310.
80. Ibid., 311.

PART II

Ecological Engagements

CHAPTER 4

Intertidal Poetry: Making Our Way Through Change

Bridie McGreavy

By all accounts, we are living in a time of amplified, intensified change in Earth's material flows. These changes are due, in part, to high impact agricultural land uses and the burning of fossil fuels that mark our transition into the era now commonly, and contentiously, referred to as the Anthropocene.[1] Across Earth's major life support systems we are at or have crossed critical thresholds beyond which everyday life as we know it may not be sustained. We have blown past the proposed limit of 350 parts per million (ppm) of atmospheric carbon dioxide and have, in the course of writing the essay, now exceeded 400 ppm. This gaseous material layer traps excessive heat energy, accelerating global ice melt and sea level rise, causing marked changes in ocean chemistry and circulation. We are flushing massive amounts of nitrogen and phosphorus into waterbodies, polluting aquatic ecosystems at the same time we pump, transport, and consume water in ways that disrupt how it has cycled on this planet for eons.[2] By looking at the 540 million-year-old fossil

The original version of this book was revised: Editor affiliation has been updated. The erratum to this chapter is available at https://doi.org/10.1007/978-3-319-65711-0_14

B. McGreavy (✉)
Department of Communication and Journalism,
University of Maine, Orono, ME, USA
e-mail: bridie.mcgreavy@maine.edu

B. McGreavy et al. (eds.), *Tracing Rhetoric and Material Life*,
Palgrave Studies in Media and Environmental Communication,
https://doi.org/10.1007/978-3-319-65711-0_4

record, we also know that the current rate of species extinction is 100–1000 times higher than trends that date back to the time when clams' distant cousins first formed hard shells and long before human beings were a glimmer in the evolutionary eye.[3]

We can take a critical look at all of these trends, identifying the potential consequences of Anthropocene as discourse, how a metaphor like "threshold" may mask complex and unequal systems of power that reproduce our current unsustainable moment, and how research conducted by the Stockholm Resilience Center enacts modes of control that may limit how we identify and implement more equitable ways of relating to each other and the world.[4] With a healthy dose of critique mixed in, we can accept scientific evidence that the world is pressing in.[5] The material world calls our attention to these changes and demands that we respond.

Rhetoric provides orientations to shape how we respond to change. With its sustained focus on adaptation and memory, rhetoric has much in common with a foundational concept in the science of ecology, namely resilience. Resilience was first used in a landmark paper by C.S. "Buzz" Holling to describe predator–prey dynamics in lake ecosystems and how systems respond to change to maintain structures, functions, and overall identity.[6] Social–ecological systems literature has since refined the concept to explain outcomes from human–environment interactions in contexts as diverse as coral reefs, urban water systems, sub-Saharan grasslands, and high mountain streams.[7] More recently, *resilience thinking* offers a way to analyze how individuals and communities anticipate, respond to, and absorb change.[8] Resilience thinking attends to the role of memory in shaping how people make sense of and respond to changes they observe. In popular accounts, the goal of resilience is to "bounce back" and return to normal after disturbances like natural disasters, disease outbreaks, or economic collapse.[9] Approached in this way, memory is a social resource on which people draw as they attempt to find ways to cope and return to normal following change.[10] However, as ecologists and rhetoricians would likely agree, bouncing back or any form of return to a previous state is, in fact, untenable because perpetual change is ontological. More simply, we know we can't ever go back. What, then, is memory and how do we remember our way through change?

This chapter addresses this question by weaving together concepts from rhetoric and resilience to articulate a poetic approach to memory and responsiveness to environmental change. I draw from a rhetorical ethnography with clammers on the coast of Maine who are already

Fig. 4.1 This clammer uses his hand to pull clams from the soft mud. Here he searches for clam holes as the tide comes in. Photo by Bridie McGreavy

grappling with climate-induced changes in intertidal ecosystems, such as warming ocean temperatures and species shifts.[11] Clammers inhabit liminal territories where land and sea meet and where their everyday practices of digging in mud and water condition their awareness of material change (Fig. 4.1). They work in a zone that Rachel Carson described in *The Edge of the Sea* as a place where "there are other tidal effects, mysterious and intangible," and where "the present is linked with past and future, and each living thing with all that surrounds it."[12]

Working with clammers, I have learned from how they make sense of change. Clammers follow the cyclical rhythm of the tides every day. With previous days' digs and tide charts as guides, they sense the right time to go into the intertidal mudflat to dig for clams. As they do, they observe the variation in the fine grained texture of the mud and how this composition and variation affects the clams. They describe, in rich detail, the smell of the mudflat and the sounds that the mud makes as they wiggle their hoes into the mixture of rock, clay, sand, silt, shell, and organic matter. Some note changes in the numbers of clams, the thickness of the

Fig. 4.2 Baby clam shells, some dead, scattered just below the surface of a mudflat. Photo by Bridie McGreavy

shells, and the taste and texture of clam meat. Others have seen entire mudflats die, baby clams the size of fingernails littered on the surface of gray mud (Fig. 4.2).

In these daily habits that, for many, stretch across a lifetime, a clammer "thinks with his body, not in the abstract but in lived temporality" where he "observes and retains smells as traces that mark out rhythms. He garbs himself in this tissue of the lived, of the everyday."[13] By observing clammers' sense-making, I have learned how metaphors trace these rhythms as *poesis*, allowing them to make the world through memory, a process of remembering and forgetting. I analyze an archive of texts produced within rhetorical ethnographic fieldwork from 2011 through 2016 to show that what and how we remember shapes what the world becomes.[14] The poetic work of metaphors is multifaceted, so I begin by exploring how biological metaphors trace organic rhythms in ways that articulate bodies and ecologies in the present. I then turn to cycles as a specific type of metaphor, a recognizable trope that folds multiple realities in a single figuration.[15] Cycles help us remember ourselves as ecological in ways that extend far beyond immediate sense impressions. I conclude with reflections on how poetry and biological metaphors may shape how we remember resilience, making our way with cycles of grief and hope through these *troubled times*.[16]

Crisis, Ecology, and Poetry

Crisis is a feeling of disruption in patterns and rhythms, marking turning points and places for crossing over to new orderings and histories. Donna Haraway argues that the crises of climate change, species loss, and systemic injustices that serve as material-discursive markers of the Anthropocene are such a disruptive transition. For her, the Anthropocene is not an era but a boundary event where "what comes after will not be like what came before," where the aim "is to make the Anthropocene as short/thin as possible to cultivate with each other in every way imaginable epochs to come that can replenish refuge."[17] She proposes Chthulucene as one name among many possible for this era. The impulse in naming is to remind us that "It matters which stories tell stories, which concepts think concepts" and "we need stories (and theories) that are just big enough to gather up the complexities and keep the edges open and greedy for surprising new and old connections."[18] She advocates for sympoeisis, a standpoint for making refuges—a concept in many ways analogous to resilience[19]—in ways that trace and (re)make chthonic rhythms.

Rhetoric can help trace a path out of the Anthropocene and into refuge by remembering, in new ways, the relationships between ecology as *oikos*, home; crisis as *krisis*, turning; and poetry as *poesis*, making together. Rhetorical scholars have slow danced with poetics in ways that connect with and enrich Haraway's sympoeisis and may help with this work of making refuge in a world in flux. As a rhetorical scholar with a distinctly ecological sensibility, Kenneth Burke offers an initial foothold for making a new, more liveable home at such a turning point.[20] Burke's writing was deeply influenced by the science of ecology and the crippling drought that gave rise to the 1930s' Dust Bowl.[21] In response to the ecological imbalances caused by rationalism and the drive for efficiency characteristic of the growing neoliberal paradigm, Burke turned to poetry because, for him, poetry offers a corrective by helping us feel ecological imbalances in new ways.[22] Poetry heightens a sense of the "revisions made necessary by the nature of the world itself" so that "all universe building is ethical"[23] in an ontological sense.[24] Working from the example of a potter molding her clay, Burke argues that poetry enacts a "loyalty to the sources of our being."[25] The potter and the clay become themselves through their relationship to each other and the world. This loyalty, expressed through poetry, is an orientation to becoming together where life is created "like the writing of a poem, though some people write their poems on paper, and others carve theirs out of jugular veins" and still others shape theirs from mud in various ways.[26]

In this approach, symbols are modes of action where associated meanings "serve to 'expand' the symbol." Susanne Langer treats such symbol use as an imaginative art, whereby artistic expression occurs in how the artist creates a "virtual experience."[27] The virtual experience forms when symbols, such as metaphors, (re)create "the rhythms of life, organic, emotional and mental...which are not simply periodic, but endlessly complex, and sensitive to every sort of influence. All together, they compose the dynamic pattern of feeling."[28] The creation of a virtual experience, ecologically shaped, is "the principle of poesis."[29] The critical questions for poetry then go beyond the intended meanings or the feelings a poet wanted to produce to ask instead:

> What has the poet made, and how did he [sic] make it? He has made an illusion, as complete and immediate as the illusion of space created by a few strokes on paper, the time dimension in a melody, the play of powers set up by a dancer's first gesture...But what he creates is not an arrangement of words, for words are only his materials, *out of which* he makes his poetic elements. The elements are what he deploys and balances, spreads out or intensifies or builds up, to make a poem.[30]

The emphasis in poesis shifts from the symbols themselves as discrete entities to the relationships between the symbols and to what the symbols create for those who interact with them. This is why it matters what stories tell stories and what thoughts think thoughts, in Haraway's sense. In poetry, the symbols create a virtual experience in the ways that they are strung together as a whole resembling the "*intricacy of a group dance*"[31] or, possibly, the materialization of a world with a distinctly ecological pulse.

Seeking to inspire such a creative dance with language in ways that attend to ecological meaning-making, Burke advocates for the use of biological metaphors because "a point of view biologically rooted seems to be as near to 'rock bottom' as human thought could take us."[32] This rock bottom exists somewhere beyond cognition and inside affect. It is an intuitive place where we register the syn-aesthetic connections between body and world. Biological metaphors enhance our ability to feel the "liquidity in the description of an event."[33] Yet, using biological metaphors is more than turning to a simple trope to acknowledge and understand inherent connectedness in the universe. Through poetics, metaphors express a patterned virtual experience, as the ways in which biological metaphors trace organic patterns provides relative consistency in apprehension and directed motion.[34] For example, notice what you apprehend when you read or speak this phrase: The heartbeat of the Earth drums in our bones. Though different for every instantiation, there is a consistency in the affective response to this phrase due to the rhythms that we as ecological, transcorporeal entities perceive.

Langer explains this phenomenon, saying that as "organisms, all our actions develop in organic fashion, and our feelings as well as our physical acts have an essentially metabolic pattern. Systole, diastole; making, unmaking; crescendo, diminuendo. Sustaining, sometimes, but never for indefinite lengths; life, death."[35] As poetry, biological metaphors create ways of relating to the world shaped by these rhythms that are necessary and available to sustain organic and rhythmic patterns. Biological metaphors are forms that interrelate with our sense perceptions through which we are drawn to actual experiences within myriad relations. For Alfred North Whitehead, biological metaphors as distinct forms of qualitative and spatio-temporal patterning "are indifferent to emotion, being just themselves, namely the vivid realization of things capable of abstraction from that instance of actuality with its cargo of emotion."[36] Approached in this way, language, like the written word "heartbeat" is held as distinct from the thought "heartbeat." Language is the systemization of expression and thought, the metronome through which rhythm connects experience with pattern and expression.[37] Poetry then weaves thought and language together to create the space where they co-mingle as virtual, ecological experience.

Thus, biological metaphor orients movement in ways that heighten rhythmic and chthonic coherence. Biological metaphors turn our attention to ecology as ontology and ethic and, when paired with poetics, can guide how we make the world so attuned. As a performance of this orientation to metaphor and poetry I invite readers to take a moment to try the following imaginative exercise:

> Imagine, for a moment, the oceans breathing with deep watery inhales and exhales. Pause, and take a breath as you do. Breathe again and this time the oceans inhale tiny silvery fry (baby salmon) from the thin, fibrous branches of its river veins into vast plankton-filled lungs. In a long, slow, rushing exhale, they release fattened adult salmon back into rivers to spawn. This fish-river-ocean breath-rhythm pulses, pulses, and has done so for millennia. Inhale. Pause. Exhale. We step onto a mudflat adjacent to the river, straining our eyes to see into the dark empty water, waiting for the breath to rush in. No fish. Our lungs are empty. We try to draw oxygen into our lungs but the air is heavy and we labor in its emptiness. We pull an inhaler out of our coat pocket and suck wind on the few fish that pass by in a spawning season. Breath-less river, we gasp.

This narrative illustrates affective rhythm that can accompany biological metaphors and an ecological ethic shaped by what the world itself conditions. Writing and then reading this passage, I feel the extinction of Atlantic salmon as a bodily register. This recognition occurs by attending to the synesthetic rhythms connecting earthly bodies, like the steady flow of breathing, river runs, and fish migration. In this strategic use of symbols, I want to call forth specific meanings about rivers, fish, and the consequences of choices about how we relate to the environment. But more to the point, this poetic narrative intends a shift from meaning to affect in a way that "reopens the interval between sense and sense-making, and multiplies the sensory dimensions of acting in the world and the milieux of inter-corporeal movement."[38] The sensing and sense-making, the virtual experiences created in the interaction with these symbols, depend on one's own ecological entanglements. It matters if you, as a reader, have stood on a river bank, looking into the dark water, waiting for movement. It matters if air quality or genetic legacy gives you a corporeal awareness of asthma. It matters, too, how you performed the above reading. Did you pause for breath or not, and how did this action come to matter?

Remembering Intertidal Rhythms

The ocean breath above and the biological metaphors I describe in depth below approach metaphor as movement, a connotation that connects to metaphor's lunar etymology of passing phases of the moon and to the turning action that tropes have long performed.[39] Working with metaphors as movement requires that we consider the material-discursive configurations by which such metaphors become "consolidated by the disciplinary processes and tactics that help us sediment [it] into the soft tissue of cultural life."[40] For example, Atlantic salmon are an endangered species. As metaphor, the salmon swims into policy realms, where standing on the banks of an empty river searching for salmon becomes a different kind of a thing for this discursive fact. The metaphor "salmon" is not the word salmon but the movement that guides action. It is a movement that articulates, moment by moment, a host of materialities into a recognizable pattern, one that may result in more salmon in more rivers as the threads tangle together through time.[41] Or it might not, depending on the specific configurations of discourses and materials as they come together. Salmon swim at the edge of memory, that tenuous rivulet that runs between remembering and forgetting; the couple of thousand that remain in the Penobscot River show how fragile and finite that edge can be.

Biological metaphors can attune us to rhythms and patterned movements with a unique potential for directing poetic awareness of *kairotic* opportunities to maintain or change habitual ways of living. Jennifer Lin LeMesurier demonstrates how *somatic metaphors*, or specific types of patterned movements through bodies, express materiality within a complex territory that emerges at the confluence of materials and discursive practices. She explores how dancers remember dance as the repetition of their previous training, called forth to shape their performance of the present moment. Metaphors participate in shaping the present, helping to gather threads into a recognizable pattern or set of actions. A dance teacher in a studio instructing a student to open her arm like a rusty door on a hinge is creating a world in which dancers remember movement at the same time they forget themselves as sedentary, untrained, or otherwise stationary beings.[42]

There is thus unique potential in poetry and biological metaphor for ecological attunement; yet the question of how we remember our way through change remains. A recursive or, for clammers, tidal approach to rhetoric offers one response.[43] Metaphor, and by extension rhetoric, "expresses materiality; it is a way of incarnating ourselves in and with the world through practices we call 'discourse'."[44] Memory as conceived as the practices through which we bring the past into the present becomes a performance shaped by complex relationships between matter and bodies, remembering and forgetting. As memory practice, metaphor is not an empty vessel waiting to be filled with meaning. Instead metaphors are dynamic movements for emplacing the present.[45] Nathan Stormer refers to this dynamism as recursion: "a process in which the present recurrently encounters the past. Rhetoric's materiality, be it speech, spectatorship, custom, research, you name it, has discursive *and* recursive qualities that establish adaptable relations between places then, now, and later."[46] Remembering occurs through a set of entangled practices that shape the present into a recognizable form, retaining some patterns while letting go of others. Poetic approaches to metaphor and memory strive to create "a symbol of feeling, not by recalling objects which would elicit the feeling itself, but by weaving a pattern of words—words charged with meaning and colored by literary [and rhythmic] associations akin to the dynamic pattern of feeling."[47] Poesis is an imaginative exercise in which metaphors can connect with patterned, organic, and remembered occasions in ways that open up new articulations of the present.[48]

In sum, this chapter brings together poesis, metaphor, and memory to shape a critical analysis of how clammers working within intertidal ecosystems respond to change. In scientific approaches to resilience thinking, clammers' memories allow them to draw on the past to guide how they recognize and act in the present. A close look at the metaphors they are using to remember the past shows that memory is more than storage and symbols are more than vessels. A poetic approach to metaphor and memory creates standpoints for considering rhythmic relationships between experience, symbols, and identification of and responsiveness to, in this case, environmental change. As I demonstrate in the sections that follow, my work with clammers has helped me find poesis at the intertidal, where ocean and land intermingle, and where metaphors guide us to remember in complex, unexpected, and potentially transformative ways.

An Ecological Rhetorical Ethnography

My rhetorical ethnographic research with clammers began in the middle of things.[49] I was in the middle of another research project with a group called the Frenchman Bay Partners. Through this engaged research with the Partners, I began to work in 2011 with the Frenchman Bay Regional Shellfish committee, a group of about 80 commercial clammers who had organized into a seven-town cooperative to manage the intertidal mudflat ecosystem. Initially, my role with this group was to strengthen the relationship between the Partners and the shellfish committee. I did this by showing up at and participating in their committee meetings, writing grants with them to support their conservation and management efforts, providing recommendations from the clammers to inform the conservation action plan the Partners were advancing, and volunteering when they needed help with mudflat monitoring and restoration.

At the specific moment when this current project emerged, I was helping move clams from one mudflat to another where they would presumably grow better. In addition to being in the middle of another research project, I was in the middle of a tide as it was flowing back into Raccoon Cove. I was also sitting in the middle of a canoe. As we paddled through the middle of the cove, one of the clammers described a night when he was paddling in Raccoon Cove during a full moon. He described how every time he dipped his paddle, the water exploded with shimmering phosphorescent plankton. He went on to note that these types of experiences were what he loved most about clamming. Up to that point, I had been asking questions about municipal management, conservation issues, and information needs. The phosphorescent tide made me want to know more about digging, materiality, and memories, for what I might also learn about resilience and sustainability.

This research began with the commitment to occupy "the space of live rhetorical action" and participate as "potential actors in emancipatory practice, not simply critics of interventions already performed."[50] I attended monthly town hall meetings, where fishermen persuade each other about how to manage intertidal mudflats for sustainability.[51] I also go out on the mudflat with clammers to observe and help count baby clams in their annual resource surveys, plant clams, and install nets to keep predators like green crabs and gulls from eating what we sow (Fig. 4.3). As a rhetorical ethnographer, I seek to "engage in direct participation inside invention and advocacy."[52] This research is a form of rhetorical action that can only be accessed through direct participation in Frenchman Bay as a particular location.[53] Being in the town hall and going out on the mudflat, again and again, matters for what I notice and describe through embodied critical analysis.

My fieldwork was based on showing up at a host of events as often as I could and actively and critically reflecting on how my participation shaped the context. I also interviewed twenty-three clammers between December 2014 and October 2016, asking about how they got into clamming, how it makes them feel, the types of tools they use, memories about digging, what they think about when they dig, and their perceptions of change. My time in the field created different types of *texts*

Fig. 4.3 College students volunteer to help with the clam recruitment census in Bar Harbor, Maine as part of the town's marine resources committee mudflat monitoring and restoration program. Photo by Bridie McGreavy

as artifacts. From audio recorded interviews, I produced and analyzed verbatim transcripts. I took extensive notes during shellfish committee meetings, and written reflections following fieldwork and interviews. I took photos and recorded audio and video of clamming. My interpretation of these texts was shaped by the mud, the tides, the clammers, and more, in ways that exceed full awareness but that rely on belonging to a field site and showing up sensitive to ecology, metaphor, and kairos.[54] In the following sections, I begin with a narrative drawn from my fieldwork that demonstrates how clammers use biological metaphors to shape their movements in response to change, creating unique patterns of remembering and forgetting. The opening narrative helps us ask of this clammer-poet: what does he make with metaphor? I then elaborate his poesis to show how cycles move clammers in ways that may help guide how we all cross through this transition zone.

Feeling Edges, Counting Lifelines, and Tracing Change

We stood in the dark, small clam shop on a cold December evening. Tom, a clammer who has been working the tides in coastal Maine for 45 years, began to describe the changes he is noticing in the relative "scarcity" and size of the clams.[55] With the help of hand gestures and actual clam shells, he remembered the biggest clam he ever pulled: it was seven and a quarter inches long and weighed eleven ounces. He describes the moment he pulled it out of the mud, how his finger wrapped around the bottom half of the shell but did not touch the top. Feeling the clam before seeing it, he said to himself:

> "It cannot be that big."... It was at the fifteen-foot level. The thirteen-foot tide that went down into a valley of a channel and it was on the edge of that mud ravine. I saw this hole. It was like this [shows opening with hands]. I said, "Oh no. It's a lobster hole." I found lobsters in holes before, pulling clams. I said, "It's gotta be a lobster hole." But it wasn't.

He was digging on a thirteen-foot tide, an unusually big tide for this part of the coast where the average difference between high and low tide is eleven feet. He saw the hole and waded into the water to slip his hand into the mud, feeling the massive clam as he did.

He went on to wonder about the age of the clam and then showed me two quahogs he dated to about 120 years, found in a similar way (Fig. 4.4). He knew they were that old by counting the lifelines. Much like counting the rings of a tree, the raised ridges on the shell tell the age of a clam. He pulled out a hand lens and we counted them as we traced the raised ridges on the shell with our fingernails, edge by edge. I asked where he found them and pointing to a distant cove outside the clam shop he replied:

> There in Goose Cove. Normally what I've heard the quahogs. Of course the Arctic quahogs have been dated in excess of 400 years old, but far as I knew these [quahogs] could get to 50. But the lines don't lie...I found them bigger, but not with that amount of lifelines. I've got a monster in the house that's bigger than those, but it's only dated around 38 years. These ones were in the rocks where they're safer and they're impervious to anything at that point. They have no predators. No crab's gonna put a hole in that baby. Ice isn't going to affect them. So with their low metabolism, they can live indefinitely.

Fig. 4.4 The dense lifelines of a quahog that may have been more than 100 years old when it was harvested in Goose Cove, Frenchman Bay. Photo by Bridie McGreavy

Tom's description of being drawn to the low tide margin, slipping his hand into the mud, feeling the hefty weight of a clam, and tracing the ridged lifelines to determine the age illustrates how clammers live at the intertidal edge, as they find and feel clams and make sense of changes they witness. This brief but thick description illustrates how clamming as a practice and livelihood is shaped with and within intertidal ecosystems.[56] Clamming occurs in that ecotonal space Rachel Carson characterized as fluid and mysterious, and where the past and present coalesce in immediate sense impressions that are materially, discursively shaped.[57] In his remembering, Tom was drawn to the watery edge just beyond the intertidal margin. Searching with his hand, he found a clam whose lifeline was much longer than his own. He traced this pattern by running his finger over the raised ridges of the shell counting the cycles of seasons, tides, and calcium carbonate fluxes as he did.

The way in which Tom remembers his encounter with the massive clam shows how metaphors as symbols can connect with systems of meaning, as slippery, unstable, and diffuse as those meanings may be. For example, the set of possible meanings that articulate *shell* is shaped within the constraints of this chapter, the English language, childhood experiences of listening to shellbound oceans, and so on. This range may be more narrowly circumscribed when the written symbol is articulated with a photograph (Fig. 4.4). The spoken and then written narrative are certainly part of the performance, but a poetic approach to biological metaphor emphasizes how meaning and memories are made recursively and ecologically. It matters that, as part of this instantiation of the present, Tom left me alone in the cold dark clam shop to go into his house to retrieve the clam. It matters that he brought a hand lens and then showed me how to count the thin lifelines by tracing his fingernail along the shell's ridges. It matters that we moved together and with a host of materialities to create this present, one in which the life of clams is articulated with ours but where their life cycles extend far beyond ours too. Tracing the lines of the shell with his hand creates a virtual experience where this present carries traces of mudflats that once contained monster clams.

A contemporary neologism in ecology, shifting baselines, describes how our memory of past ecosystems is inaccurate because we don't have good records of species size and abundance.[58] From a rhetorical perspective, we only ever have shifting baselines, as memory is fluid and the past is never recaptured but instead recursively instantiated, always different for how it is shaped ecologically. In this account, Tom creates a sense of mudflats that contain hidden clams lurking in their depths. In this performance, he remembers clamflats by selecting from a range of possible elements to create a sense of the present. Turns out, there were no lobsters in that hole. More to the point, he tells this story to create a relationship to clams lurking in the future, forgotten before they have had a chance to form. In the next section, I more fully consider this poetic shaping of past, present, and future with biological metaphor.

Sensing Change: Rhythm, Memory, and Metaphor

Tom is an active member of the clamming community in Frenchman Bay and holds licenses in multiple towns. Though he does a lot of other things, like caretaking properties, fishing for baby eels during spring migrations into streams, wild mussel harvesting, hunting, and foraging for mushrooms, he primarily identifies himself as a clammer with

Mic Mac and Penobscot ancestry. The island on which he now lives was inhabited by Native Americans for centuries before European colonization. The heaping piles of clam shells they left are evidence of their sustained dwelling in this *clam digging place*.[59] In the two hundred years since the Penobscot Nation was displaced from its ancestral territory and coastal, riverine, and forest ecosystems across the United States succumbed to the aggressive material forgetting that marks the Industrial Revolution, the pH of ocean surface waters has fallen 0.1 units, from approximately 8.2 to 8.1.[60] This may not sound like a big change but because the pH scale is logarithmic, this represents approximately a 30% increase in acidity. Acidification is the oceanic manifestation of the consequences of pumping massive quantities of carbon dioxide into the atmosphere, which the ocean absorbs and transforms into carbonic acid. This is a weak acid that can have a dramatic impact on organisms, like shellfish, plankton, and coral that draw calcium carbonate from the water to produce their skeletons or shells.

Having heard about this trend on the news and through his participation in his town's shellfish committee, Tom described the acid issue as "a canary in a gold mine." He went on:

> That's the one that's killing the clams and I have no control over. None. Zero. Unless we did drastic changes to pollution on this planet, it is not gonna go backwards. So the clam population, I don't see coming back. I've seen a half a million baby clams just drop dead in one three-month period. I watched it physically happen, with no sign of any damage to the clam shells. None. There was no reason for those clams to die. There had to have been something else going on.

When he first heard about ocean acidification he felt he had found the, "smoking gun. This is why my clams—Thomas Bay out here right behind us—went dead. I pulled clams there all my life, pulling the clams one at a time in the mud; soupy, best mud around. Just dropped dead."

After describing the death of a clam flat, Tom compared changes in pH to our body temperature:

> [Our bodies] can run at 98.6. That's where it should be, but you rise up to 99 for a full year, you're gonna get sick. It's that sensitive. People who aren't out there, they don't observe like that. It's so sensitive.

Here, biological metaphor guides the remembered resonance between bodily temperature regulation and an Earth that also relies on dynamic thermal and material balances. This practice recurs throughout the clamming community, as clammers invoke biological metaphors to remember corporeality and change in particular ways. As clammers encounter these differences, they talk about noticing changes that are, as one says, "tough for the eye to handle." Tom's friend Ron demonstrates this in a related way when he describes how he notices change in the mudflats from being out there every day, "When your head's in that mud, and if you are aware. And it's like Tom and I. He said 'Ron, what's going on with this mud? I said, 'I don't know. Somebody cooked up a bad pea soup today.'"

Similarly concerned about ecological balance and limits, Matt, a clammer who works the tides on the other side of Frenchman Bay, shows this in the following passage where he attempts to convey a sense of the dire consequences for going beyond limits, especially as they relate to climate change. He says:

Matt: Bottom line is I worry about global warming. If the sea rises too high or acid rain makes it so they can't form shell, I'm out of business. I see that coming. Probably, maybe not in my lifetime, but they're saying that the ice caps ain't gonna melt in my lifetime, and I don't believe that for a minute. I think this place is gonna heat up a whole lot quicker than they say, and they're gonna all be playing musical chairs here in probably another 20 years.

Bridie: What do you mean by playing musical chairs?

Matt: Seven billion people playing musical chairs, maybe eight billion by then, and that ain't gonna be pretty. And as far as the clams go, well they won't ... Let's say if the northern hemisphere, if all the ice melts, you've got 20–30 feet of sea rise, and if the Antarctic melts, you've got 200 feet. So at 20, 30 feet, that's like half the world's population lives here, if not more. And if it melts the Antarctic ice cap, then you're talking the whole place is screwed.

Approaching this passage as poetry recognizes that the description of playing musical chairs, at the level of language, matters for the specific material-discursive configuration of meaning. But the movement extends far beyond the symbol. Have you, the reader, played musical chairs? What virtual experience did this evoke when Matt described our future through this childhood game? My virtual experience includes a musty church basement. I'm five or six and I am watching a group of children, but I don't know who they are. The light is dim, coming in from windows high on the walls. I am standing to the front and center of the memory frame and the chairs are wooden. I feel left out, but I am also not playing and I don't lose a chair. This scene as I remember it with Matt never happened. It is a recursive performance of memory that is ecologically shaped by previous experiences that coalesce from immediate sense impressions that are further figured by where and how I conducted this interview and the relationship I have with Matt. I am not seeking causal explanations for why he told this story to me in this way nor for the virtual experience it evokes for me. I am more interested in ecological attributions of this dynamic shaping, its immediate effects, and how the pattern of remembering ourselves ecological (which also means remembering that, as biological beings, the music will eventually stop) recurs through time and space.

Much like the narrative of the river and salmon that introduced how biological metaphor conjures fluidity in the description of the event, it matters if and how you have played musical chairs, or felt that tug when you no longer have a place to sit. It matters if you have smelled bad pea soup or an analogous mixture of cooking gone wrong that hangs on the inside of your nostrils and settles at the back of your throat. These feelings are conditioned by repeat action that dynamic forms, such as biological metaphors, have carved out in time and space. This tracing through metaphor offers a distinct approach to rhythm. Debra Hawhee, channeling Heraclitus, describes such rhythm as a relationship between a river's path, "with its interruptive rocks and sediment, on the one hand, and the force of the water's current, on the other, that produces the eddies and swirls, the sudden shifts in direction within the general flow—herein lies the rhythm." In this sense rhythm, "produces distinctive movements within a generalized direction; it combines fixity with variability. Put simply, rhythm emerges from difference."[61] It is not just that the mud smells like someone has cooked bad pea soup; it is that it smells that way and that smell is different from what has been perceived before. In this material tracing, the recursive performance occurs in how the present is felt, placed, and remembered differently.[62]

Clearly, biological metaphors are not simply the words we use to characterize change. They are instead apparatuses between body and world, movements, and phenomena that attune us to rhythm, difference, and patterns.[63] The next section takes this further by looking at how cycles, as a particular type of apparatus, turn between body and world, complicating immediate sense impressions as we forget bodies and remember ourselves as ecological.

Cycles as Poetic Crossing: Forgetting Bodies and Remembering Worlds

Shellfishing is a patterned relationship between the moon, earth, oceans, tides, mud, clams, and clammers' bodies. Frequently, clammers describe the connection between the physical practices of harvesting, their sense impressions, and their co-habitation in ecosystems as what they like most about this livelihood. They say, "It's nice to just get to the shore and just leave it behind, you know? Hot day, there's usually cool breezes. The smell of that moist, fresh, salty air. Seagulls screaming in the background. It's very therapeutic, you know?" Clamming connects bodies within ecologies in visceral, mud-covered ways where, "You're stretching your body all over" and "the main part is just being outside. It makes you feel good." There is an opening, a letting go that occurs in the mudflat, as another clammer describes: "[It is] the only place I feel calm. Inside. I don't like being inside. Too many thoughts go on. When I'm out here, I'm just blank. Everything goes away."

Clammers routinely describe how, when they are on the mudflat, the dichotomies between body and mind become porous and a distinct rhythm takes over. It starts as the boots slip across the mud and when the hoe makes its first flip:

> First you start to think about clamming and the tide. You pay attention to what you have to do with work, but once you get in a *rhythm*, you think about life. You solve your problems out, and hopefully things are going well. But I've had some times in the flats when life wasn't going so good. Just try to figure out what's going on. Keep working and just a lot of times you get into clams and start thinking about, "Where are these clams going? Do I go left? Do I go right?" Just try to put it all together. The tide's the big thing that you think about. "How much time do I have left?" And you're always watching because that tide is fairly inconsistent sometimes.

Another describes this dynamic interplay as his:

> peace of mind. Because when I'm out there—to attempt to describe it, I mean in the summertime when that sun is just coming up and it's the heat and the sun and it warms up 15 degrees within 10 seconds and nobody's around—because in the summer the big tides are in the morning.

These quotes show the ecological, rhythmic quality of clammers' lives within the intertidal where the practices of digging physically articulate a connection between body, mud, and tide. The *peace of mind* occurs through movement as clammers arrive in the intertidal and start digging in the mud, searching for clams, remembering themselves as clammers with each soft thunk as the clams fill up their hods. I argue that this peace of mind is a mode of equanimity that creates the space for remembering change in ways that open up movements across and into new realities; I return to this point in the conclusion.

This practiced way of living uniquely situates clammers for noticing patterns and making sense of change in ways that are loyal to how they become themselves in relation to the world. Taking up Burke's notion of metabiology, Bryan Crable explains that "As embodied language-users, we draw linkages and develop orientations, learn to socialize our actions (or learn a vocabulary of motivations), and select means to combat problematic situations."[64] Recall the clammer who described how grounding himself in the mud and finding that space between thought and action became the means through which he dealt with his problems. This orientation transcends, though of course still includes, a focus on symbolicity to also attend more fully to ecological relations that connect embodied, observable experiences and systems of expression. Our system that guides our selections of means, interpretations, and resulting choices about how to act "is therefore not a tissue of lies and mystifications that unconsciously works to reproduce a mode of production, but the symbol-systems and orientations that are grounded in our metabiology."[65]

This analysis has, so far, described how biological metaphors shape clammers' movements between past and present to create a sense of change by tracing, through language, difference in pattern and rhythm. Yet there is a distinct form of biological metaphor that shapes how and what they remember in complex and unexpected ways, as evident in the following narrative in which Greg, a clammer who has been working the tides his entire life, orients to change through the metaphor of cycles:

Greg: Yeah. I don't see it. I see Mother Nature just doing cycles. We have good cold winters and warm winters. We don't see the effect around here, like Iceland and them places, I'm sure.

Bridie: So you think climate change may be happening in other places?

Greg: Oh, I'm sure it is. We're losing a lot of the Arctic and all that. So, yeah, I'm sure. Mother Nature's changing.

Bridie: And do you think we're to blame for that? Do you think it's [human caused]?

Greg: Really, no. I'm sure we ain't helping our environment, but I think Mother Nature just has a general cycle, and she's running through one. Population control. *[Laughs]*

The move to cycles marks a shift from bodily knowing, witnessing difference and patterns on the mudflat, to ecological knowing, tracing knowledge away from and outside of immediate sense impression. This move is a means for expressing conflicted feelings about uncertainty, grief, and hope. From a poetic standpoint, cycle metaphors connect with Susanne Langer's view that "Every good work of art has, I think, something that may be said to come from the world, and that bespeaks the artist's own feeling about life."[66] For Greg, this metaphor traces uncertainty and a movement beyond immediate sense impressions. For other clammers, cycles appear to create a virtual experience of hope, a feeling that may accompany not knowing to help maintain a commitment to a different future: "To me that was like the beginning of really seeing all that mud and everything just totally disappear... And that's why I'm hoping that that cycle's over, and after seeing the re-growth of the new clams that's set in this summer and this fall."

Cycle metaphors trace rhythms between bodies, ecosystems, and the physical universe in distinct ways, as Henri Lefebvre shows:

> movements, undulations, vibrations, returns and rotations are innumerable, from the microscopic to the astronomical, from molecules to galaxies, passing through the beatings of the heart, the blinking of the eyelids and breathing, the alternations of days and nights, months and seasons and so on.[67]

We live by cycles as material movements, these recursive patterns and repetitions that break up discursive constructions of linear time at the moment when day transitions into night, when the first snow falls, or when we feel the complete and final exhale of another living being. These cycles operate at the scale of the body, on an earthbound register. The etymology of cycle refers to periodicity and recurrence, and the use of cycle metaphors is influenced by clammers' daily habits of digging clams, where they sense and live by cyclical rhythms in seasons, tides, in their bodies, and by scientific discourses. When cycle is articulated to bodies it guides how we recognize the periodic and recurrent patterns we experience in every moment of our biological being. Our breath moves in a cycle, in and out of lungs. Our blood cycles continuously in and out of our hearts and uteruses. Our food and nutrients cycle from soil, to plate, to stomachs, and back into the Earth. At the same time, we forget these cycles. In our *practices of everyday living*, these cycles are at once intimate and obscured, remembered and forgotten, as two sides of the same figure.[68] We remember the movement of oxygen through trees and lungs and water, through rivers and veins, not directly through bodily sensing but instead through ecology as an organized set of scientific practices. Remembering distant cycles through ecology obscures and may encourage us to forget the more proximate cycles of the body.

Stormer's genealogy of bodily pathology demonstrates how tropes, such as metaphors, can guide these complex and contradictory movements, for example, "transforming knowledge *in* the body to knowledge *of* the body in the same motion. A combined trope like this is called a metalepsis in the Greek" and it is "the means by which tropes that work conjointly, a reversal and a substitution in this case, creat[ing] a passageway from one place to another even though the conceptual distance may seem far."[69] Metalepsis describes the ontological turns that occur through combined tropes and how metaphors guide kairotic and ecological attunements to multiple and different realities. Coming at this symbolic phenomenon through the study of poetry, Edward Hirsch similarly describes this move as a poetic crossing where, within a poem, it becomes possible to move:

> from one plane of reality to another…A poetic crossing, which follows the arc from physical motion to spiritual action, requires the blacking out of the quotidian world and the entrance into another type of consciousness, a more heightened reality. It is a move beyond the temporal, a visionary passage.[70]

Poet Octavio Paz, according to Hirsch, is particularly adept at this kind of crossing. Paz's poetry creates new spaces for thought, feeling and action and reconfigures a sense of the body in relation to the world. What matters, for Paz and Hirsch, as well as for Langer, Burke, and Lefebvre, is not the specific symbol or intended meaning when a symbol is invoked, but conditioned rhythm and what rhythm does for organizing the present. Rhythm is attunement to myriad presences, similar to how memory occurs as recursive articulations that connect with and expand patterns of material-discursive meanings. Referring to the organismal desire to maintain particular patterns,[71] like those that occur at intertidal edges, Paz says "In order to save *it* we must *convert it* to rhythm."[72]

Cycles comprise a complex metaphor that may guide this conversion and, as poesis, shape how we remember and forget. Let us return again to the breath. Our breath cycles in and out of our body, a rhythm we easily forget until we are asked to direct attention to it, or until we struggle to remain loyal to it at the edge of death. It is arguably the most proximate pattern of repetition for all breathing bodies. At a distinct and yet immeasurable distance from breath in our lungs, salmon similarly pulse in and out of rivers in migratory cycles, and clammers pulse in and out of mudflats at regular tidal cycles. Approaching cycle as biological metaphor, we can begin to recognize the affective sense of loyalty to cycles of breath and water. We experience the proximity of these cycles with every inhale. Attempting to feel rivers and salmon through the cycles of breath demonstrates this affective proximity and how memory trembles at the edge of remembering and forgetting.

However, our loyalty necessarily extends far beyond our immediate bodily rhythms and sense impressions because the materials that compose the air in our lungs and the water in our veins are loyal to us for brief moments before they travel other and more distant paths. The science of ecology, as a set of rule-bound practices, is an attempt to follow air and water on these paths and organizes our sense of the patterns in distinct ways.[73] Religion and spirituality comprise other ways of tracing these more distant relations between body and world. Are these paths water cycle or reincarnation? Ecology as poetic science organizes our loyalty and ability to remain faithful to these more distant places beyond the fluid boundaries of our biological corporeality. As metalepses, cycles open a passageway from the body into broader spatiality and, in doing so, create ontological shifts where body, science, and spirituality comprise different realities organized around a shared trope.

Pausing at the Intertidal Edge: Finding Refuge in Grief

Returning to the opening discussion of amplified change, crisis, and how the Anthropocene marks disruption and not an era, Haraway notes that "Change on earth is not the problem; rates and distributions of change are very much the problem."[74] The problem is thus, in part, how we are living in this time of rapid and wide-ranging loss. We are currently forgetting on a scale we have not experienced as earthly bodies for 60 million years, when an asteroid forced a forgetting of much of life while it simultaneously remembered and gave rise to mammals. We may feel a sense of pain, physical and virtual, in this rapid dissolution because we encounter the loss as discordant rhythm materialized in diverse ways that pull at our lived and imagined realities. So, though this analysis of bodily metaphor may be read as a missive of hope, and may also provide rhetorical tactics to, for example, communicate about change in ways that tune an affective and ecological register, this is not the meaning I intend to make. This conclusion comes during a moment of deliberate political forgetting of climate change science and of refuge for kin of many kinds. In this intensification of current practices of forgetting "I find myself in the middle of an eye, / watching myself in its blank stare. / The moment scatters. Motionless /I stay and go: I am a pause."[75] In this moment of pause in the watery eye of despair, poetry and metaphor may open up collective grieving. Standing here, I am not alone.

When I met with Tom, the clammer searching at the intertidal edge, to share this chapter and get his feedback on it, he had just survived a two-month long shellfish closure due to a bloom of a phytoplankton species previously unknown to the Gulf of Maine. This plankton produces a biotoxin known as domoic acid that causes amnesic shellfish poisoning (ASP), so named because people who eat clams that have eaten the plankton can experience permanent memory loss. I mark the trope of irony, turning this turn, folding forgetting back on itself. I asked him how he was doing. He paused, turned to look at me, and quietly said: "Clamming's *dead*."

Staying with this kind of trouble means living with the hollowness carved out by presence made absent.[76] Finding and making refuge through poetry is not about feeling better or, as resilience thinking would have it, finding ways to cope, bounce back, and otherwise return to normal. Finding refuge in poetry is about tracing grief into a future where we don't yet know how to dwell. Grief creates "a path to understanding entangled shared living and dying; human beings must grieve *with*, because we are in and of this fabric of undoing. Without sustained remembrance, we cannot learn to live with ghosts and so cannot think."[77] Edward Hirsch, who gave us the poetic crossing above, expresses such grief in a book-length poem for Gabriel, the son he lost to a drug overdose. He describes how mourning is "like carrying a bag of cement / Up a mountain at night" where we take no comfort as we witness "Almost everyone carrying bags / Of cement on their shoulders."[78] We remember our corporeal attachments to each other and to life, feeling the weight of cement bearing down. Forgetting, the moment of letting go and leaning into indeterminacy, is what compels us to take another step in that darkness.

This path out of the Anthropocene, to find a home in more earthly, ecological, liveable terrain requires changing our relationship to hope in favor of equanimity: a radical, ecological homecoming that distributes our sense of self through space and time and that will, moment by moment, teach us how to grieve in ways that also teach us how to dwell. Remembering poetry at the intertidal edge expresses this radical, dynamic becoming, where grief, equanimity, and collective striving for a different future commingle. We, too, can find this place at the intertidal edge, dig our hands into the muddy unknown, and feel, in the words of Patti Ann Rogers, how the song of the world emerges at intertidal edges. We may tune to a single "molecule on the tip of one / tentacle of an anemone attached /to the rim of a low-tide rock, / and briefly balance there."[79] We can pause at this edge, feeling this trembling indeterminacy. So balanced, we find ways to stay with grief and learn to trust in the rhythm of the tides that run thick and slow in our chthonic veins.

Notes

1. Paul Crutzen, "Geology of Mankind," *Nature* 415, no. 6867 (2002): 23; Donna J. Haraway, *Staying With the Trouble: Making Kin in the Chthulucene*, (Durham: Duke University Press, 2016).
2. Johann Rockström, Will Steffen, Kevin Noone, Asa Persson, F. Stuart Chapin III, Eric F. Lambin, Timothy M. Lenton, Marten Scheffer, Carl Folke, Hans Joachim Schellnhuber, Björn Nykvist, Cynthia A. de Wit, Terry Hughes, Sander van der Leeuw, Henning Rodhe, Sverker Sörlin, Peter K. Snyder, Robert Costanza, Uno Svedin, Malin Falkenmark, Louise Karlberg, Robert W. Corell, Victoria J. Fabry, James Hansen,

Brian Walker, Diana Liverman, Katherine Richardson, Paul Crutzen, Jonathan A. Foley, "A Safe Operating Space for Humanity," *Nature* 461, no. 7263 (2009).

3. Elizabeth Kolbert, *The Sixth Extinction: An Unnatural History*, (A&C Black, 2014), 92–110.
4. Bridie McGreavy, "Resilience as Discourse," *Environmental Communication: A Journal of Nature and Culture* 10, no. 1 (2016).
5. Judith Butler, "What is Critique: An Essay on Foucault's Virtue," in *The Political: Readings in Continental Philosophy*, ed. D. Ingram. (London: Basil Blackwell, 2002); Steve Schwarze, "Environmental Communication as a Discipline *of* Crisis," *Environmental Communication: A Journal of Nature and Culture* 1, no. 1 (2007).
6. C.S. "Buzz" Holling, "Resilience and Stability of Ecological Systems," *Annual Review of Ecology and Systematics* 4 (1973).
7. Brian Walker and David Salt, *Resilience Practice: Building Capacity to Absorb Disturbance and Maintain Function* (Washington, DC: Island Press, 2012).
8. Carl Folke, Stephen Carpenter, Brian Walker, M. Scheffer, Terry Chapin, and Johann Rockström, "Resilience Thinking: Integrating Resilience, Adaptability and Transformability," *Ecology and Society* 15, no. 4 (2010).
9. Andew Zolli and Anne Healy, *Resilience: Why Things Bounce Back* (New York: Free Press, 2012), 1–23, 211–238.
10. F. Stuart Chapin, Carl Folke, and Gary P. Kofinas, "A Framework for Understanding Change," in *Principles of Ecosystem Stewardship*, eds. F. Stuart Chapin, Carl Folke and G.P. Kofinas (New York: Springer Science + Business Media, 2009), 3–28.
11. Aaron Hess, "Critical-Rhetorical Ethnography: Rethinking the Place and Process of Rhetoric," *Communication Studies*, 62, no. 2 (2011); Michael K. Middleton, Aaron Hess, Danielle Endres, and Samantha Senda-Cook, *Participatory Critical Rhetoric: Theoretical and Methodological Foundations for Studying Rhetoric* In Situ (Lanhman: Lexington Books, 2015), 1–210.
12. Rachel Carson, *Edge of the Sea* (New York: Houghton Mifflin, 1955), 36–39, 195.
13. Henri Lefebvre, *Rhythmanalysis: Space, Time and Everyday Life* (London: Bloomsbury Academic, 2004), 31.
14. Nathan Stormer, "Recursivity: A Working Paper on Rhetoric and Mnesis," *Quarterly Journal of Speech* 99, no. 1 (2013).
15. Ibid.
16. Haraway, *Staying*.
17. Ibid., 100.

18. Ibid., 101. Haraway notes that she learned about needing stories that remain open from Jim Clifford.
19. McGreavy, "Resilience," 104–121. Haraway's concept of seeking refuge responds to some of the issues with resilience as a discourse which strives to bounce back and maintain a status quo as opposed to finding refuge, which is about acknowledging the broader ecological material shaping of worlds.
20. Kenneth Burke, *Permanence and Change: An Anatomy of Purpose* (London: University of California Press, 1954).
21. Marika A. Seigel, "'One little fellow named Ecology': Ecological Rhetoric in Kenneth Burke's *Attitudes toward History*," *Rhetoric Review* 23, no. 4 (2004).
22. Ibid., 399.
23. Ibid., 250.
24. Burke, *Permanence*, 257, 250.
25. Ibid., 71.
26. Ibid., 76.
27. Susanne K. Langer, *Feeling and Form: A Theory of Art Developed from a Philosophy in a New Key* (London: Routledge, 1953).
28. Ibid., 241.
29. Ibid., 253.
30. Ibid., 211.
31. Ibid, 231.
32. Burke, *Permanence*, 261.
33. Ibid.
34. Alfred North Whitehead, *Process and Reality*, eds. David Ray Griffin and Donald W. Sherburne (New York: The Free Press, 1978), 18–20.
35. Langer, *Feeling*, 99.
36. Alfred North Whitehead, *Modes of Thought*, (New York: The Free Press, 1968), 73.
37. Ibid., 34.
38. Sarah Whatmore, "Materialist Returns: Practising Cultural Geography in and for a More-Than-Human World," *Cultural Geographies* 13, no. 4 (2006): 604.
39. Debra Hawhee, *Bodily Arts: Rhetoric and Athletics in Ancient Greece* (Austin: University of Texas Press, 2004), 82.
40. William E. Connolly, *The Fragility of Things: Self-organizing Processes, Neoliberal Fantasies, and Democratic Activism* (Durham: Duke University Press, 2013), 74.
41. Nathan Stormer, "Articulation: A Working Paper on Rhetoric and Taxis," *Quarterly Journal of Speech* 90, no. 3 (2004).
42. Jennifer Lin LeMesurier, "Somatic Metaphors: Embodied Recognition of Rhetorical Opportunities," *Rhetoric Review* 33, no. 4 (2014).

43. Stormer, "Recursivity," 50–55.
44. Ibid., 29.
45. Ibid., 54.
46. Ibid., 31.
47. Langer, *Feeling*, 230.
48. Whitehead, *Process,* 18–30.
49. Gilles Deleuze and Felix Guattari, *A Thousand Plateaus*, trans. Brian Massumi (London: University of Minnesota, 1987), 19–21.
50. George F. McHendry, Michael Middleton, Danielle Endres, Samantha Senda-Cook, M. O'Byrne, "Rhetorical Critic(ism)'s Body: Affect and Fieldwork on a Plane of Immanence," *Southern Communication Journal* 79, no. 4 (2014): 295.
51. Here I use the broader term fishermen to include mussel harvesters, worm diggers, and lobster fishermen, as some fishermen work in multiple fisheries or play an active role in managing the clam fishery even if they are not themselves clammers.
52. Hess, "Critical-Rhetorical," 129.
53. Danielle Endres and Samantha Senda-Cook, "Location Matters: The Rhetoric of Place in Protest," *Quarterly Journal of Speech* 97, no. 3, (2011).
54. Hess, "Critical-Rhetorical," 129, 133.
55. Names have been changed to maintain confidentiality.
56. Clifford Geertz, *The Interpretation of Cultures: Selected Essays* (New York: Basic Books, Inc., 1973), 6. Geertz notes that the concept of thick description is borrowed from Gilbert Ryle.
57. Carson, *Edge*, 37.
58. Daniel Pauley, "Anecdotes and the Shifting Baseline Syndrome of Fisheries," *Trends in Ecology and Evolution* 10, no. 10, (1995): 430.
59. Julia Clark and George Neptune, "People of the Dawnland," *Friends of Acadia Journal*, Spring 2014, accessed April 16, 2017. https://friendsofacadia.org/.
60. Peter M. Haugan and Helge Drange, "Effects of CO_2 on the Ocean Environment," *Energy Conversion and Management* 37, no. 6 (1996): 1019–1022.
61. Hawhee, *Bodily*, 142.
62. Stormer, "Recursivity," 54.
63. Karen Barad, *Meeting the Universe Halfway: Quantum Physics and the Entanglement of Matter and Meaning* (Durham: Duke University Press, 2007), 140–153.
64. Bryan Crable, "Ideology as 'Metabiology': Rereading Burke's Permanence and Change," *Quarterly Journal of Speech* 84, no. 3 (1998): 313.
65. Ibid., 313.
66. Langer, *Feeling*, 253

67. Lefebvre, *Rhythmanalysis*, 84.
68. Stormer, *Sign*, 51–52; Michel de Certeau, *The Capture of Speech and Other Political Writings*, trans. Tom Conley (Minneapolis: University of Minnesota Press, 1997), 129–132.
69. Stormer, *Sign*, 88.
70. Edward Hirsch, *A Poet's Glossary*, (New York: Houghton Miffflin Harcout, 2004), 471. I originally came across this reference to Edward Hirsch's poetic crossing in an interview with Terry Tempest Williams, *Yes Magazine*, accessed October 6, 2017. http://www.yesmagazine.org/issues/the-debt-issue/terry-tempest-williams-survival-becomes-a-spiritual-practice-20151005.
71. Alfred North Whitehead, *Process*, 18–344.
72. Edward Hirsch, "Octavio Paz: Words are Bridges," *Poet's Choice* (Orlando: Harcourt Books, 2006), 180. Hirsch is quoting Paz's *Alternating Current.*
73. Michel Foucault, *The Order of Things: An Archeology of the Human Sciences*, (New York: Random House, Inc., 1970), 128–138.
74. Haraway, *Staying*, 73.
75. Hirsch, "Poet's Choice," 182.
76. Haraway, *Staying*, 76.
77. Ibid., 39.
78. David Greene, "A Poet On Losing His Son: 'Before You Heal, You Have To Mourn'," National Public Radio, September 5, 2014, accessed October 6, 2017. http://www.npr.org/2014/09/05/345796530/a-poet-on-losing-his-son-before-you-heal-you-have-to-mourn and http://www.npr.org/books/titles/345796185/gabriel-a-poem#excerpt.
79. Patti Ann Rogers, "Song of the Oceans of the World Becoming," *Song of the World Becoming: New and Collected Poems, 1981–2001* (Minneapolis: Milkweed Editions, 2001): 67–68.

CHAPTER 5

Walking in the City: The Arrival of the Rhetorical Subject

John M. Ackerman

The everyday is a "halfway house, a halting place and springboard, a moment made of moments...the inevitable starting point for the realization of the possible."[1]

"The body surges. Out of necessity, or for the love of movement."[2]

This chapter locates rhetorical capacity within the rhythms of everyday life and, as such, within the agencies and recurrences that populate vital ecologies.[3] For scholars, it is perhaps less common to fuse the everyday with ecological vitality because, as a scholarly trope, everyday life has been overwhelmed by commodification, and is therefore a scene of modern and postmodern alienation. This critique is led by Henri Lefebvre, who maintains that quotidian spaces and routines have been plundered across modernity through the expression of economic capital and control. As such, the banal gestures and orchestrations of ordinary existence, certainly ecological by virtue of proximity and exposure to natural and built environs, have been invaded by the modern capitalist enterprises,

J.M. Ackerman (✉)
University of Colorado, Boulder, CO, USA
e-mail: john.ackerman@colorado.edu

B. McGreavy et al. (eds.), *Tracing Rhetoric and Material Life*,
Palgrave Studies in Media and Environmental Communication,
https://doi.org/10.1007/978-3-319-65711-0_5

a conquest both in the historical material sense of commodification of most realms of daily living and through the tyranny of representation.[4] By his teachings, we who occupy the everyday would disdain any separation of ourselves or our environs by machine, by design, or by symbolic representation—any philosophical or structural move that divides vitality, complicity, and critical reflection into discrete parts and orders of time.[5]

Yet, therein lies the critical potential of the everyday for ecological reflection and action. If ecologies are understood to be inclusively vital, then all forms of agency share "recurrence" in "linear and cyclical repetitions across natural and rational time."[6] Thus, the everyday could be, on occasion and as Lefebvre proposes, "the meeting place of all repetitions" and as such a field of encounter for human and nonhuman agencies (as is more common to say today in speculative theory).[7] Sometimes these pulsations rise to the level of conscious attention by people but mostly they do not, percolating well below human awareness, and for good reason. Rhythms and pulses within vital ecologies help to hold in place an "equilibrium" that balances (in the best of times) matter, bodies, routines, nature, and human need. In the worst of times, in moments of ecological precarity, the equilibria tilt toward inequity if not outright catastrophe.[8] Throbbing ecologies, rich with movement and interaction, earmark rhythms as an irreducible condition for living, a plane of common existence.[9] After all, rivers run on their time and so do buses (we hope); tides move in an out and so do people from houses; climates warm and so do geopolitical desires for conquest—no one would argue such rhythms are homogeneous but neither would one say, upon reflection, they are unrelated.

Were we to take rhythm, circulation, recurrence, and movement to be conditional for regional everyday ecologies, how would one enlist vitality in mind, place, and body to *punch* through the everydayness that Lefebvre disdained; can a rhythmed body beside a rhythmed stream or traffic way usurp the quotidian tableaus of systematic framings, scheduled events, and archives of pre-ordained artifacts? My inquiry into these questions is framed by a now lifelong study of residential life in post-industrial neighborhoods, a study that took root years ago in the theocratic urban spaces of Salt Lake City, in the ethnic borderlands of Flagstaff, and then most pointedly in a multi-year, multi-sited study of economic redevelopment in two communities in Northeast Ohio from 2000 to 2007. I now live elsewhere, but I cannot let these places go—and, why would I? Everything of value that I propose here benefits from situational and interpretive retrospection born out of living in residence in neighborhoods adjacent to other biological and urban orders.

To live *in residence* could mean much more, as an event space, than a physical address. Were it to be a vital residence, it would counteract the closures of the everyday noted by Lefebvre. Living in residence, in this way, accumulates as exposure to rhythmic vitality, and such that *residence* exceeds in catalytic potential the economic, geographic, or legal boundaries given to *real estate* as in "the house that I own" or the "state to which I owe taxes." To be in residence, as an event space, grants agency to someone's "sense of place," recognizing its experiential volume, history, and geography, even though much of the traffic through one's sensed place remains semiconscious. People are entangled in everyday ecologies without the need for conscious attention, yet over time we acquire, by necessity and then sometimes with peril, a "cognitive map" of experience, difference, security, and threat.[10] The urban orders I return to (mostly in mind, partly in body) for this chapter depend upon socio-geographic familiarity, yet I am, frankly, humbled by what I do not know about these places. The desire to know more, to pay closer attention, is an homage to past residences, but it also motivates Timothy Morton's advocacy of "ecological thinking" to discern the basis and roles, rhythms and textures of ecological inclusion as "interconnectedness in the fullest and deepest sense" so that neither nature, nor built worlds, nor ourselves are alienated from daily registers of contact and coherence.[11] His other motive is that without such "thinking," we further our planetary peril.

My specific concern in this chapter is to identify those moments of exposure, through sensorial travel, that puncture the ordinary to connect bodies to matter in space and time and that might, upon reflection, lead to ecological thinking and equitable, inclusive action. Kathleen Stewart identifies these moments as "jumps" in visceral attention wed to movement through ordinary spaces. The jump works the terrain of the touched, the perceived, and the newly imagined all within residential places, and its energetic stickiness depends on the lively role of sensuous bodies in place. The jump is "ordinary affect in the textured, roughed surface of the everyday.... People might be touched by it, or hardened to its obnoxious demands."[12] The jump on certain occasions may hover above routine to become "hyper-vigilance," as proposed by Lauren Berlant to generate a particular kind of insight, well beyond melancholy or regret, an insight that "refuses a return to the ordinary" and instead gives way to "exemplary laboratories for sensing contemporary life in new idioms of affective realisms."[13]

The jump is a commonplace event without political motive until it rises to hyper-vigilance (or some lesser form of discursive insight) to aid in the struggle to comprehend a vital urban ecology and its preferable states of equilibrium. It might guide the divination of affective exchanges amid ordinary spaces that percolate outward to punch through the commodified ordinary and tyrannies of representation. The jump could serve to honor socially accepted narratives, objects, and routines—a reminder of what a community has accomplished. But the jump could also index moments of disjuncture, instances of lost time and objects. As a tactile, reflexive moment in qualitative study, the jump pinpoints an *oscillation* between presence in the world and, then, meanings derived from that presence adjacent to other more typical fieldwork methods, such as interviews, public meetings, scheduled observations, object analysis, and archive retrieval.

When Stewart situates the jump in the "vagueness of the ordinary," she honors its capricious attitude and stickiness. Jumps occur as she "walks the neighborhood with Ariana in the very early morning, laying down imaginaries."[14] The jump that eventually becomes a textual referent is partially minted in a moment of occurrence. Yet (as I consider both Stewart and Lefebvre), Austin and its thereabouts are Stewart's halfway house, and her walking gives way to critical inscriptions that do not terrorize. The rambling territories and textures common to cities and neighborhoods, to rivers and plains, to estuaries and industrial parks, archive an ensemble of residential memory and affect-laden expressions.[15] If the *post-industrial neighborhood* generally names an ebb and flow of economic conditions, in places where I have lived, then residual memories may appear sporadically but always interdependent with that label. Residual memory is the ecology's archive, the "stones of the city."[16] The retrieval of memories by walking or driving, by taking a bus or a plane, gestures to a peripatetic rhetorical potential, a lively practice of ecological movement that gives rise to a more affective realism in place.

Across this chapter, oscillation names the catalytic exchange between the jump as the "charge" that passes through the body that, then, is incorporated into an "order of meaning."[17] Oscillation challenges the tyranny of representation by highlighting the hermeneutic potential given to vitality itself. Oscillation occurs for Hans Gumbrecht at capricious moments of synapse between *presence* in the world, oft times inarticulate, and always pulsing through the body, and *meaning* enlivened through the aesthetics of thought and prose.[18] The failure to recognize the ecological precursors to representation are, for Gumbrecht, regrettable with the kind of regret that haunts a lifetime, due in hindsight to lopsided investments in an overly textual comprehension of the world.

Were you to be driving or walking to a field-study interview, or on the way to work on location, or in a facility, travel restocks the capacity for oscillations between presence and meaning. If you were to leave a worksite to press on with the obligations of family or self, these moments compile in effect because you, like the informants and places in our studies, are moving through the rituals and rhythms of daily living. Oscillation names a synaptic potential available to the observer and the observed. Most of what passes people by as a "fundamental structure of exposure," to use Diane Davis' phrase, occurs well below the threshold of intelligibility within a locale, and that is why it carries so much potential.[19] The rhetorical subject continuously *arrives* through physically and geographically assembled subjectivities. It arrives enabled by jumps in the sensorium of daily existence. The rhetorical subject gathers vitality and conceptual power within the "residue" of daily existence, taking the form of "derisory, minuscule decisions in which freedom is rediscovered and experienced: taking the bus to this or that stop; speaking or not speaking to a particular person."[20]

As I consider the rhetorical subject, she arrives endowed with the grace and wit of quotidian exposure such that oscillation divines a sense-world opened by passage through the everyday. The everyday, then, must be amplified ecologically and differentiated from economic alienation; it needs weight, dimension, and recuperative capacity. In *Everyday Life in the Modern World*, Lefebvre grants the *resident* a phenomenology of recurrence: "gestures of labor and leisure, mechanical movements but human and properly mechanic, hours, days, weeks, months, years, linear and cyclical repetitions, natural and rational time, etc., the study of creative activity (of production in the widest sense)."[21] Habitation exposes one to the simultaneity and interconnection of modernity and everydayness, a reciprocity not as signifier and signified but as the "crown" and the "veil"; "The quotidian is what is humble and solid, what is taken for granted and that of which all the parts follow each other in such a regular, unvarying succession"; a pulsing life world that affects "objects and beings" with a "praxis" and "poiesis" in the "*residuum* (of all the possible specific and specialized activities outside social experience) and the *product* of society in general"; an escape route from the tyranny of banality and exhaustive consumption as "adaptation" of "body, time, space and desire" from "recurrent gestures of a world of sensory experience"; a conceptual domain and finally a publicity based not only in consumption, conquest, and control but also "the imaginary existence of things," a "rhetoric and poetry superimposed on the art of consuming and inherent in its image; a rhetoric that is not restricted to language but invades experience."[22]

I collocate Lefebvre's writing on the everyday to begin the demonstration that everyday life in vital ecologies need not be downcast or futile. Rather, that to be that alive and aware suggests a virtuous life in the everyday. We want to meet this self- and place-aware person, this resident in full bounty, do we not? We want to share the common boulevard or river bank or work site. We welcome the reassurance that recurrence and vitality pulse through our banal routines as well, so that the arriving subject anticipates our arrival as we do hers. And should such consubstantiations gain political, recuperative traction, would we not try to open our pores to other moments of oscillations elsewhere, to revisit different locales and experiential histories to gain a fuller comprehension of the world's residences?

Walking in the City

There is no more likely starting place to consider the figurative powers oscillating between the body in everyday life and inscription than Michel de Certeau's classic essay, "Walking in the City." For those in rhetoric, and communication more broadly, *The Practice of Everyday Life* and this chapter are probably the most read, taught, and adapted statements on the micropolitics of daily living because the rhetorical subject is granted nobility through mobility and scribal power.[23] The essay provides a luminous depiction of the difference between *tactical* and *strategic* practices—that is, between, on the one hand, ordinary gestures of walking along the "thicks and thins of an urban 'text'" and "jaywalking," and, on the other hand, the gestures and structures of corporeal control built and coded into a city environment.[24] In this highly speculative essay, walking enunciates an immense "texturology" that is anthropological, poetic, and mythical in nature. Walking comes equipped with an "unlimited diversity" of tactical, perambulatory movements and gestures; yet those figurations echo silently within the "nowhen" of late modernity that flattens out "all of the data in a plane of projection" into a single narrative of progress and achievement.[25]

The allure of this essay to rhetorical study and environmental communication is easy to spot because mercurial movements in everyday life are analogical to lyrical, narrative, and poetic invention called "walking rhetorics."[26] Walking is rhetorical because phatic movements rise to the level of emphatic gestures and figures of speech; for de Certeau, "The act of walking is to the urban system what the speech act is to language." The strategic urban text is transgressed, although the movements and routes of bodies remain condemned to invisibility—"spaces that cannot be seen; their knowledge of them is as blind as that of lovers." Walking opens up a world of occupancy and habitation, a "poetic geography on top of the literal, forbidden or permitted meaning," but with a cost—the actions of bodies enunciate most dynamically as (merely) figures of speech—the designation of the part to the whole (synecdoche) and the suppression of explicit connections (asyndeton).[27] The walker is written to become a lyric when she "skips over links and whole parts…every walk constantly leaps, or skips like a child, hopping on one foot and is therefore less performative or persuasive by other means. [She] practices the ellipsis of conjunctive *loci*."[28] Rhetoric, for de Certeau, vivifies within its own traditions of figuration (as *loci*) and those in rhetoric or other interpretive fields may well celebrate the discovery of figurative practices in everyday life. After all, these practices help to valorize an "anti-text" of disruption and escape within the subterranean world of the street. Yet the ecological counter movements to textual figuration are left silent. The walker tells a good story, it seems, but the story never gets published, never incorporated into master narratives or mythic designs. Stories conquer the rumor mills of mass mediation only momentarily, a chance encounter, a blind corner without sustained consequence.

Pedestrian speech acts oscillate through the streets but the oscillations set forth in this classic essay never gain traction as "hyper-vigilance." There is no jump between bodily presence to interpretive meaning that claws upward for more strategic value. Recalling the vitality given to the everyday (and thus ecologies) by Lefebvre, his critical project is to give voice to the walker (and to any participant) and never to leave them "silent" or moving in "secret."[29] The tactical walker for de Certeau deftly maneuvers the concept city, but it is that city's figurative dominion that establishes the terrain of movement across the "thicks and thins of the urban text." The walker enlists an "ancient catalogue of tropes" that compares "pedestrian process to linguistic formations" to leave the walker, ironically for my purposes, to "lack a place" and "in search of the proper."[30] De Certeau's gift to ecological communication is one frame of illumination, a textual frame, yet to do so without caution ignores the propensity in late modernity to rely on tropes of textuality to capture through representation otherwise complex social and technological ecologies, a logical habit of mind that Mark Hansen calls "technesis" or the "putting of technology into discourse."[31]

Technesis works in complementary ways with Lefebvre's specter of representational tyranny, because a singular reliance on textuality as the root metaphor and diagrammatic framing of the technical, social, and material dimensions of everyday life constitutes a sin of ecological omission. And, it is a sin in times of extreme technological density and ecological fragility. Stewart's project in *Ordinary Affects* is partly to "slow the quick jump to representational thinking" for the broader project of putting fragility and precarity into analytic relief.[32] She announces no grand salvation through movement in her book; were we to consider her lessons, we are taught to look for the poignancy of the jump to expose the opportunistic textures in banal circumstances and thereby increase the potential for critical reflection. The walker for de Certeau may hail the ancient catalog of tropes, but Hansen points out that metalepsis is one of them, a figurative collapsing of one complexity into another: "As the rhetorical operation that collapses (linguistic) text and (nonlinguistic) context, *metalepsis* produces 'materiality' within the space of language" and therefore a different materiality than a city block, a river's eddy, or hurried walk to catch the bus.[33] Whenever we find technesis at work in descriptions and interpretations of technologically or biologically dense environments—essentially places where we work and live—we might consider the "domestication of what lies beyond the text"[34] and with that recognition, a counterstatement is possible to reclaim a "phenomenological (or corporeal) embodiment…to restore the unconstrained domain of embodied complexity as the locus where technologies impact us immediately and practically."[35]

Oscillation undergirds the powers of inscription in "Walking in the City," but as a walking rhetoric it is not, yet, a fully equipped *techne*. Were it to be so, walking, or movement more generally, would relax the grip of technesis upon presence to foster the discovery of the more-right words to capture the more-right vitality in an ecological scene and event. Oscillation might well engender an ethically more inclusive registry of the exposures in everyday life, one jump, skip, or hop at a time. Were one to look down from tall buildings in New York City as de Certeau did, or in any other city even 30 years ago, one would see not only walkers but a dynamic grid of concrete and steel pulsing with vehicular traffic and economic pursuit that constitutes the walkability of street life, for better or for worse. Were oscillations to co-depend upon presence in the world and meaning making of those worlds, the walker would learn to trust her physical inscriptions as more than a pedestrian speech act. She would leave a more lasting imprint than a phatic gesture in the concept city because vital ecologies have, according to their catalog of figurations, left a mark on her.

Lucia's Arrival

If walking on occasion opens up a poetics of exposure, transcendent of linguistic configuration, what physical and geographic textures along the way elevate the jump to sustained critical awareness? Ash Amin and Nigel Thrift in *Arts of the Political* present affectability as an enabling condition yet awash in limitations. Affection is a "semi-conscious phenomenon" because of limitations of the body, limitations of time, and constrictions on movement through different sorts of milieu.[36] Yet, it is rhetoric, they propose, that transcends these limitations, because rhetoric in relation to affectability and movement connects "passion" to signification, motive, and influence. Passion, it would seem to follow, is not so much structured emotion but rather an intensity, and then reflections of those intensities (physical as well as mental) derived from newly discovered abilities to move through otherwise familiar, otherwise challenging, social and technical domains.

Lucia takes the bus and walks with passion as a first-generation student at the university. She is introduced both as friend and character in *Lives on the Boundary*, an extraordinary book about the struggles of non-traditional students to succeed at a public research university.[37] By Mike Rose's telling, Lucia is 28, a single parent, a working mother, a returning student, and without childcare on the day she commutes from East Los Angeles to the sprawling campus and intellectual terrain at the University of California at Los Angeles. She declared psychology as her major in part because her brother suffers from mental illness, and as Rose accounts for her motives, because "she had seen that look in her brother's eyes, felt drawn to help people whose mind had betrayed them."[38] In an upper-division abnormal psychology class, she encounters the elevated logic and prose in *The Myth of Mental Illness* by Thomas

Szasz, who denies that mental illness exists. The disjuncture between a hermeneutical treatment of mental illness as a *texturology* and a lived exposure to her brother's mental illness in her everyday life eventually leads Lucia to Mike's office for help with this book and her schooling more generally.

Those of the academy might well recognize the liberal humanist move to treat the science of mental health as a myth, with Szasz questioning its existence because "human sign-using behavior does not lend itself to exploration and understanding in these terms. We remain shackled to the wrong conceptual framework and terminology."[39] Such a move is not yet so familiar to Lucia, who turns to Rose for an explanation. In order for Lucia to make workable sense of this passage, she must articulate two related hermeneutic circles, one born through her residential history of living and working in LA, and the other born through a shorter-lived exposure to academic prose. With Rose's tutorial guidance, she has to entertain a world not of the flesh but of the sign, and a world that divides the profane from the sacred; a world sociologically and discursively constructed that countermands both the principles of science (her brother's diagnosis) and her experiential dependencies born from position and juxtaposition in everyday life, a different but related hermeneutical plane. When Szasz writes, "Man thus creates a heavenly father and an imaginary replica of the protected childhood," Lucia's working-class Catholic background denies the possibility of a common hierarchical, authorial plane for "God and the priests, the totalitarian leader and his apologists, and Freud and the psychoanalysts."[40]

By recounting this tutoring session in this way, Rose challenges the cultural myth of deficiency tied to a lack of exposure to the annals of higher education, a different kind of myth common to discussions of success in post-secondary education. I recall this episode because it exposes a different kind of deficiency on the part of "higher" education, one that disavows the quotidian analytics of travel to and from campus in relation to the analytics required by elevated prose. I need to quote Rose at length to take full measure of the poignancy of this episode:

> We left Campbell Hall and headed southeast, me toward a sandwich, Lucia toward the buses that ran up and down Hilgard on UCLA's east boundary. It was a beautiful California day…and Lucia talked about her baby's little discoveries, about a cousin who worried her, about her growing familiarity with this sprawling campus. *I'm beginning to know where things are*, she said pursing her lips. *You know, the other day some guy stopped me and asked*

> *me where Murphy Hall was...and I could tell him...It felt pretty good.* We walked on like this, her dress hiked up where the baby rode her hip, her books in a bag slung over her shoulder, and I began to think about how many pieces had to fall into place each day in order for her to be a student: The baby couldn't wake up sick, no colic or rashes, the cousin or a neighbor had to be available to watch him, the three buses she took from East LA had to be on time—no accidents or breakdowns or strikes—for travel alone took up almost three hours of her school day. Only if all these dropped into smooth alignment could her full attention shift the complex and allusive prose of Thomas Szasz. "Man thus creates a heavenly father and an imaginary replica of the protected childhood situation to replace the real or longed for father and family."[41]

Lucia's intellectual journey coincides with her urban journeys, suggesting the generative powers of oscillation when textual interpretation depends directly upon lived experience. In this single story of personal and intellectual achievement we get closer to the textures of exposure necessary to connect two hermeneutic circles: one derived from reinterpreting a text; and one derived through the physical repetition of travel (the bus, the street, the walk to campus) and arrival. These knowledges are unequal in their origin and substance, yet they are irrevocably present in this narrative of discovery and achievement. They are starkly unequal in cultural value in the political economy of the university, but what they share is a spiraling capacity to *jump* between presence to meaning and toward hypervigilance, in this case of the travails of learning to arrive in person and in mind in an intellectual milieu. "It felt pretty good" identifies the discovery of personal authority bound to territorial familiarity and coincidental to facility with academic discourse. Precisely when and where Lucia jumps between travel and awareness is unknown, but in Rose's telling we see the spark of political potential in its most pragmatic and embodied instance as she maneuvers through position and juxtaposition in a technologically dense urban ecology. *It does feel pretty good.* Learning jumps from familiar to foreign terrain as much as one lexicon to another. This has happened to me, and it has happened to you.

Fieldwork as Travel in Ohio

Up to this point, I have proposed movement and travel as uniquely poised to sponsor the jump from ecological presence to interpretive meaning, but without guarantee that jumps will occur or that oscillations will hover in consciousness. "Walking in the City" portrays a lively rhetorical subject with plenty of tactical moves but without a breakthrough into the dominant narratives of the "concept" city. For Lucia, who lives in a concept city, Los Angeles, I infer that her jumps from presence to meaning might well increase in frequency and consequence because of how one locality infects another. Her movements are along a city street, but they also oscillate between home and university, and between her personal history and a history of thought. There exists, then, a medial space of occurrence, an in between-ness, that elevates bodily movement and sensation to the level of hermeneutic insight open to everyone. If technesis identifies a societal overdependence upon textual frames, these medial spaces help to dethrone the alphabetic text by elevating diverse exposures in everyday life. Stewart's jumps celebrate a medial occurrence independent of digital milieux to exonerate the body from its linguistic imprisonment, though never quite leaving the text-world behind. The rhetorical subject oscillates through the flesh, blood, sensed embodiment emplaced in social, technical, earthly, and textual environments, a selfhood, and as Brian Rotman explains, "coming into being to the side of the written form, what might be termed a *para-self* (italics his)" interiorized, as well as emancipated within, a "trans-alphabetic ecology of ubiquitous and interactive, networked media."[42] His media are more digital than streets in cities but the parallel is striking: as we discover a para-self to the side of the alphabetically induced body, so still thrives a para-self by virtue of habitation in densely textured built and natural ecologies. The discovery of rhetoric through the body and at the edge of language, as Debra Hawhee has discussed, grants the body metronomic capacity, situates it as a divining device that reveals the "pulsing, moving and vibrating" processes and agencies that constitute everyday life as a vital political platform for common concerns.[43]

In all of the neighborhoods I have studied I have encountered some form of bridging apparatus that sparks the jump and feeds oscillation. The centrality of bridges in Northeast Ohio has much to do with the structuring influences of rivers that carve out valleys and passageways, as well as bisecting communities.[44] My route to one site, the Orchard Community on the near west side of Cleveland, begins in Kent Ohio, crossing the Haymaker Bridge to leave campus to drive north along the Cuyahoga, then splitting the Twin Lakes neighborhood before reaching the I-480 and I-80 interchange in Streetsboro. I-80 heads West to cross the sprawling Cuyahoga river valley where trains run north to the industrial center of Cleveland and where a national park preserves hardwood forests, wetlands, and agriculture from a bygone era. In driving this route, I thrice crossed the Cuyahoga. Once in Kent, and then again far above the valley floor below before turning north along I-77 to tour the remains of industrial centers in Cleveland. I crossed the Cuyahoga a third time before turning east on I-90 to Exit 169, which dropped down to rail corridors and another small bridge that served as the gateway to the Orchard neighborhood at 41st and Bailey Avenue, the southeast corner of the former Orchard Middle School.[45]

This school was scrapped and rebuilt after the great recession years (2008–2011) to become a secure, modern building devoted to science and technology. The former Orchard Middle School was an aging structure of yellow brick facades and cracked linoleum hallways, its green spaces and playgrounds long ago paved over, the school serving as a civic, community center because of its proximity to houses and its physical potential. Around 2000 a "Green-School Initiative" was launched by the city of Cleveland that brought an avalanche of city officials, grant-seekers, local advocates, and economic redevelopers that ultimately tore back the concrete to design a multi-use living and learning laboratory on school grounds and to augment one of the best science programs in the city and in one of its most overlooked neighborhoods.

Along this school's north wall, adjacent to Orchard Avenue, a different kind of socio-technical innovation took place, invented by people who reside in the Orchard community and with knowledge preceding and transcending the design of the school. A small, abandoned parking lot abutted the avenue and provided local children, adolescents, and adults a common site for play, typically after school hours, in the evenings, and on the weekends. It centered the community by localizing talk and play; it galvanized a community that had no civic structure, no meeting hall, no town square. It challenged the proprietary instincts of financiers and landscape designers, and altered the pace and the form of the proposed park to preserve the public record of expropriated spaces—a handball court that helped to blur the bounds between schooling and living, learning and playing, the young and the old, and recreation and publicity.

I cannot account for the recuperative value of oscillation for those who frequented the handball court on the north wall of school because across all of my observations and interviews, I did not know to ask or look for movement, much less jumps or oscillation. Yet, the critical turn in the design processes of a highly publicized redevelopment project was provided by a high-school aged young man who spoke in a public forum about the heritage and value of ordinary spaces and repurposed surfaces (the north wall of the school and adjacent street), with everyone who lived in or near Orchard murmuring in assent. That speech act was granted authorial force though a history of movement and thus oscillation because meaning making was conditionally dependent upon routine exposure to the physicality of invented civic spaces. My first impulse as observer, of course, was to listen dumbly for the effects of language, yet I witnessed how others were listening with a visceral attention to the young man's spoken history of games and laughter, affiliation and resistance. There had to be a temporal, physical, and geographic basis for him to reset the horizon of intelligibility, just as Lucia found analytic credibility through bus rides from East LA to the UCLA campus. How many jumps must have occurred for the neighbors in Orchard to recognize his argument—how many jumps does it take to find the courage to speak—I cannot answer those questions, but jumps that oscillate were retrospectively on display within a local political, economic, and ecological moment of discovery and control. The unsanctioned territories of publicity were pulled into strategic alignment through an invented space, a handball court that preceded its identification through language. Locality, it appears, a-synchronically pushes through a collective unconscious oriented to the technical apparatus of the street when provoked by a community's dissolution or when one's status in the world is challenged.

I can report with confidence that the longer I visited the Orchard community, the more times I showed up for local events like handball, or movies in the park, or ward meetings, the more I became aware of reasons why the locus of design shifted from parking lots and learning centers to the handball court and repurposed streets. Orchard as a community was chosen for redesign to gentrify the community property values and to recruit a more selective student population. When pressed to remember, new idioms of affective realism were clearly in play, for this observer, a provisional geography of civic commemoration. I was directed by both residents and redesign advocates to look away from the formal aspects of design, keeping those designs at the periphery of my attention, while listening and watching the rhythm of physical and scriptural enactment of emplaced civility. I can say now that I had little of Lucia's compounded awareness of the interrelationship between formal learning at the university and the temporal and spatial knowledge of commuting. I crossed a bridge to leave Kent to drive north to Orchard; I crossed a bridge near the tracks to leave I-90 to enter this community. Walls and streets were bridges of another kind for residents in this community (as were buses and campus boundaries for Lucia). Walls as bridges, bridges as walls—by recollection these were the *topoi* and *loci* of my jumps. The remap pulled my commute across valleys and railways into a hermeneutic relation to the daily rhythms and rituals of play, unseen by architects and planners, unseen nearly by everyone encumbered by distance and advanced learning, yet pulsing, moving, and vibrating at the invented center of Orchard.

And then in time the same bridge that I crossed as I left Kent, Ohio, to drive north to Ohio City and Orchard, began to oscillate differently, provoking a conditional relationship between transportation routes between Orchard and Kent and the structuring forces at work in economic vulnerability and public tragedy. Haymaker Bridge was proposed right before the tragic events of May 4, 1970 and was commissioned shortly thereafter. The final planning decisions and reports were published just months after National Guard troops and vehicles entered the community along the same river that flows north to the steel mills and automobile plants in Cleveland. The bridge was built to move traffic from the west side of Kent to the east so that the city would expand its economic districts in the decades after 1970. What is now clear by tracking the cycles of unfulfilled redevelopment planning[46] is that once the shooting occurred and the deaths and injuries were recorded, the bridge also made its jump[47]: instead of serving its intended purpose to move people more into the area, the bridge and the new passage through the city severed the campus community from the central business district. This separation calcified the loss of trust from those who blamed students for political agitation and protest, and the distrust of those who blamed government for violent intervention. It took forty years to reconcile the two affected territories and to alter the resulting debris fields of abjection, forty years to figure out how to rebuild a downtown business district known for underutilized streets and buildings. The fortitudes given to movement in Orchard, ones that oscillate outward upon provocation to express pride and strength; such fortitudes were missing in Kent. Oscillations of the kind that jump from scenes of tragic loss and distrust would be, it seems, necessary for both social and economic development to occur.

Undermining Regret by Traveling the Everyday

I do not mean to suggest that living in Kent or Ohio City or anywhere else is regrettable because redevelopment happens or does not. If there is regret to confront, it belongs to critics and observers, as well as those who may reside in a community, where a needed collective response to regret, remorse, or melancholy has not materialized. Kent had become since 1970, in the eyes of some, a notorious site of economic incapacity, disparaged for not dealing with its local tragedy and bouncing back to its former economic strength and coherence. Orchard and Kent thrive in the same post-industrial region of the US; both communities suffered through cycles of economic downturn. If there is regret, it may foremost belong to those who foster narratives of decline and dissolution. Hans Gumbrecht in *The Production of Presence* marks the unequal yet reciprocal territories of oscillation as the substrata of a non-hermeneutic alterity for the critic as resident. Presence refers to spatial and temporal domains of awareness, the space of the "world and its objects" in retrospection. Presence becomes tangible though the senses and rises to the threshold of language and brings objects and spaces forward from their recessive positioning. What I suggest through my vignettes and narratives from the field is the recompense presented by those who have lost, those who are in need, and achieved through locality and exposure, in contrast with the recompenses presented by textual interpretation in its highly trained, academic forms. Gumbrecht's motive for bracketing textual interpretation and to allowing oscillations to thrive speaks to a tired bias in the modern history of hermeneutical interpretation. After years of study, and at the highest level of achievement in his field, his scholarly project enunciates a "pledge against the tendency in contemporary culture to abandon and even forget the possibility of a presence-based relationship to the world" because metaphysics has become "both an everyday attitude and an academic perspective that gives higher value to the meaning of phenomena than to their material presence: the word thus points to a worldview that always wants to go 'beyond' (or 'below') that which is 'physical.'"[48]

Gumbrecht and Lucia share a fragile common ground as she traverses campus and he recounts his intellectual journey, although their timelines for discovery are quite different. As he recounts a colloquium more than forty years ago in Croatia, he returns to a lasting memory of the "beauty and liveliness" of Dubrovnik, Yugoslavia, that sponsors his intellectual pledge to re-establish presence in relation to meaning. Over the course of an extensive intellectual review of literary and linguistic events (shared in his book) that demarcate the interpretive domain of academic disciplines, he found himself with his colleagues in a moment of geographic and philosophical stasis, a profound disappointment with intellectual history that led to encouragement to look elsewhere for what would be discovered to be a commitment to the "materialities of communication." The decisive moment of intuition came on a Sunday morning in 1985, "when we were walking down Stradun, the marble-paved street that runs through the old city of Dubrovnik" when Gumbrecht with his colleagues chose a different intellectual path instead of a long "detour" through intellectual history.[49]

His jumps are retrospective; Lucia's more prospective, and it is worth considering whether the difference is more than simply by age or degrees of edification. Bodily movement within a technological apparatus constitutes a particular nexus for discovery, a material hermeneutic that "gives things voices where there had been silence and brings to sight that which was invisible."[50] A "loss of the world" is, in fact, regrettable not only when the critic discovers what has been lost from sight, but also those occasions where criticism is slow to recapture its political edge.[51] A loss of the world, is indeed regrettable but it is also retrievable. Charting and naming the jumps that lead to interpretive possibility as oscillation is one approach; although Stewart shares little comment on methodology; she is a practicing cultural anthropologist who takes notes, talks with people, and moves between home and field. The semi-conscious nature of moving through common spaces can be mitigated by recognizing that subjects neither tower over their ecological environments, nor succumb to technological determinism.

Nigel Thrift has written extensively on a "technological unconscious" that would appear to riddle the practice of walking or moving by other means.[52] Pulsing through ecologies, however, is a partial antidote to the unconscious: within any ecological circumstance, built, found, or otherwise, resides a "prepersonal substrate of guaranteed correlation, assured encounters, and therefore unconsidered anticipations."[53] Quoting Tim Ingold, Thrift would find any ecological terrain and field of possibility irreducible to its more elemental forms, certainly its personages or inscriptions, because "it exists and takes on meaning in relation to the beings that inhabit it." These relations constitute the "properties of dynamic self-organization of relational fields."[54]

To name the technological as (routinely) unconscious means that most exposures in everyday life—and for reasons of efficiency and practicality—circulate well below the thresholds of cognition but closer to the surface when encountered in residence. Instead of a deficit of attention, ecological vitality provides a currency of exchange that enables the jump with extraordinary potential through residence. Requisites for the jump are less the formal knowledge of structures, design, and history than "knowledges of position and juxtaposition" that bend our bodies in one direction or another "without the benefit of any cognitive inputs, a pre-personal substrate of guaranteed correlations, assured encounters, and therefore unconsidered anticipations."[55]

When Thrift reads "Walking in the City," the walker colludes with automobiles, therefore streets, therefore the urban apparatus to travel and to make the city, and the walker shoulders more complicity in the orchestration of an urban ecology with the benefit of displaying more vitality than a "representational metaphysics of presence."[56] The walker consubstantiates with other people, things, and territories in everyday life, and modulates the technological unconscious—in the most basic sense, one walks perhaps to avoid taking the bus or to grab a private moment but the technological dependencies are still there; the car enables the body to go further and faster, while walking is a reprieve from speed; the car opens up a different field of affection through status, style, and economic distribution. The walker and the car consubstantiate ecological responsibility. They have multiplied to the point of ecological exhaustion but have also multiplied the opportunities for oscillation.

I end with optimism: that people productively mired in residential requirements for living in vital ecologies are closer to recognizing everyday life as a halfway house, as a repository for shared logical and biological potential. Oscillations may prove to be indifferent to all the distances and boundaries built into tall buildings and bordered communities—if we know to look, listen, and touch. Nathan Stormer proposes that a "given order of discourse will consist of a specific articulation of bodies and language as well as a given order of things."[57] However true that may be, in this chapter I suggest that reversing the authorial temporality of that claim is useful as well because any given order of things carries within it the residential potential to jump to a given order of discourse. The oscillations that enable that reversal are distributed widely in everyday, ecological consortia. It would then be possible, and advisable, to go out and "lay down the imaginaries" as Stewart would say and in as many different ecological settings as possible. In a more recent essay, Stormer conspires with Bridie McGreavy, aided by the coastlines of Maine, to grant the powers of oscillation to clams, mud, tides, and economic tricks made of wood by harvesters. The ecological frame they offer exists with little regret, by critic or resident, when the harvesters are "working the tides, [and] the tides work them too."[58] As I began this chapter, once more we find rhythm at work in everyday life, giving structure to exposure and opportunity for rhetoric, dialog and ecological attunement within residence. I suspect we will find it here, there, and everywhere should we care to look.

Notes

1. Henri Lefebvre, *Everyday Life in the Modern World*, trans. Sacha Rabinovitch (New Brunswick: Translation Publishers, 2007), 14.
2. Kathleen Stewart, *Ordinary Affects* (Durham, NC: Duke University Press, 2007), 113.
3. I have not found a sufficient definition of ecology to fit the scenes and motives in this chapter. Ecologies have quotidian content because they occur daily, but a "lively" ecology (an unnecessary redundancy except that ecologies are presented often as static and bounded) would need to extend well beyond the legacies of biological science. It would need to consider the geographies and interrelations given to technological advances to include social accumulations of property, law, ethnicity, economy, governance, and so on. Ecologies are "entangled" and they "entrap" people in dependent relations with things, as Ian Hodder explains. See Ian Hodder, *Entangled: An Archaeology of the Relationships between Humans and Things* (Malden, MA: Wiley-Blackwell, 2012). Ecologies are recently understood to be rhetorical, as Jenny Edbauer has argued but that, as Edbauer admits, is a starting point. See Jenny Edbauer, "Unframing Models of Public Distribution: From Rhetorical Situation to Rhetorical Ecologies," *Rhetoric Society Quarterly* 35, no. 4 (2005). I must then work implicitly to build a provisional theory of ecology, one that is inclusive of multiple agencies, both material and immaterial, and I am, therefore, in league with a number of scholars in and out of rhetoric, communication, and composition studies, who share that goal.
4. If "tyranny of representation" is too strident an idea for readers of this collection, it would not be for Lefebvre, who likens representation to a form of "terror" worthy of critical disdain and subversion. This motif is found across his collected works. See Henri Lefebvre, *The Critique of Everyday Life, Volume 1* (London: Verso, 1991).

5. It is important to understand that the everyday and its critique substantiates Lefebvre's later proposals for the vitality of social space (e.g., *The Production of Space*), the poly-rhythmic nature of liveliness (e.g., *Rhythmanalysis*), and the human "right" to equity in city environments (*Writings on Cities*). See Lefebvre *Critique of Everyday Life: Volume 1*; Lefebvre, *Critique of Everyday Life: From Modernity to Modernism, Volume III*, trans. Gregory Elliott (London: Verso, 2005); Lefebvre, *Everyday Life in the Modern World.*
6. Lefebvre, *Everyday Life in the Modern World*, 18.
7. Ibid., 19.
8. The principle of equilibrium comes from Gregory Bateson and the connection to the everyday by Ezio Manzini. See Gregory Bateson, *Steps To an Ecology of Mind: Collected Essays in Anthropology, Psychiatry, Evolution, and Epistemology* (University of Chicago Press, 1972); Enzio Manzini "Prometheus of the Everyday: The Ecology of the Artificial and the Designer's Responsibility," *Design Issues* 9, no. 1 (1992).
9. A common plane of existence as a speculative frame is found in Elizabeth Grosz and Jacques Rancière. See Elizabeth Grosz, *Chaos, Territory, Art: Deleuze and the Framing of the Earth* (New York: Columbia University Press, 2008); Jacques Rancière, *The Future of the Image* (London: Verso Press, 2007).
10. The critical and interpretive value of a "cognitive" map of embodied experience inscribed through exposures to cultural environments is elucidated by Fredric Jameson and based on the foundational architectural criticism of Kevin Lynch. An illustration of its usefulness to critical race and gender scholars can found by Ruth Frankenburg. See Fredric Jameson, *Postmodernism, Or, The Cultural Logic of Late Capitalism* (Durham, NH: Duke University Press, 1991); Kevin Lynch, *The Image of the City* (Cambridge. MA: MIT Press, 1960); Ruth Frankenburg, *White Women, Race Matters: The Social Construction of Whiteness* (University of Minnesota Press, 1993).
11. Timothy Morton, *The Ecological Thought* (Cambridge: Harvard University Press, 2010), 7. My use of "nature" here is shorthand because I agree with his goal to erase "nature" from the ecological imaginary because the word freights a legacy of domination by people over the biosphere.
12. Stewart, *Ordinary Affects*, 38–40, 127.
13. Lauren Berlant, *Cruel Optimism* (Durham: Duke University Press, 2011), 54.
14. Stewart, *Ordinary Affects*, 127.
15. I find it useful to keep in tension "places of memory" that tend to be designated by society as sites of commemoration and those places undesignated but rich in quotidian exposure. The distinction deserves more detail than I can muster here because oscillations will certainly be triggered by

attending to dedicated memorial sites. The vantage point in this chapter has more to do with oscillations that occur in ordinary, unsanctioned, or previously unrecognized, sites and spaces. See Greg Dickinson, Carole Blair, and Brian Ott, eds., *Places of Public Memory: The Rhetoric of Museums and Memorials* (Tuscaloosa: University of Alabama Press, 2010).
16. Max Pensky, "Three Kinds of Ruin: Heidegger, Benjamin, Sebald," *Poligrafi* 12, no. 61 (2011): 65–90.
17. Stewart, *Ordinary Affects*, 39.
18. Hans Gumbrecht, *Production of Presence: What Meaning Cannot Convey* (Stanford: Stanford University Press, 2008).
19. Diane Davis advances the pre-hermeneutic space of occurrence throughout her book. See Diane Davis, *Inessential Solidarity: Rhetoric and Foreigner Relations* (Pittsburgh: University of Pittsburgh Press, 2010), 3.
20. Lefebvre, *Critique of Everyday Life, Volume III*, 127.
21. Lefebvre, *Everyday Life in the Modern World*, 18.
22. This compendium of quotations was excerpted from *Everyday Life in the Modern World*, 24–90. I compressed them, as such, to try to capture the vitality in everyday life for ecological purposes. I am trying, therefore, to be architectonic in the sense of a figurative gesture sculpted to imitate an ecological event.
23. Michel de Certeau, "Walking in the City," *The Practice of Everyday Life*, trans. S. Randall (Berkeley: University of California Press, 1984).
24. Because most of the bodily movements in this chapter belong to the able-bodied—those who are able to walk or move freely—I apologize to those who may read this chapter or my sources and find no images of bodies that cannot move so easily, for whatever reason, through complex ecologies. In fact, it could be shown that those who are disabled or are detoured because of threat or loss of capacity have a heightened readiness for jumps and oscillations because of the struggle to achieve mobility.
25. De Certeau, "Walking in the City," 94.
26. For an account of "mercurian" rhetoric, see Peter Simonson, "The Streets of Laredo: Mercurian Rhetoric and the Obama Campaign," *Western Journal of Communication* 74, no. 1 (2010).
27. De Certeau, "Walking in the City," 93–101.
28. Ibid., 101. De Certeau identifies the subject in this passage as rhetoric and not a body per se, yet across this essay the embodiment of the walker is presumed to move across materiality (the street) and figuration (the text). I want to return the subject of this passage to its embodied figuration. That said, the slippage between a body that walks and a body that narrates provides a classic illustration of oscillation though favoring—that is elevating—inscription over embodiment.

29. Ibid., 98. The walk remains blind, silenced, secretive, forgettable, illegible, and so forth throughout this essay. To be fair, the distinction made by de Certeau is invaluable for the critique of the concept city or any overly deterministic framing by words or by numbers of life within the city. Yet to make that argument, de Certeau condemns the walk to the fate of inscription without a vital ecological counterpoint, as will be discussed. It is through the technology of abstraction by virtue of theoretical representation that the image, the body, the city as technological form is reduced to a figure of speech.
30. Ibid., 102–103.
31. Mark Hansen, *Embodying Technesis: Technology Beyond Writing* (Ann Arbor: University of Michigan Press, 2004), 4.
32. Stewart, *Ordinary Affects*, 4.
33. Hansen, *Embodying Technesis*, 91.
34. Ibid., 92. Elsewhere in *Embodying Technesis*, Hansen compares technesis to the "naturalizing" of the image by Barthes and before that "gynesis," the "putting of women into discourse that Alice Jardine discovers at the heart of French feminism and poststructuralism" (86).
35. Ibid., 27–29. To "restore the unconstrained domain of embodied complexity" could also be extended into the literature of "cognitive mapping" underwritten by critical architects, post-structural critics, and feminists. The body, down below, by their account does leave a trace, more than a fading textual reference.
36. Ash Amin and Nigel Thrift, *Arts of the Political: New Openings for the Left* (Durham: Duke University Press, 2013), 161.
37. I have turned to Lucia and the following vignette before to close that essay with a poignant example of physical and intellectual arrival. I have wanted to give her story a more central role in locating hermeneutic potential in everyday life. See Lucia, "The Space for Rhetoric in Everyday Life."
38. Mike Rose, *Lives on the Boundary* (New York: Penguin Books, 1989), 183.
39. Ibid., 183.
40. Ibid., 184.
41. Ibid.
42. Brian Rotman, *Becoming beside ourselves: The alphabet, ghosts, and distributed human being* (Durham: Duke University Press, 2008), 5.
43. Debra Hawhee, *Moving Bodies: Kenneth Burke at the Edges of Language* (Columbia: University of South Carolina Press, 2012), 17.
44. See John Ackerman, Caroline Gottschalk Druschke, Bridie McGreavy, and Leah Sprain, "The Skunkwork of Ecological Engagement," *Reflections* 16, no. 1 (2016).

45. The recollection of the three bridges is new to this author even though I have traveled back and forth between Kent and Orchard numerous times. In recounting those journeys, I am at a loss for how to convey the navigation of the volume of the Cuyahoga valley and the progression of scenery from an otherwise sleepy university town to a once-vital industrial region. Pictures and maps would not help, and the struggle to recapture moments of oscillation, on demand to write this chapter, is not only difficult; it makes the point of how limited we are in recovering the potential once removed in time and place. Yet the catalytic value of that travel and the exposure and challenges it involved haunt my professional and residential identity to this day.
46. John Ackerman, "Rhetorical Engagement in the Cultural Economy of Cities," in *The Public Work of Rhetoric: Citizen-Scholars and Civic Engagement*, eds. John Ackerman and David Coogan (Columbia: University of South Carolina Press, 2010).
47. Without going too far into arguments for new materialism by scholars such as Levi Bryant, Jane Bennett, and Thomas Rickert and many others, bodies that move through built/natural ecologies, and that sometimes "jump" to other kinds of awareness and interpretations might well be more (not less) attuned to the archaeological fact that "things" continue to evolve, their affordances, their symbolic value, their practical purposes (c.f., Hodder, *Entangled*). See Jane Bennett, *Vibrant matter: A political ecology of things* (Durham: Duke University Press, 2009); Levi Bryant, *The Democracy of Objects* (Open Humanities Press, 2011); Thomas Rickert, *Ambient Rhetoric: The Attunements of Rhetorical Being* (Pittsburgh: University of Pittsburgh Press, 2013).
48. Gumbrecht, *Production of Presence*, xv.
49. Gumbrecht, "Materialities/Non-hermeneutic/Presence," 6.
50. For a short introduction to Ihde's rethinking of hermeneutics, see Don Idhe, "Material Hermeneutics: Symmetrical Archaeology," accessed January 18, 2016, http://humanitieslab.stanford.edu/23/746. See also Don Idhe, *Expanding Hermeneutics: Visualism in Science* (Evanston: Northwestern University Press, 1998).
51. Gumbrecht, *Production of Presence*, 49.
52. See Nigel Thrift, *Knowing Capitalism* (London: Sage Press, 2005); Patricia Clough, *Auto Affection: Unconscious Thoughts in the Age of Tele-Technology* (Minneapolis: University of Minnesota Press, 2000).
53. Thrift, *Knowing Capitalism*, 213.
54. Timothy Ingold, "From Complementarity to Obviation: On Dissolving the Boundaries between Sociology and Biological Anthropology, Archaeology, and Psychology," in *Cycles of Contingency: Developmental Systems and Evolution*, eds. Russell Gray and Paul Griffiths (Cambridge: MIT Press, 2001).

55. Thrift, *Knowing Capitalism*, 213.
56. Nigel Thrift, "Driving in the City," in *Non-representational Theory: Space, Politics, Affect* (London: Routledge Press, 2008) 77.
57. Nathan Stormer, "Articulation: A Working Paper on Rhetoric and Taxis," *Quarterly Journal of Speech* 90, no. 3 (2004): 261.
58. Nathan Stormer and Bridie McGreavy, "Thinking Ecologically," 19.

CHAPTER 6

(Re)Arranging Regional Rhetorics

Joshua P. Ewalt

And that precisely would be…one of the truly productive characteristics of material spatiality—its potential for the happenstance juxtaposition of previously unrelated trajectories, that business of walking round a corner and bumping into alterity, of having (somehow, and well or badly) to get on with neighbors who have got "here" (this block of flats, this neighborhood, this country—this meeting-up) by different routes than you; your being here together is, in that sense, quite uncoordinated. This is an aspect of the productiveness of spatiality which may enable "something new" to happen.

Doreen Massey, *For Space*

We live in a world populated by structures—a complex mixture of geological, biological, social, and linguistic constructions that are nothing but accumulations of materials shaped and hardened by history. Immersed as we are in this mixture, we cannot help but interact in a variety of ways with the other historical constructions that surround us, and in these interactions we generate novel combinations, some of which possess emergent properties. In turn, these synergistic combinations, whether of human origin or not, become the raw material for further mixtures. This is how the population of structures inhabiting our planet has acquired its rich variety, as the entry of novel materials into the mix triggers wild proliferations of new forms.

Manuel De Landa, *A Thousand Years of Nonlinear History*

J.P. Ewalt (✉)
University of Utah, Salt Lake City, UT, USA
e-mail: joshua.ewalt@utah.edu

B. McGreavy et al. (eds.), *Tracing Rhetoric and Material Life*,
Palgrave Studies in Media and Environmental Communication,
https://doi.org/10.1007/978-3-319-65711-0_6

Nebraska in the early days of June. Although quite hot outside, it is a nice day for driving down a highway through the middle of the United States. As I drive the distinct modes of matter that constitute central Nebraska, and the larger Great Plains region, pulsate, vibrate, and move around me: the cows eat green grasses in a crowd up against a fence; the brown and yellow prairie grasses sway above the shorter, green grasses and the broken down wooden fences; the hot sun, blue sky, and white clouds move slowly over the landscape; my fellow travelers make occasional eye contact with me as we collectively pass that dull brown land that slowly becomes a vibrant green, returns to a dull brown, and then becomes golden. Symbolic, technological, and (not-just)human flows also move among and intermingle with my moving body. Signs and billboards reference maintaining "traditions" and "roots." My quickly moving car passes anti-abortion discourse: "Abortion stops a beating heart"; "If you're pregnant, it's a baby!" A semi-truck drives by, carrying large, muddy pigs. The harsh wind slaps against a pig's nostrils as it peers out of the truck's carrier. Inside an old gas station, I see a lighter with an image of the *Duck Dynasty* family.[1] A car drives by with a bumper sticker for local conservative politician Ben Saase.[2] While moving with/in the moving matter of this dynamic space-time, I become critically attuned not only to the prairie grasses of the Great Plains but also to food politics, anti-abortion sentiments, and heteronormative articulations of class and sexuality.

Central Nebraska is, indeed, a complicated collection of matter, a unique arrangement of human, plant, animal, technological, and symbolic flows, always intertwining in novel and well-worn ways. The cows eating grasses by the broken down fence intertwine with the cars that drive down the state roads, the exhaust that moves towards the cows, the people that feed them as they carry *Duck Dynasty* lighters in the front pocket of their shirts, the hot sun, and the prairie grasses that struggle for survival against the encroaching monocultures of corn. Together, the intermingling of these moving materials forms the contours of space. In Doreen Massey's words, this space is a product of the articulation of matter, a "sometimes happenstance, sometimes not…*arrangement-in-relation-to-each-other* that is the result of there being a multiplicity of trajectories."[3] When driving through central Nebraska and the larger Great Plains, one endures, is shaped by, and contributes to this particular arrangement of material phenomena.

Currently, I am moving with/in this space to attend a particular event: the 2014 Willa Cather Spring Conference in Red Cloud, Nebraska. While the entire conference, as a celebration of the life, works, and times of Willa Cather in her Nebraskan hometown, maintains an epideictic character, I am especially interested in one session that seems geared toward affecting change through the form of public address, a prairie foraging demonstration on the Cather Memorial Prairie. Occurring multiple times over the course of the three-day conference, the demonstration aims to cultivate more sensitive relations between plant and human bodies and instill more sustainable eating habits. More specifically, its goal is to develop a new orientation to the prairie among human attendees; many of the plant bodies that we identify as weeds actually have the potential to become-food or become-medicine if human bodies were to choreograph their dance-with-plants differently.[4] The demonstration, moreover, seems heuristic from the standpoint of rhetoric's materiality: stylized in the form of a prairie walk, it is highly mobile and incorporates a variety of non-human participants (plants, winds, sunlight, cameras, books, and brochures) into the rhetorical encounter. Thus, as I move toward Red Cloud, I carry a question with me: what can the foraging demonstration teach us about the event of rhetoric?[5]

In the following chapter, I offer an answer to this question. I encourage an understanding of rhetorical events as *regional arrangements*, or configurations of moving materials which have come from, and are on their way to, elsewhere. Arrangement refers to the organization of simultaneously existing material histories into a spatial configuration.[6] Such arrangements are *regional* insofar as their constitutive flows are drawn from larger networks and their inventions also make their way elsewhere, moving variously into local, national, and international ecologies. Thus, rhetorical events are regional articulations—spatio-temporal events embedded within a wider topography of material flow—and the extent to which their inventions circulate within those multiscalar ecologies shapes their ability to affect change.[7] By bringing into contact different modalities of moving matter—human and non-human, discursive and non-discursive, digital and analogical—with the ultimate goal of altering the complex, intra-twined systems that inform food cultures on the Great Plains, I argue that the prairie foraging demonstration invented utterances and orientations attentive to the aesthetics and utility of wildflowers, and also enhanced the wildflowers' "thing-power."[8] Elements of the arrangement, however, also reproduced notions of authenticity that *selectively* attended to the event's fundamental foreignness—though the rhetorical arrangement included multiple bodies with decidedly non-local histories, certain non-local mobilities were presented as "invasive," while others, notably those affiliated with European settler colonialism, became markers of spatial authenticity.

The following chapter develops two considerations. First, by connecting scholarship on materiality with rhetorical theories of arrangement, I conceptualize *arrangement's inventiveness*. I argue various materialist theories and their uptake in rhetorical studies centralize the canon of arrangement. They present arrangement as a process that transforms natural-cultural ecologies by bringing simultaneously existing material histories into new arrangements, altering those histories through the inventiveness of those arrangements, and dispersing those trajectories back out into the larger set of interlocking systems within which the arrangement is embedded.[9] Thus, I argue that engaging with the world as a rhetor involves cultivating arrangements out of multiple material phenomena in order to disperse outward a set of emergent mobile inventions.[10] Moreover, by unpacking the event of the prairie foraging demonstration, I provide a field-based account of this process at work.[11] Second, I ground the case in scholarship concerning regional theory in order to direct attention to the *regional* character of inventive arrangements: they are dynamic configurations composed of materials coming from, and on their way to, other space-times, and thus are energized by their heterogeneous and ubiquitous foreignness. Through attention to how the foraging demonstration reveals and conceals its foreignness in inconsistent ways, I evidence the critical necessity of attending to rhetoric's fundamentally regional character.[12]

In the remainder of this chapter, I perform the following: first, I overview the relationship between rhetoric, materiality, and arrangement, concluding with an account of how the canon of arrangement has appeared in theories of feminist new materialism, and proposing the regional character of rhetorical arrangements. I show that arrangement offers a unique opportunity to connect rhetoric with discussions of materiality and ecology, particularly insofar as the canon has remained tied to another: invention.[13] Second, I offer an analysis of the prairie foraging demonstration. Grounded in the language of a mobile, materialist orientation to rhetoric's arrangements, I account for the rhetorical event as a space of "relation in which all kinds of unlike things can knock up against each other in all kinds of ways."[14] In doing so, I evidence rhetorical moments as regional arrangements inventive of subsequent mobilities. I conclude with the concept of cultivating arrangements, elucidating this chapter's approach to the relationship between arrangement, materiality, and the production of rhetorical moments.[15]

On Rhetorical Events as Regional Arrangements

Arrangement: A Brief History

To outline an ecological and materialist sense of rhetorical events as regional arrangements, a brief history of arrangement is needed, in particular emphasizing its relationship to invention. As one of the five canons of rhetoric in the classical period arrangement concerned the division of speech into parts and the ordering of arguments within a composition.[16] Rhetorical handbooks typically offered four basic divisions of speech: introduction (*exordium*); narration (*narratio*); proof (*probatio*); and conclusion (*peroratio*).[17] As Cicero explained, of these four parts, "two of them, the statement of the facts and the proof, serve to establish the case and two, the exordium and the peroration, to influence the mind of the audience."[18] Arrangement, moreover, remained inextricably tied to invention. An orator or logographer, when inventing rhetorical works, necessarily considered each part of speech as well as contextual factors such as the class of the case, ends sought, and type of audience, factors which determined the arrangement appropriate to the subject matter.[19] Cicero, in *De Partitione Oratoria*, explains that a suitable arrangement differs according to whether the orator seeks judgment (decisions about the past), deliberation (decisions about the future), or embellishment (evoking pleasure).[20] And, as Richard Leo Enos demonstrates, although the rhetorician's advice seems rather strict for each class of case, Cicero permitted students a degree of creativity when discovering a suitable arrangement appropriate to the content and context.[21] This perspective on the inextricable relationship between invention and arrangement continues into the modern period of rhetorical theory, influencing, to offer one example, Francois Fénelon for whom, "implicit in the configuration of the subject" was "the pattern of arrangement best suited to it."[22]

As rhetorical theory turns to the topic of materiality, the relationship between arrangement and invention becomes configured quite differently from its classical predecessor: arrangement becomes affiliated with articulation, and invention with articulation's emergent effects.[23] Initiated by Michael Calvin McGee's call to address rhetoric as a social phenomenon as real as air or water—a phenomenon which results in the formation of consciousness—materialist approaches to rhetoric, written around the turn of the century, privilege a logic of articulation over representation.[24] Rather than analyzing rhetoric as symbolic material merely representative of cognitive activity, rhetorical discourse becomes studied as a material force that orders social relations.[25] Grounded in (this version of) the materiality of rhetoric, critics become concerned less with what a rhetorical operation or discursive fragment *means* than *what it does* when it arranges, or is articulated with, the other natural-cultural materials of socio-cultural life. Developing the direct connection to the canon of arrangement, Nathan Stormer encourages critics to address the historical emergence of any given rhetoric through attention to articulatory processes, or through the process of spatializing "culture and nature by arranging diverse material-semiotic elements into recognizable bodies and languages."[26] Thus, what appears as "rhetoric" and "not-rhetoric," such as the difference between the natural-cultural configuration of a prairie foraging demonstration at an event celebrating Willa Cather, on the one hand, and thunder on the Great Plains, which is articulated differently in varying social/historical contexts, on the other, are the product of historical practices that arrange material-discursive elements into meaningful configurations.[27] Arrangement refers less to the parts of speech, or the ordering of arguments, and more to, following Stormer, the ways a "heterogeneous network of material-semiotic elements" encounter each other and materialize into intelligible rhetorics.[28]

Articulation, moreover, has emergent effects. As Stormer argues, "articulation is transformational and emergent, creating new possibilities where none existed before, augmenting by factors rather than by linear accretion…diminishing possibilities in ways that defy rudimentary calculation."[29] Following Doreen Massey, to spatialize is not only to order the boundaries of a rhetoric, it is also to create the conditions of possibility from which "something new" might emerge: new meanings, subjectivities, bodies, and practices.[30] Indeed, in the tradition of materialist rhetoric, a multitude of semi-stable phenomena, including "meaning and subjectivity," become understood as "contingent effects of discursive arrangement."[31] As such, we might theorize a different relationship between arrangement and invention. When material-discursive fragments encounter each other, they create the conditions of possibility for invention: the emergence of subjectivities, meanings, bodies, and arguments that form relations.

Rhetorical scholarship that takes an "ecological" orientation to invention, and focuses on the work of objects and technologies in complex assemblages with humans, further establishes arrangement's inventiveness. These cases work from an emerging interest in posthumanisms in rhetorical studies, pointing to a broader investment in the ways in which arrangements of humans and non-humans collectively produce emergent effects. Through a reading of the history of *Kairos*, Rickert argues "invention becomes an emergent process" that works through ambient environments rather than "an autonomous, self-willing, and clearly individualized subject."[32] Following both Rickert and Edbauer Rice, Rivers and Derksen present deception as a phenomenon emergent from ambient rhetorical ecologies, or a "complex of agents that collectively produce effects," rather than as the product of an agential subject separated from their environment.[33] Simonson, moreover, argues invention occurs "through an *interlocking and dynamic array* of…inventional media."[34] Although Simonson argues inventional media may be conceptually distinguished into eleven categories, such as technologies, space/place, bodies, and regimes and relations of power, in practice the generation of rhetorical materials involves a "meshwork," or an arrangement of interrelating media.[35] Finally, working from Stormer and from materialist theories of performativity, Hallenbeck examines nineteenth-century "object arrangements," including particular relations between women, men, and bicycles, among other things, that invented a reinforcement and alteration of gender norms and practices.[36]

Feminist New Materialisms and Inventive Arrangements

An emphasis on articulation's emergent effects can be found in scholarship on new materialisms, particularly that which has gone under the title "feminist new materialisms."[37] Although characterized by vital lines of distinction, feminist new materialisms collectively insist upon materialization as a process involving not just human actors, or discourse, but rather the agential activity of material phenomena, discursive and not, human and not, categories which are, themselves, overlapping and always sliding into each other (the human, for instance, is composed of many non-humans).[38] Under feminist new materialisms, traditional distinctions between subject and object become replaced with *entangled material processes that find their agency not in autonomous freedom, but in their inevitable arrangements with simultaneously existing phenomena*.[39] As such, conversations surrounding the topic present a number of concepts—entanglement, the agency of the assemblage, intertwined complex systems—that, with different emphases, center the inventiveness of arrangements.[40]

Jane Bennett's explication of the agency of the assemblage is illustrative. Exploring a North American power outage, Bennett illustrates how the emergence of a blackout cannot be located in any one cause but as a result of the communicative capacity of the grid-as-assemblage; electricity, power plants, transmission wires, Enron, consumer demand, brush fire, and economic theory (among many other histories) all participated in the emergent moment.[41] As Bennett explains, "an actant never really acts alone. Its efficacy or agency always depends on the collaboration, cooperation, or interactive interference of many bodies and forces."[42] Thus, participation in an assemblage ultimately alters the "thing-power" of any material actant.[43] Food, to use one of Bennett's examples, gains different capacities when it enters into an assemblage with human bodies: "a carrot as it first enters the eater's mouth is a full-blown entity, with a distinctive taste, color, texture; once swallowed, however, its coherence gradually dissipates until, if one were to continue to observe it via a tiny camera inserted into the gut, the difference between carrot and eater vanishes altogether."[44] As the carrot becomes-eater, moreover, it gains the power to alter the body, shaping its mood, energies, speech, and, ultimately, appearance. Relational arrangement becomes central to the materialization of phenomena.

Karen Barad's "agential realism" offers another account of arrangement's role in materialization, emphasizing the material-discursive entanglements that shape the formation of new, or reiteration of previously existing, boundaries and bodies.[45] As Barad explains:

> To restrict power's productivity to the limited domain of the social, for example, or to figure matter as merely an end product rather than an active factor in further materializations is to cheat matter out of the fullness of its capacity...Surely it is the case—even when the focus is restricted to the materiality of "human" bodies (and how can we stop there?)—that there are "natural," not merely "social," forces that matter. Indeed, there is a host of material-discursive forces—including ones that get labeled "social," "cultural," "psychic," "economic," "natural," "physical," "biological," "geopolitical," and "geological"—that may be important to particular (entangled) processes of materialization.[46]

This perspective displaces materialization, or the ordering of materials into intelligible configurations, from the work of the human being, or even cultural-discursive processes as clearly defined categories of practice, to the complex natural-cultural assemblages of matter that move around, with, and through the (not-just)human. If we consider the production of speech to be one of these emergent, intelligible configurations, then we can say that invention, from its narrowest conceptualizations as the discovery of persuasive means to its wider application as the generation of all the "symbolic and physical elements that enter into or are gathered for the purpose of communicative address," results from multimaterial, agential arrangements.[47]

Regional Arrangements

It is from these traditions that I posit the event of rhetoric as a regional arrangement. I encourage attention to rhetorical moments as productive events, "complex, pluralistic, relatively open" arrangements composed of a constellation of moving materials, out of which something new is invented to circulate into natural-cultural ecologies on multiple scales.[48] From this perspective, any body (human, plant, technological) participating in the rhetorical moment is composed of a variety of material histories, and those histories exhibit a responsiveness to the affective vibrancies of the other bodies with which they form a complex mixture.[49] This responsiveness-through-arrangement results in the generation of new materializations (ideas, subjectivities, utterances, feelings) to circulate in natural-cultural ecologies, finding their own arrangements, and inventing and transforming again into differently articulated materials.

Moreover, it is in reference to the fact that the constitutive materials come from and make their way into subsequent ecologies that I speak of their *regional* character. As Ash Amin, Doreen Massey, and Nigel Thrift explain, contemporary regional theory focuses less on bounded territories and more on regions as "sites where diversity and difference are juxtaposed and where flows and networks of different spatial reach overlap."[50] For the rhetorical event to have regional characteristics, it has to be an articulation connecting up to multiple sites within a broader spatial topography. Thus, given that both the materials forming any rhetorical moment come from elsewhere, and its inventions also make their way into arrangements distant or near, *all rhetorical moments are regional phenomena, while the regional cartography is different to each.* It is also through attention to their regional character that we can attend more critically to the differences *and distances*, the constitutive foreignness, characterizing inventive arrangements, as their fundamental materials have come from somewhere else, in a process of becoming. A celebration of Willa Cather in her "hometown" of Red Cloud, Nebraska, features for instance: a wildflower grown from a seed that migrated to the Great Plains through the wind or with the travels of an animal; the European migration that brought Willa Cather's ancestors to the United States, and contributed to the genocide of Native American Tribes throughout the Great Plains, including the Oglala Lakota and Chief Red Cloud, after whom the town is named; the humans who arrived to work for the Willa Cather Foundation, and the books, lectures, habits, and college seminars they bring with them; and the artifacts from Cather's travels, brought to Red Cloud, to provide a sample. Afterward, a bevy of materials found their way into a number of natural-cultural sites. The word "region" captures these constituent travels, and offers a framework for critically attending to the ways they are revealed (and concealed) in the unfolding of a rhetorical event.

Thus, in what follows, I provide an example of rhetoric as a regional arrangement. I examine the movements and relations constitutive of the prairie foraging demonstration that occurred as part of the 2014 Willa Cather Spring Conference. Upon accounting for the inventive arrangement of the rhetorical event, I will turn to what it might mean to produce such rhetorics.

Learning to Eat the Prairie

After eating lunch in town I join a group of about fifteen conference attendees gathered around the entrance to the Willa Cather Memorial Prairie. An unintimidating barbed wire fence with dark grey wooden posts and a "No Hunting" sign on a swinging wooden door separates the dirt and grass parking lot from the 621 acres of restored grassland prairie. Some attendees mingle among themselves, but most are calm, silently taking in the magnitude of the prairie. Large and very professional-looking cameras move with some of the human bodies; most wear tennis shoes. In the center of the group, a young man with black hair, a slightly bearded chin, a large nose, an orange T-shirt, and light brown corduroy pants explains the purpose of the upcoming walk. The point, he says, is to learn about wildflowers we did not know were edible or had other kinds of medicinal value, wildflowers we thought were weeds. His name is Evan, the foraging guide, and he is leading the group through the swinging door with the "No Hunting" sign onto the 621 acres of vast and immensely sublime prairie.

Indeed, the Willa Cather Memorial Prairie in Red Cloud, Nebraska, is quite the place. Maintained by the Willa Cather Foundation, a "501(c)(3) not-for-profit organization that works to perpetuate an interest throughout the world in the works, life, and times of Willa Cather through education, historic preservation, and the arts," it is an ecosystem of moving matter meant to resemble the prairies Cather encountered when she first moved to Nebraska.[51] Soft hills, sometimes light and sometimes dark green, occasionally rise into a patch of small trees jutting out from the plateaus above old creek beds. Today, the sky is the lightest of light blues and the clouds the fluffiest of white-tinted-with-grey clouds, the kind of clouds that children admire because they always look like something else. Discarded tree limbs are stacked in piles on the side of short hills that head straight down into valleys below. Mown trails cut through green and brownish-yellow grasses and scattered patches of colorful wildflowers. As we pass through the wooden gate to start the walk, we enter into a more intimate arrangement with this assemblage of moving phenomena: the wildflowers, grasses, dirt, sky, bugs, birds, and air that gather here.

An Arrangement of Mobile Phenomena

A diverse array of mobile, material phenomena arrange together to form this rhetorical moment. After we pass through the entrance gate, we continue walking along the mowed trails. The constant chirping of birds and the rustle of feet walking on dry grasses intermingles with the sounds of human speech moving between walkers, and between walkers and the foraging guide. The movement of my body and the relations between my body and the other phenomena composing this place—the slow movement of the joints, the releasing of endorphins, and the incredibly soft brush of the wind against my arm hairs—help constitute an environment of relaxation. The aroma of grass and wildflower and slight tinges of human sweat completes the enveloping sensory atmosphere. While attending closely to the flora and fauna of the prairie, the human beings undertake different bodily movements. Arriving at wildflowers, we slow down and coordinate our bodies to become physically closer to the grasses, patches of dirt, and wild plants; in order to hear the human speech sounds moving with, and often upstaged by, the sometimes quickly moving wind, we lean our ears in the direction of the forager; we touch specific wildflowers that have their own internally complex and heterogeneous movements, emerging from the movements of soil. Touch, sound, smell, sight: the human senses become activated through encounters with the moving, material histories composing this arrangement.[52]

Throughout the walk the foraging guide interlaces descriptions of prairie flowers with jokes and polemics about the importance of foraging as a way of life. The speech emitted from his body intermingles with the plants, and the vibrations of affect emitted from them, with the sky that serves as the visual backdrop, and with the wind that breaks up any clear apprehension of the flowing sentences. Next to a few piles of cow excrement, speech sounds direct attention to a tall, green thistle: "Oh, here we go....So you look at this and you think it doesn't look incredibly appetizing [audience mumbles: 'that's for sure'], but once they get nice and tall, you can cut off the top and...you kind of peel off the outside and you can cook the inside of the plant like asparagus." The forager's speech continues to refer to this particular type of thistle as a "bully" and thus a good plant to eat. Moving on, audience members engage in conversation and the flows from the foraging guide now move through their bodies: "That plant's really a bully," a fellow walker exclaims. Further into the foraging walk, the pauses in the walking in order to identify wildflowers become more rapid. The guide approaches a thin and tall plant with small yellow leaves and identifies it as a "yellow sweet clover, a leguminous plant which means it actually fixes nitrogen into the soil." The forager explains that we can find these all over the side of the road and that they can be used as a "kind of vanilla-like flavoring." Throughout the walk, audience members ask questions, write notations in their notebooks, and take pictures of the prairie and the wildflowers that grow among it. The audience intermingles through lines of affect, becomes a collective body, and relates more intimately with the forager, leading to jovial questions—like "Okay, but here's the question: how gooood do all of these plants actually taste?"—pointing to a persistent anthropocentrism flowing through this arrangement.

Everything is moving here: bodies, smells, human speech sounds, birds, bird chirps, and cameras. Even the seemingly immobile entities—the wildflowers, signs, dead trees, and dirt—are really "mobile, internally heterogeneous materials whose rate of speed and pace of change are *slow* compared to the duration and velocity of the human bodies participating in and perceiving them."[53] This is a meeting-up of mobile processes temporarily articulated into a regional arrangement, and the mobile phenomena carry with them networks of varying spatial reach: some plants emerge from seeds, and thus plants, very near to this prairie; others arrive from distant places and times. Audience members arrive after participating in the program, and, more broadly, from places as close as Eastern Nebraska and as far as New York. A collection of places, texts, conversations, foods, and symbols encountered while moving as part of foraging culture arrive through, or rather as part of, the body of the foraging guide. One text, in particular, *The Edible Plants of the Prairie,* seems to be ventriloquizing his body, speaking through him as we move.[54] During the walk, the continued influence of the text becomes explicit as he says, "What's cool about doing these sorts of things is that you always learn new things….I didn't know that purple poppy plants were edible, but through finding that book, that pink book, *The Edible Plants of the Prairie,* it pinpoints all of the edible plants that are on this prairie." This constellation of mobilities connects the arrangement to many sites, to distant spaces and times.

Arrangement and Invention

One significance of this regional arrangement, this rhetorical configuration of multiple, mobile processes, is its fundamental *inventiveness.* In the utterances and attitudes emergent during this walk, we find the inventive capacities of this multimodal arrangement at work. As we near the end of the walk, the foraging guide asks us if we want to head back or continue on to a valley of cattails. The audience responds enthusiastically; a bald man with black-rimmed glasses yells, "Cattails! Ho!" Before we arrive, the forager explains the politics of foraging as a way of life: "I do this 'cause I love it. For me, it's about being one with the land…seeing the land this way is about being locked into the great human history. I think we gain resiliency as a species by understanding this stuff. That's why I'm here. It's a love." As we walk, he describes cattails as "the most useful plant on the prairie." A few minutes later, we arrive at a hill that looks upon the valley of cattails. I've never thought of these tall skinny plants with a brown flower that grow in watery areas as particularly visually stunning, but as we look at the valley, I hear a fellow walker loudly exclaim an aesthetic judgment: "They're beautiful!"

Although the utterance was emitted from one person's body, it is not only that body's product but an articulation born from the collective force of this particular arrangement of mobile, material histories: walking on the prairie; the human speech sounds which identified cattails as "the most useful plant on the prairie"; the fact that the forager asked whether we should return to Red Cloud or move toward the valley of cattails; the speech about the importance of foraging; the prairie grasses, the cattails themselves, and spring weather; the set of histories each body brings to the encounter. An inventive arrangement. To phrase it in Doreen Massey's words, this regional arrangement, or this "coming together of trajectories," is a "locus of the generation of new trajectories and new configurations."[55] Or, to refer back to the epigraph written by Manuel De Landa, this "complex mixture of geological, biological, social, and linguistic constructions that are nothing but accumulations of materials shaped and hardened by history" generates materials to circulate after the event desists in its concentrated form.[56] The activity of arranging into regional articulations results in the invention of new materials.

This process, however, is non-linear; sometimes the inventing arrangement becomes influenced by the very inventions it produced. In other words, some of the histories that form an arrangement are the result of feedback loops.[57] For instance, during the foraging demonstration, the foraging guide temporarily halts the motion of the audience as we arrive at a purple flower on a green five-leaved plant: the "purple poppy molly," which has "roots that are very carrot-like." The forager bends down and his fingers graze the tips of the flower petals. His speech explains that the plant has a carrot-shaped root in order to get down to the water that is lying below the land. Moreover, fire will not necessarily destroy these "non-invasive" plants. This is because, he explains, "their hearts are underground." Immediately this figure moves into the body of the person next to me. "Oooh...their hearts are underground," her body exclaims. Now mingling with multiple bodies, the figure produces collective affects in the group of walkers as they mutter and mumble and pay great attention to the interaction. Finally, the figure, and the network of affects generated by its movements, circulates back into arrangement with the body of the foraging guide: "Yeah...yeah...so when it burns off, it doesn't hurt them...Cause a lot of these plants: their hearts are underground. And the woody invaders: their hearts are above ground." Thus, this figure, emergent from the arrangement of histories that makes up the foraging guide-moving-among-the-moving-matter-of-this-prairie (and the plethora of histories that makes up that particular arrangement), arranges with the bodies of audience members, producing affects, utterances, and *mimesis*, inventions which then re-arrange with the body of the foraging guide, altering it, and inventing further explanation of the plant through the perpetuation of this figure.

Critical Analysis of the Arrangement

As a rhetorical event, the prairie foraging demonstration consisted of a temporary meeting-up of trajectories that invented new configurations and new material processes. A new sense of orienting to the beauty of the prairie, for instance, emerged out of the relations of human and non-human phenomena. While there is reason to question the value of beauty as an aesthetic sensibility that promotes consumption, as opposed to mere preservation, of prairie wildflowers, there is reason to believe that such an orientation could have some utility. Indeed, human beings waste a great deal of food because it does not meet the beauty standards inherent in contemporary late-capitalist consumptive practices.[58] Moreover, this was only one of many inventions. Another participant, later in the conference, admitted she now feels the need to make her inherited prairie open to the public. This discloses a *mimetic* impulse as, if the inventions of the experience remain powerful enough, it will result in a body working to replicate certain components of the foraging demonstration. These material re-orientations can shape how bodies move after having experienced this "juxtaposition of previously unrelated trajectories."[59]

The rhetorical arrangement also enhanced the "thing-power" of the wildflowers with regards to their ability to act on human bodies.[60] The prairie flowers gained unique properties when participating in this regional arrangement, rendering sensible (to humans) a future state of the prairie wildflower as food, spice, or medicine. While always latent in the wildflower, this future-state remains muted in natural-cultural arrangements such as that of travel, in which the slowly moving roadside wildflowers, sometimes suffering from their arrangement with deadly chemicals, encounter swiftly moving human bodies, often arranged with automobiles. However, if the event is to disclose this potential state of wildflowers, a greater emphasis on one modality of arrangement and sensation could have potentially enhanced the wildflowers' thing-power: activating eating and the sense of taste. Imagine, for instance, if the audience were to taste the wildflowers, to experience their virtual futures as food and spice, during a group dinner at the end of the demonstration. Such an encounter would have allowed the wildflower to actually enter into that future state, producing memory, and becoming a part of the human bodies as they move elsewhere. What, I wonder, could the wildflower have done, what subsequent influence might it have affected, after becoming-human through the passage of taste? To incorporate this strategy, of course, would require an anthropocentric commitment: that, in order to enhance the rhetorical effectivity of this event on human bodies, a plant's current, living state would be sacrificed.

Finally, particular discursive materials moving through the arrangement selectively revealed the fundamental foreignness that constituted the event as a regional rhetorical arrangement. While some mobilities became revealed as inauthentic, others were concealed by being positioned as markers of spatial authenticity. The Cather Memorial Prairie Wildflower Guide, which walkers carried, explains the preservation of this land in the following terms:

> The Willa Cather Foundation acquired the area in 2006 with the assistance of The Nature Conservancy. With grant funds from The Nebraska Environmental Trust, we have removed thousands of non-native and invasive trees that inhibit the growth of desirable native species such as purple prairie clover, big bluestem, showy milkweed, echinacea and baptisia. The Willa Cather Foundation continues to dedicate significant time and effort to restoring the Prairie to its pre-1900 condition. Responsible grazing practices and the inclusion of fire in our management rotation has also allowed us to control invasive shrubs, and to re-establish natural springs that provide crucial habitat for grassland birds, amphibians, and small mammals. The preservation of the Cather Prairie is part of a holistic approach to the study of American art, history, and culture through the works of Willa Cather, who championed the prairie and its "fierce strength, its peculiar, savage kind of beauty, its uninterrupted mournfulness." We aim to be part of this land's struggle back to itself.[61]

The brochure utilizes a discourse of belonging and, interestingly, while some plants are positioned as not belonging to the prairie proper, Cather becomes that prairie's fiercest champion. Of course, Willa Cather's family migrated to Nebraska from Virginia, and their ancestors came to the United States from Wales.[62] The brochure presents the prairie as moving towards a more authentic state of ecological life but ties that authenticity to settler colonialism.

A regional emphasis accounts for all of the inherent heterogeneous mobilities of a rhetorical encounter: all plants, and soil, and rocks, and people, and pamphlets made from trees, and discourses have migrated here from someplace else, as something else. As Massey argues:

> The rocks of Skiddaw are immigrant rocks, just passing through here, like my sister and me only rather more slowly, and changing all the while. Places as heterogeneous associations. If we can't go "back" home, in the sense that it will have moved on from where we left it, then no more, and in the same sense, can we, on a weekend in a country, go back to nature. It too is moving on.[63]

Regionalism seeks a consistent emphasis on all the mobilities that come into the rhetorical arrangement from somewhere else. Doing so means admitting that the rhetorical arrangement is *thoroughly* saturated with foreignness and that the praise of Cather, far from being a way to access spatial authenticity, belongs to the mobilities of settler colonialism, and, thus, to some degree, this arrangement of semiotic-material elements at Red Cloud celebrates such movements.

Conclusion: On Cultivating Arrangements

Conceptualizing rhetorical events, like that of the prairie foraging demonstration, as regional arrangements means thinking about them as inventive articulations wherein a wide variety of mobile, material phenomena, coming from places of various distances, encounter, affect and are affected by, each other. These complex mixtures, moreover, result in the generation of new material phenomena to circulate into ecologies of varying spatial reach. Thus, if we return to our opening epigraphs, we get a sense of a major goal of this chapter: to think through rhetorical practices in a way responsive to the recent insights of spatial studies and new materialisms in which emergent phenomena, or the production of something new (in this case, vernacular utterances, aesthetic judgments, and the simultaneous celebration of authentic place and settler colonialism) are the result of arrangement. The event of rhetoric always consists of something more than humans or discourse, even if it is significantly characterized by those modes of organization; a site where multiple historical accumulations meet and generate a series of utterances and habits, enhance the "thing-power" of constituent mobilities, and/or reiterate old habits and familiar ways of seeing. In overviewing the history of arrangement with particular emphasis on rhetorical materialisms, and the canon's relation to invention, and inscribing an account of a prairie foraging demonstration, I aimed to build and evidence this perspective.

This perspective on arrangement and invention, moreover, contributes to our understanding of, and justification for, the practice of rhetorical field methods. From this perspective, the invention of theory, or a scholarly article, *also results from an arrangement of material histories.* As Bennett writes, "the sentences of this book also emerged from the confederate agency of many striving macro- and micro-actants: from 'my' memories, intentions, contentions, intestinal bacteria, eyeglasses, and blood sugar, as well as from the plastic computer keyboard, the bird song from the open window, or the air or particulates in the room, to name only a few participants."[64] This necessitates a critic's attention to the field in which they are entangled; studying rhetoric *in situ* matters as the modes of moving matter that make up a live rhetorical moment in which the scholar's body is immersed are active participants in the production of theory. As Barad explains, "the point is not simply to put the observer or knower back into the world (as if the world were a container and we needed merely to acknowledge our situatedness in it) but to understand and take account of the fact that we too are part of the world's differential becoming."[65] Thus, the "rhetorical scenes" that emerge after a rhetorical moment are inventions emergent from the multitude of material histories, the complex mixture that comes into being when the researcher is *immanently* entangled in a rhetorical encounter.[66] Such scenes are not only the critic's inventions. At the same time, the researcher still brings a particular set of histories to the encounter and, thus, the event becomes "diffracted," to use Barad's term, through the body of the researcher; the written account that the event produces will differ depending upon which body, and corresponding histories, becomes the writing apparatus.[67]

If rhetorics are regional arrangements, then the process of building a rhetorical event involves *cultivating arrangements* out of the moving materials in an ecology.[68] From a social change perspective, the goal might be to bring simultaneously existing phenomena into contact—such as plants, humans, soil, and discourses—with the ultimate goal of inventing new materials, or disallowing articulatory processes. Indeed, some materials might become undone by a rhetorical moment.[69] If the prairie foraging demonstration were to incorporate eating, for instance, the wildflower would be undone in its current form. I have focused on invention, but other work might devote attention to arrangement's capacity to "wrench apart."[70] As Connolly notes, the world is full of self-organizing systems which, upon encountering a jolting experience, will restlessly seek "a new resting point." "Such activity," Connolly argues, "may periodically bring something new into the world."[71] That is, as he argues, "self-organization often involves a rhythmic interaction between entities, when one or both has been disrupted. It subsides, to a degree at least, if and when a new equilibrium is established."[72] To cultivate arrangements is to develop micro-habitats wherein multiple material trajectories are brought into more immediate contact in order to generate new materials that then disperse back out into those self-organizing systems to jolt them into a new stasis, or to reinforce pre-existing systemic practices.

Finally, if arrangements invent and humans are irremediably entangled in creative complex systems—if we are, in fact, made up of the stuff of those systems—then the invention of an arrangement does not happen from a position of exteriority. We are all already arranged. As Debra Hawhee states in regards to invention, we operate from "the middle," which is to say that "one invents and is invented, one writes and is written, constitutes and is constituted."[73] At the same time that we are arranging, we are being arranged; other forces are working on us, through us. The prairie foraging demonstration is organized as much by the foundation's workers as it is by the lingering presence of Cather through her writings and her writings' influence on the worker's bodies. What we must do is figure out how to move and operate from within the inside, an action that necessitates both an "attunement to and negotiation of assemblages."[74] It also means maintaining a sense of humility and remaining open to the "slight surprise of action" as other things also do their work through us.[75] Nathaniel Rivers notes, "We each act in time and each act participates in the agonistic composition of a world. There cannot, finally, be a path laid out ahead of us that allows us to save and fix the world.…There is only ever a series of agonistic encounters between wild objects…we cannot fully know what our responsibility will look like."[76] What we can do, I have argued, is become more attuned to arrangement: to the ways we are arranged; the possibilities of organizing arrangements; and the life of arrangement's inventions as they circulate through ecologies into another unfolding rhetorical moment.

Notes

1. *Duck Dynasty* refers to a television series chronicling the activities of the Robertson family, who became wealthy via the production of duck calls for duck hunting. A member of the series, Phil Robertson, became the subject of public controversy after he made derogatory remarks toward LGBTQ publics. He became even more popular among some communities after *A & E*, the station that airs the series, suspended him. See Brian Stelter, "A & E: 'Duck Dynasty' Resuming 'with the Entire Robertson Family,' Including Phil," *CNN.com*, December 28, 2013, accessed November 10, 2014, http://www.cnn.com/.
2. Ben Saase is a US senator from Nebraska, elected during the 2014 mid-term elections.
3. Doreen Massey, *For Space* (Thousand Oaks: Sage, 2005), 111.
4. On ontological choreographies, see Donna Haraway, *Companion Species Manifesto: Dogs, People, and Significant Otherness* (Chicago: Prickly Paradigm Press, 2003).
5. As will become more clearly articulated throughout, the "event of rhetoric" refers to an emergent moment composed of human and non-human materials arranged into a momentary configuration. In this chapter, I am particularly focused on explaining relatively traditional epideictic and deliberative moments as natural-cultural emergent events composed of moving human and non-human materials that find agency in their arrangement. Other works, while still focused on regional arrangements, would likely find other uses of the rhetorical modifier.
6. Much of my take on spatial configuration, and specifically regional arrangements, as a product of the articulation of material phenomena is indebted to Doreen Massey, as, for Massey, spatial configurations are arrangements of simultaneously existing material histories. See Massey, *For Space*.
7. Stormer's use of articulation explicitly ties the concept to spatiality. See Nathan Stormer, "Articulation: A Working Paper on Rhetoric and *Taxis*," *Quarterly Journal of Speech* 90, no. 3 (2004): 270.
8. Jane Bennett, *Vibrant Matter: A Political Ecology of Things* (Durham: Duke University Press, 2010), 3.
9. William Connolly, *The Fragility of Things: Self-Organizing Processes, Neoliberal Fantasies, and Democratic Activism* (Durham: Duke University Press, 2013).
10. Connolly, *The Fragility of Things*.
11. This chapter works from the tradition of rhetorical field methods. See Michael Middleton, Samantha Senda-Cook, and Danielle Endres, "Articulating Rhetorical Field Methods: Challenges and Tensions,"

Western Journal of Communication 75, no. 4 (2011): 388. See also Aaron Hess, "Critical-Rhetorical Ethnography: Re-thinking the Place and Process of Rhetoric," *Communication Studies* 62, no. 2 (2011): 127–152; Michael Middleton, Aaron Hess, Daniel Endres, and Samantha Senda-Cook, *Participatory Critical Rhetoric: Theoretical and Methodological Foundations for Studying Rhetoric In Situ* (Lanham, MD: Lexington Books, 2015). In particular, I worked from a tradition of generating "rhetorical scenes." See Peter Simonson, "The Streets of Laredo: Mercurian Rhetoric and the Obama Campaign," *Western Journal of Communication* 74, no. 1 (2010): 94–126; Joshua P. Ewalt, Jessy J. Ohl, and Damien Smith Pfister, "Activism, Deliberation, and Networked Public Screens: Rhetorical Scenes from the Occupy Moment in Lincoln, Nebraska (Part 1)," *Cultural Studies Critical Methodologies* 13, no. 3 (2013): 173–190. The creation of rhetorical scenes refers to a process of entering into a field in order to create performative imitations of rhetoric. Scenes are not meant to be exact replications, but rather differential iterations of the originary scene of action; there is always a recognition of the participant-observer's role in shaping the inscriptions. The goal of writing rhetorical scenes is to account for rhetorics that might otherwise not be documented, analyzed, or imitated. I will complicate the production of scenes from the perspective of new materialisms later.

12. For an introduction to relational regionalism, see Ash Amin, Doreen Massey, and Nigel Thrift, *Decentering the Nation: A Radical Approach to Regional Inequality* (London: Catalyst, 2003): 5.
13. See also Stormer, "Articulation." My goal in this chapter is to connect discussions of arrangement's materiality with recent conversations in new materialisms, regional theory, and to exemplify theoretical arguments about arrangement by using fieldwork to unpack a specific vernacular rhetorical moment.
14. Amin, Massey, and Thrift, *Decentering the Nation*, 5.
15. On rhetorical production, see also Jodie Nicotra, "Assemblage Rhetorics: Creating New Frameworks for Rhetorical Action," in *Rhetoric, Through Everyday Things,* eds. Scot Barnett and Casey Boyle (Tuscaloosa: University of Alabama Press, 2016): 185–196.
16. George Kennedy, *A New History of Classical Rhetoric* (Princeton: Princeton University Press, 1994).
17. Ibid.
18. Cicero, *De Partitione Oratoria,* trans. H. Rackham (Cambridge: Harvard University Press, 1942): 4.
19. Kennedy, *A New History.*
20. Cicero, *De Partitione Oratoria.*

21. Richard Leo Enos, "Ciceronian Dispositio as an Architecture for Creativity in Composition: A Note for the Affirmative," *Rhetoric Review,* 4, no. 1 (1985): 108–110.
22. Barbara Warnick, *The Sixth Canon: Belletristic Rhetorical Theory and its French Antecedents* (Columbia: University of South Carolina Press, 1993): 55.
23. There are multiple versions of rhetorical materialism. On the one hand, rhetorical materialism has traditionally been a way of theorizing rhetoric that collapses the ideal/real (signified/signifier) distinction at the heart of traditional rhetorical theorizing, and which takes rhetorical activity as an object that moves through the world. In doing so, it also combines the analysis of the signification or "meaning" of a discourse with consideration of its force and function. For an overview, see Michael Calvin McGee, "A Materialist's Conception of Rhetoric," in *Explorations in Rhetoric: Studies in Honor of Douglas Ehninger,* ed. Raymie McKerrow (Glenview, IL: Scott, Foresman, 1982): 23–48; Barbara Biesecker and John Lucaites, "Introduction," in *Rhetoric, Materiality, and Politics,* eds. Barbara Biesecker and John Lucaites (Tuscaloosa, AL: University of Alabama Press, 2009), 1–16; Ronald Greene, "Another Materialist Rhetoric," *Critical Studies in Media Communication* (1998): 21–40; Barbara Biesecker, "Rethinking the Rhetorical Situation from within the Thematic of Différance," *Philosophy and Rhetoric* 22, no. 2 (1989): 110–130. It is important to note, however, that others have contested the materiality of discourse hypothesis in favor of more traditional materialist analyses of the production of ideology. See Dana Cloud, "The Materiality of Discourse as Oxymoron: A Challenge to Critical Rhetoric," *Western Journal of Communication* 58, no. 3 (1994), 141–163. While highly related to this effort to theorize "rhetorical materialism," I locate "new material rhetorics" in its own paragraph. On the theorization of the materiality of discourse through the concept of articulation in particular, see Stormer, "Articulation"; Kevin DeLuca, "Articulation Theory: A Discursive Grounding for Rhetorical Practice," *Philosophy & Rhetoric* 32, no. 4 (1999): 334–348.
24. McGee, "A Materialist's Conception."
25. Green, "Another Materialist Rhetoric."
26. Stormer, "Articulation," 261. For another explicit connection between materiality, rhetoric, and arrangement, wherein arrangement is positioned at an ontological level, see Celeste Condit, "Race and Genetics from a Modal Materialist Perspective," *Quarterly Journal of Speech* 94, no. 4 (2008): 383–406.
27. This is not to say that thunder is inherently non-rhetorical. In particular cultural configurations, thunder becomes affiliated with meanings more

associated with rhetoric. Moreover, following Plugfelder, micro-rhetorics could be argued to permeate this process. See Ehren Helmut Pflugfelder, "Rhetoric's New Materialism: From Micro-Rhetoric to Microbrew," *Rhetoric Society Quarterly* 45, no. 5: 441–461. Rather, this is to say that "thunder on the plains" and "a foraging demonstration" are two different constellations of (non-)human materials, human materials, and semiotic materials, and one is rather easily folded into a larger configuration known as "rhetorical theory." It is precisely this fact—that thunder is disarticulated with the set of meanings that makes up the history of rhetorical theory—that scholarship related to new materialisms aims to address by rethinking our received understanding of rhetoric's ontologies. See Scot Barnett and Casey Boyle, "Introduction: Rhetorical Ontology, or, How to Do Things with Things," in *Rhetoric, Through Everyday Things,* eds. Scot Barnett and Casey Boyle (Tuscaloosa: University of Alabama Press): 1–14.

28. Stormer, "Articulation," 270.
29. Ibid., 264.
30. Massey, *For Space*, 94.
31. Davi Johnson, "Mapping the Meme: A Geographical Approach to Materialist Rhetorical Criticism," *Communication and Critical/Cultural Studies* 4, no. 1 (2007): 32.
32. Thomas Rickert, *Ambient Rhetoric: The Attunements of Rhetorical Being* (Pittsburgh: University of Pittsburgh Press, 2013), 77. See also Rickert's discussion of *chora*.
33. Nathaniel A. Rivers and Maarten Derksen, "Ecologies of Deception in Psychology and Rhetoric," *Quarterly Journal of Speech* 101, no. 4 (2016): 647. It is important to note that Rivers and Derksen, drawing from Edbauer Rice, Rickert, and Barad, make distinctions between ambience and ecologies, with ambience referring to "the irreducible and indeterminate whole from which cuts are made," and an ecology as a "cut made in the name of inquiry." Rivers and Derksen, "Ecologies of Deception," 640. See also Rickert, *Ambient Rhetoric*; Jenny Edbauer, "Unframing Models of Public Distribution: From Rhetorical Situation to Rhetorical Ecologies," *Rhetoric Society Quarterly* 35, no. 4 (2005): 5–24.
34. Peter Simonson, "Reinventing Invention, Again," *Rhetoric Society Quarterly* 44, no. 4 (2014): 313.
35. Ibid., 318.
36. Sarah Hallenbeck, "Objects, Material Commonplaces, and the Invention of the 'New Woman'," in *Rhetoric, Through Everyday Things*, eds. Scot Barnett and Casey Boyle (Tuscaloosa: University of Alabama Press, 2016), 210.
37. See Karen Barad, *Meeting the Universe Halfway: Quantum Physics and the Entanglement of Matter and Meaning* (Durham: Duke University Press, 2007); Bennett, *Vibrant Matter.* A tendency that distinguishes feminist

new materialism from some of the other inquiries into materiality, science, and technology, is its intensive emphasis on relational ontologies over autonomous ontological being. On the distinction between new materialisms and object-oriented ontologies, see Nathaniel A. Rivers, "Deep Ambivalence and Wild Objects: Toward a Strange Environmental Rhetoric," *Rhetoric Society Quarterly* 45, no. 5 (2015): 420–440.

38. Diana Coole and Samantha Frost, "Introducing the New Materialisms," in *New Materialisms: Ontology, Agency, and Politics,* eds. Diana Coole and Samantha Frost (Durham: Duke University Press, 2010), 7.
39. Ibid.
40. On entanglement, see Barad, *Meeting the Universe;* on the agency of the assemblage, see Bennett, *Vibrant Matter;* on intertwined complex systems, see Connolly, *The Fragility of Things.*
41. Bennett, *Vibrant Matter,* 24.
42. Ibid., 21.
43. Ibid., 3.
44. Ibid., 49.
45. Barad, *Meeting the Universe Halfway.*
46. Ibid., 66.
47. Simonson, "Reinventing Invention, Again," 313.
48. Coole and Frost, "Introducing," 7.
49. See Manuel De Landa, *A Thousand Years of Nonlinear History* (Cambridge: Zone Books, 1997).
50. Amin, Massey, and Thrift, *Decentering,* 37.
51. "About the Willa Cather Foundation," *The Willa Cather Foundation,* accessed March 14, 2015, https://www.willacather.org/.
52. See Debra Hawhee, "Rhetoric's Sensorium," *Quarterly Journal of Speech* 101, no. 1 (2015): 2–17.
53. Bennett, *Vibrant Matter,* 58.
54. On ventriloquism, see François Cooren, "Communication Theory at the Center: Ventriloquism and the Communicative Constitution of Reality," *Journal of Communication* 62, no. 1 (2012): 1–20.
55. Massey, *For Space,* 141.
56. De Landa, *A Thousand Years,* 25–26.
57. On nonlinearity and new materialist philosophy, see De Landa, *A Thousand Years.*
58. See Dana Gunders, *Wasted: How America is Losing up to 40% of its Food from Farm to Fork to Landfill* (National Resources Defense Council, 2012). The contemporary adage that "you eat with your eyes first," and the establishment of "presentation" as an area of culinary expertise, popularized, among other places, on the reality television show, *Chopped,* support this point. Thus, there is reason to believe orienting to cattails

as beautiful could enhance the drive towards consumption. At the same time, as Elaine Scarry notes, when a person experiences an object as beautiful, they experience a drive to preserve it, to hold it within one's line of sight, which might counteract a drive to transform the cattail into food. See Elaine Scarry, *On Beauty and Being Just* (Princeton: Princeton University Press, 1999).

59. Massey, *For Space*, 94.
60. Bennett, *Vibrant Matter*, 3.
61. The Willa Cather Foundation, *Willa Cather Memorial Prairie* (Red Cloud, Nebraska: Willa Cather Foundation).
62. Milton Meltzer, *Willa Cather: A Biography* (Minneapolis: Twenty-First Century Books, 2008).
63. Massey, *For Space*, p. 137. I want to note that I also fell under the influence of this sense of an authentic prairie, using words, in earlier drafts, such as "this immaculately cultivated place," and I exhibited a reluctance to critique, both of which were challenged by the editors of this volume. I attribute this, in part, to the enjoyments that continued to linger in my body after walking with this piece of land, and, relatedly, my participation in the histories of settler colonialism. It is also important to note that there were moments during the conference that critiqued, quite thoroughly, the problem of celebrating Cather's writing even though she was silent about the genocide of American Indians on the Great Plains. See Clay Jenkinson, "The Sense of Primordial in Plains Literature," Presentation, 2014 Willa Cather Spring Conference, Red Cloud, Nebraska, June 5–7, 2014.
64. Bennett, *Vibrant Matter*, 23.
65. Barad, *Meeting the Universe Halfway*, 91.
66. For a strong account of fieldwork and immanence, see George F. McHendry Jr., Michael K Middleton, Danielle Endres, Samantha Senda-Cook, and Megan O'Byrne, "Rhetorical Critic(ism)'s Body: Affect and Fieldwork on a Plane of Immanence," *Southern Communication Journal* 79, no. 4 (2014): 293–310. What I am suggesting is that the field critic and the materials being studied participate in the invention of the critical essay.
67. Barad, *Meeting the Universe Halfway*.
68. It is worthwhile to briefly justify the use of the term "cultivation." As Nathaniel Rivers notes, "cultivation suggests at once both the cultivation of plants (as in agriculture or horticulture) and people (as in the cultivation of a following or a society)" and thus "works against the bifurcation of nature and culture…by reinforcing their relationship and making salient the numerous forces and agents participating in it." See Nathaniel Rivers, "Rhetorics of (Non)Symbolic Cultivation," in *Ecology, Writing*

Theory, and New Media: Writing Ecology, ed. Sidney Dobrin (London: Routledge, 2012): 40–41. As such, cultivation seems to be a particularly appropriate term for identifying and orienting to processes of building arrangements out of natural-cultural ecologies that bridges this chapter's theory and case.

69. Massey's explanation of spatiality once again becomes illustrative here: "The specifically spatial within time-space is produced by that—sometimes happenstance, sometimes not—arrangement in relation to each other that is the result of there being a multiplicity of trajectories. In spatial configurations, otherwise unconnected narratives may be brought into contact, or previously connected ones may be wrenched apart." Massey, *For Space*, 111.
70. Massey, *For Space*, 111.
71. Connolly, *The Fragility of Things*, 8.
72. Ibid.
73. Hawhee, "Kairotic Encounters," in *Perspectives on Rhetorical Invention*, eds. Janet Atwill and Janice M. Lauer (Knoxville: University of Tennessee Press, 2002), 18.
74. Nicotra, "Assemblage Rhetorics," 196
75. Bennett, *Vibrant Matter*, 103.
76. Rivers, "Deep Ambivalence," 437.

PART III

Ethical Attunements

CHAPTER 7

Better Footprints

Nathaniel A. Rivers

Prologue: Turtles All the Way Down

Two viral videos have lodged themselves in my environmental unconscious—that part of my mind ever attendant upon and shaped by environmentalism and its concomitant rhetorics. In some ways, these two videos are quite similar: Humans encounter sea turtles in various states of vulnerability, and then they intervene. Yet my responses, my affective engagements, are different.

In the first video, someone works to remove an object from the nostril of a sea turtle (Fig. 7.1).[1] The person struggles mightily. The turtle is in great pain. Blood pours from its nostrils. It takes a while. Then, the object gives up its hold. I find this moment deeply satisfying. Why shouldn't we carefully attend to our fellow creatures? That the object turned out to be a plastic straw carelessly discarded by a fellow human makes this gesture all the more just. Surely, this intervention marks an ethical environmental comportment.

The second video, shot by beachgoers witnessing the mass exodus of baby turtles from a nest, records another sort of intervention.[2] An

N.A. Rivers (✉)
Department of English, Saint Louis University, St. Louis, MO, USA
e-mail: nathaniel.rivers@gmail.com

B. McGreavy et al. (eds.), *Tracing Rhetoric and Material Life*,
Palgrave Studies in Media and Environmental Communication,
https://doi.org/10.1007/978-3-319-65711-0_7

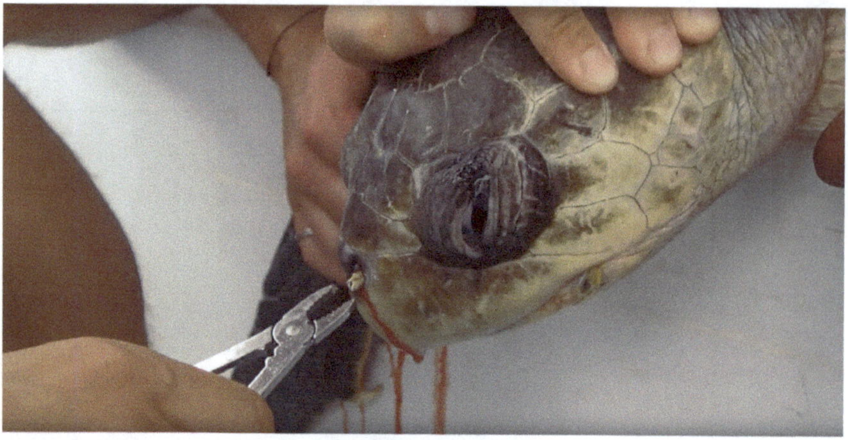

Fig. 7.1 An object is removed from the nostril of a sea turtle

individual captures a truly overwhelming scene of life's emergence. For a while, it is just baby sea turtles pulling themselves out of the nest, making awkward runs to the sea, and valiantly confronting the ocean's vanguard of surf. Except, near the end, one of the last turtles to emerge, perhaps a runt continually forced to the bottom of the pile by its more robust siblings, pokes its head from the sand and struggles to free itself from the nest. Into the frame comes the cameraman's arm to help free to the turtle (Fig. 7.2). This intervention does not sit well with me.

I find myself wanting to know why these encounters feel different and what difference this difference makes, if any. I care because I suspect this difference bears upon larger, important questions of environmental rhetoric: How do we talk, frame and move around environments? Why, in one instance, do I celebrate an intervention and, in other, decry it? Why, finally, do many commonsense environmental axioms provide me with so little guidance? Why am I left wanting for a rhetoric with which to handle this felt distinction?

Tropical Matters

The questions raised in the prologue address consequences. What do these interventions *do*? What marks do they leave upon the world? These questions, then, are about footprints: those of the turtles who emerge

Fig. 7.2 Turtle in hands, "baby sea turtle" by Franco Pecchio (CC BY 2.0)

from the nest and swim in the sea and of the humans who attend to them. The trope of the footprint has inhabited environmental discourse for some time now. When we think of environmental impact it is the *footprint* we think through. Its staying power is impressive and testifies to its potency as an argumentative trope. Since 2000, the phrase "ecological footprint/s" has appeared 5217 times in Google Book's corpus of American texts. The related phrase "carbon footprint/s" has appeared 1494 times.[3] The trope has no doubt shaped people's attitudes and actions in ways environmentalists find amelioratory. Most striking and most pervasive is the trope's symmetrical uptake by large corporations and international governing bodies. Retail giant Walmart's corporate site speaks of "reducing our collective footprint."[4] British Petroleum (BP) uses the term footprint throughout its website's "Sustainability" pages.[5] In should also be noted that footprint is used to describe the corporation's presence throughout the world. In response to penalties assessed following the Gulf of Mexico oil spill, BP sold off several assets.[6] Former Chief Executive Tony Hayward praised the move, remarking, "By disposing of assets worth more to others than to BP we can better align our strategic footprint with our global strengths."[7] The United Nations' "Greening the Blue," which concerns the environmental sustainability of the UN itself, makes use of the trope.[8] The North Atlantic Treaty Organization (NATO) uses the phrase "environmental footprint" in describing its role both in protecting energy infrastructure and in promoting energy efficiency.[9] NATO also deploys the phrase "small footprint" to describe its special forces operations.[10] The trope also emerges in many more unique, momentary usages. For example, Box Steam Brewery delivered beer during Christmas 2015 via reindeer pulling sleds. In addition to capturing the ethos of the season, the move marked a certain environmental comportment. Tony Lockyear, at the brewery, remarked, "we're lowering our carbon footprint this Christmas."[11] Footprints (ecological, carbon, water, and land, which are all terms used in various branches of climate science, each addressing either different resources or environmental contexts) move us to attend to our environmental impact in embodied as well as quantifiable ways: How does what we do discernibly impact our environment?

It should perhaps come as no surprise that the footprint has wormed its way into our discourse and so our thinking about the environment and the consequences of us in it. Parents, for instance, understand the importance of footprints. We collect and save the footprints of our children as soon as they are born. The human footprint stays the same over time, and so it has been used for the purposes of identification and even for tracking. While RFID bracelets and DNA are now embraced as better means of identifying and tracking babies, the footprint impression has become a sentimental legacy of an older system. Replaced as a means of identifying infants, the footprint nevertheless remains a significant marker of birth and presence. *You are here with us.* There is an intimacy to the footprint in its uniqueness and vulnerability. Footprints are also ways we can mark the dynamic presence of the nonhumans with which we inhabit the earth. Waking up after a snowy night reveals the innumerable mammals and birds that move around us. The opossum was back looking for fresh refuse; a pair of mourning doves, up early, has been seeking the worms; a squirrel has reaped the rewards of its autumn digging. And the footprint does what the handprint does not. Certainly a handprint might also work to capture how we mark and manipulate the earth. *Our fingerprints are surely everywhere.* I think the footprint stands out because it strikes us as unavoidable. We do not always touch and manipulate things, but we have to put our feet down.

Fig. 7.3 Banner image of a human footprint for The World Wildlife Fund website

The World Wildlife Fund (WWF) has a page devoted to the footprint, which serves as a clearinghouse for resources for understanding and measuring human impact. The banner image for the page is a human footprint in the sand, which brings to mind the humans on the beach and in the boat (Fig. 7.3). Indeed, an article on this page is about sea turtles and Australia's Great Barrier Reef. Our footprints are here understood as "threats" negatively impacting this rich ocean habitat and the turtles that inhabit it.[12] The impression of the footprint in the sand is an indelible reminder that a human was here. The footprint that adorns the WWF website is a clear, human impression made upon the earth. The particles of sand displaced and compressed to form the footprint shimmer with the sea and the sun.[13] There is though, even here, an intimacy of contact that gestures toward the fullness of the footprint trope that I wish to explore. The image works with but also against the largely quantitative approach developed in the text of the page. The footprint is frequently figured quantitatively, but there lurks an irrepressibly qualitative dimension that renders the trope fruitfully ambiguous. I return to and build from this ambiguity in the latter half of this chapter. Presently, I take aim at the more quantitative employments of the footprint.

Working from Kenneth Burke's discussion of the "Four Master Tropes," I treat the footprint as a *synecdoche*, wherein a part is taken for the whole. The footprint becomes a *representation* (Burke's term) for our impact.[14] As the WWF puts it, "Driving a car, running clothes through a dryer, turning on the air conditioning—there are activities that add up to a larger footprint."[15] Our footprints stand in for all the other ways we impact the earth. It is particularly important that in his discussion of the tropes, Burke writes the following of science (or, more accurately of scientism, which is an attitude that does not always map onto the actual practices of science):

> Science, concerned with process and "processing," is not properly concerned with substance (that is, it is not concerned with "being," as "poetic realism" is). Hence, it *need not be concerned with motivation.* All it need know is correlation.[16]

As I will argue, it is precisely motivation with which we should be terminologically concerned. To again borrow from Burke, we must attend both to means (the agency) as well as mode (the attitude).

There is, in terms of motivation, something that does not quite sit well in the synecdoche of the footprint. The footprint bears productively upon movement and place: It impresses upon us the stakes of our inhabitation. Tim Ingold, whom I turn to below, argues that places are not simply containers for movement but rather are constituted by it.[17] As I hope to show, there is much about the footprint trope, robustly employed, that is useful. Its current use, however, often comes at the cost of privileging a quantitative approach to understanding our human place in and impact upon the environment over a qualitative one. In short, a quantitative approach argues, *the one who does the least does best.* I argue that this approach to environmental ethics comes at a price. The price of the footprint, understood primarily in scientific and economic terms (as the next section traces in more details) is figured as the *amount* of marks we leave rather than the *kind*.

The footprint trope works synecdochically for much contemporary environmentalism, for instance, the quantitative approach to the footprint trope suggests an *ethics of distance* connected with typical notions of wilderness, which this chapter treats as suspect. An ethics of distance is predicated upon the hard-and-fast ontological distinction between humans and nonhumans as well as culture and nature, and on the idea that humans are therefore fundamentally *other* upon the earth. Writing on the work of shaping wilderness areas to meet human expectations of wilderness, Samantha Senda-Cook notes, "we tend to assume that nature and culture are different. Our society considers nature to be those things that are not culture or not humans."[18] This is to suggest neither that environmentalism is reducible to this trope nor that the trope perfectly represents the work of environmentalism: it assuredly does not.[19] However, the footprint's prominence suggests it is at present a concrete, strategic manifestation of environmentalism's ontology. Cronon puts it this way in his critique of the idea of wilderness:

> the trouble with wilderness is that it quietly expresses and reproduces the very values its devotees seek to reject. The flight from history that is very nearly the core of wilderness represents the false hope of an escape from responsibility, the illusion that we can somehow wipe clean the slate of our past and return to the tabula rasa that supposedly existed before we began to leave our marks on the world.[20]

What Cronon describes here, as I will show, undergirds many employments of the footprint trope, which focuses on the marks we leave upon an already formed world.

I want to think more intensely about footprints. I want to re-emphasize movement, embodiment, place, and inscription, which are all sidelined or even implicitly disparaged in the prominent employment of *footprint*. What is it to place one's feet and to inscribe with one's own body? In making environmentalism a question of a footprint's *size*, do we lose the nuance of ***kind***? Everything leaves its mark upon the world. Not all marks are the same, of course, but the simple presence of marks cannot ultimately ground critique. Privileging the quantitative over the qualitative potentially undermines other available means of persuasion. I want to think about/with the footprint metaphor beyond size and measurement—not because such approaches are wrong but because they severely limit our ability to think through the trails we make, the paths we trace as we move with/in an environment, which is something more than an already existing container.

After tracing the emergence of the footprint trope, I read it through an alternate ontology to intensify its ontological dimensions. What ways of being in the world does the footprint de/prescribe, and how does that ontology itself do rhetorical work that may run counter to the aims of its devotees? I conclude with an affirmative recovery of the footprint trope in service of a more robust environmentalism. I am not offering a full-throated critique of the footprint trope, and not only because I do not want to disparage an organization like the WWF, which is trying to save sea turtles, but because I think the trope can yet do good work. The footprint trope, like the image of the human footprint in the sand, is compelling precisely because it marks our relations within the world—our intimacy and vulnerability. It registers our existence, like the footprints we make at the occasion of a child's birth and the footprints we trace in learning about the movement of animals. Our footprints are our weight in the world and, as such, they are also the pull of the world on us. Footprints are as much something done to us as they are things we do ourselves. A footprint is the effect of relations.

Following Footprints

The footprint trope began life as the "ecological footprint," a term first used by William E. Rees in 1992 (footprint has been used earlier in other contexts to describe the landing area of a spacecraft and in sound, as in a sonic footprint). The concept of the ecological footprint was developed by Mathis Wackernagel under the supervision of Rees in the early 1990s.[21] "Ecological footprint" was their attempt at an accessible label for the concept of "appropriated carrying capacity," which is "a measure of biologically productive land and sea area required to meet the needs of human consumption on the earth".[22] The term was inspired by Rees' description of a new computer as having a smaller footprint.[23]

The ecological footprint was also developed to merge ecological and economic modeling. It was designed as a metric to account for the complex ways that people use natural resources. "Such macro ecological realities are often invisible to conventional economic analyses yet have serious implications for world development and sustainability in an era of rapid urbanization and increasing ecological uncertainty."[24] The ecological footprint seeks to model complex ecosystems: not simply geographical proximity but networked distributions of people and resources, both of which move around from place to place. Rees writes, "While we are used to thinking of cities as geographically discrete places, most of the land 'occupied' by their residents lies far beyond their borders."[25] The ecological footprint was built to quantify immense complexity. Its underlying assumptions, however, are powerfully simple:

> Like all other organisms, we survive and grow by extracting energy and materials from those ecosystems of which we are a part. Like all other

> organisms, we "consume" these resources before returning them in altered form to the ecosphere.[26]

The ecological footprint places our habitation in a complex yet finite environmental context.

To return to a previous example, the WWF, in its articulation of the footprint, employs such scientific and economic terminology. The concrete dimensions of footprints are thought to prove useful precisely because they demonstrate measurable dimensions of our uniquely human impact.

> In a typical day, you might drive to work, do a load of laundry, or watch TV in an air-conditioned room. Every one of those actions comes with a price that extends beyond the one measured in dollars and cents withdrawn from a bank—this price comes in the form of natural resources withdrawn from the Earth.[27]

The WWF website actually provides a tool to calculate one's ecological footprint, which it describes as "an indicator of human pressure on nature."[28]

Combining economics and ecology makes sense because "ecology is a science of connectivity, preoccupied with material and energy flows and their relationships to the functional integrity of ecosystems."[29] Economics unto itself ignores "physical connectedness to the ecosphere and the functional properties of exploited ecosystems."[30] Ecology *grounds* economy. This grounding is vitally important in its implications for how we might move in the future. "Failure to appreciate the systemic spatial and structural dimensions of the human ecosystem," Rees writes, "limits the substantive scope for policy responses, confines remedies to the local environment, and often results in the treatment of symptoms rather than causes."[31] The ecological footprint puts all this into play at once and with an eye toward decision-making. Lisa Deutsch reinforces this use of the ecological footprint: "There exists no pervasive recognition of the dependence of humans and our economic systems on complex adaptive ecosystems...We see a great need for ecologically founded tools with communicative power."[32] Paul Templet likewise argues that the model "should also be useful in communicating our dependence on ecosystems to decision makers and the public because it is straightforward and the concepts underlying it are easily understood."[33]

G. Cornelis van Kooten and Erwin H. Bulte write, though with less enthusiasm, that the ecological footprint "is less a scientific measure than one designed to raise public awareness and influence politics. In this regard, it must be considered a success."[34]

The ecological footprint, however, does not reign unchallenged as a model.[35] van Kooten and Bulte go on to argue that the ecological footprint as model is ultimately about the "pursuit of a political agenda."[36] Furthermore, they write,

> The EF [ecological footprint] seeks to measure human impact on nature, but public policy requires much more. It requires an understanding of human behavior and motivation, and the role of institutions, economic incentives and politics. The EF is ill-equipped to address any of these issues, so its claim is largely an empty one.[37]

Spatial welfare economics, proposed by Grazi et al., offers another critique of the ecological footprint in terms of its blindspots. "The [ecological footprint] approach," Grazi et al., argue,

> entirely omits consideration of agglomeration effects and trade advantages. The term 'agglomeration' refers to the clustering of economic activities. An agglomeration effect represents a certain type of positive externality that arises when firms share certain non-excludable inputs, such as labor and communication networks [...] Trade advantages correspond to the benefits a region receives from trading its products with another region. This includes comparative advantage, which reflects that one region has a higher relative productivity in one good than another region, while the reverse holds for another good.[38]

This critique marks the flatness of the ecological footprint model: Not all movements within an ecology are necessarily equal. van Kooten and Bulte make a similar claim, writing that the ecological footprint is "useless for policy analysis where tradeoffs at each moment in time and over time are essential."[39] While explaining Grazi et al.'s model in detail is beyond the scope of this chapter, I note that their critique, like most critiques, hinges upon the soundness of the model and its ability to reliably measure human impacts in ways that promote or prescribe sound and sustainable environmental policy. Once an economic frame has been established, it is only in economic terms that the problem can be dealt with.

Other critiques tend to address the accuracy of the trope in use. They tend to address the trope, in line with Burke's assessment of science, as simply correlational rather than motivational. They assess the technical "correctness" of the term rather than address its ethics. Wright et al. seek to consolidate a universally accepted definition of the carbon footprint specifically so that it can "become a cost-effective, practical and repeatable metric that can be adopted by all types of organizations across the globe as a 'baseline' indicator."[40] The authors remark, with a hint of dissatisfaction, "The use of the term [carbon footprint] has been driven largely by media, government, industry and nongovernmental organizations, captivating the interest of business, consumers and policy makers."[41] This proliferation of the term, the authors suggest, has hindered its pragmatic affectivity in much the same way that *sustainability* has. As Randall Teal writes, "like any phenomenon that achieves popular acceptance, sustainability runs the risk of being degraded in its aims by its very familiarity."[42] Lacking scientific precision, the argument goes, the term is less useful.

Battening down the hatches, Wright et al. argue, "To be practically useful, a carbon footprint should be an indicator of the anthropogenic contribution of a named process/product/land area (e.g., city or country) to climate change."[43] Securing a universally accepted definition of the footprint makes a great deal of sense in this regard. Wright et al. are particularly compelling because of their strong emphasis on decision making. This investment makes their critique rhetorically significant. They write,

> The carbon footprint must be recognized as a *tool for assisting decision making*, rather than a definitive answer. If decisions based on the carbon footprint indicator are correct the majority of the time, it is better than the alternative of no indicator and decision making with ignorance.[44]

Given the complexity of ecologies, which these authors as researchers daily confront, these models cannot, of course, know everything.[45] Wright et al. emphasize the need for an accurate model because of the footprint's synechdotal quality.

Their insistence that the model only "assists," however, suggests an aversion to the question of motivation. Note their use of the word proxy: "Taking these key components into consideration, the carbon footprint should act as a proxy indicator of the contributions to anthropogenic climate change made by a process, product, activity or population, accounting for the most prominent anthropogenic [gases]."[46] As Burke predicts, this assistive proxy aims largely at correlation, at measurement. But Wright et al.'s preference for accounting does not mean they are silent on implications (even if they do not explicitly address motive). "Agreement on the definition of a carbon footprint," they write, "and a method for estimating a personal carbon footprint are essential, if any decision to pursue individual carbon credits or taxation is followed."[47] For Wright et al., as for Rees and Grazi et al., securing this model is the key to sound policy. "Without a method of fair, equitable and accurate apportioning of emissions," they continue, "trading or taxation on carbon emissions will be impossible."[48] Wright et al. will get no argument from me within these confines. It would no doubt be impossible to design emissions regulations without a good accounting scheme. But the footprint, as a trope, ranges beyond this particular context.

I return to Teal's analogous treatment of sustainability. Writing as an architecture scholar, Teal's focus, much like Wright et al.'s, is on policy and the practical implications of terminology: he likewise attends to the rhetorical. Teal introduces motivational complexities with respect to sustainability. Writing critically of sustainability measures such as LEED (Leadership in Energy and Environment Design), Teal argues, "the basic juridical limitations of LEED are complicated by [among several things] its focus on scoring points … In short, LEED's value is the same as its danger; it lays a number of complicated interdependencies critical to sustainability in the built environment boldly and clearly on the table."[49] Of this phenomenon, Teal writes, informed by Martin Heidegger, "such 'solutions' for sustainability would only result in the earth being confirmed as a mere collection of resources."[50] As is perhaps to be expected, measuring things enacts certain assumptions about them. Objectivity demands objects.

The juridical features of LEED critiqued by Teal can be analogously tied to the quantitative features of footprint models. For example, there is Allan Stoekl's treatment of the carbon footprint in his contribution to *Prismatic Ecology*: "Worth stressing here is that to be calculable the earth's systems must be reformulated as finite. The calculable entails the measurable, which in turn entails eliminating the infinite, that is, all that cannot be reduced to a single number that goes into a formula."[51] Teal writes, "For if resources are all that sustainability gives us then it will have failed, as it will have forgotten such human questions as poverty, over-population, social justice, economic viability, education, affordable housing, and quality of life."[52] This is also the risk of the ecological footprint: Environmental awareness is bought at the expense of ontological complexity.

Karen Pinkus attends to the ontological complexity lost in quantitative treatments of the footprint in her exploration of the place of *ambivalence*, *ambience*, and *ambiguity* in environmental science. In particular, she emphasizes the difficulties of introducing the human into models of environmental impact—not just rates of consumption but human attitudes and action. How can these "human" factors be accounted for?

> The "human" element introduces fuzzy data or ambiguity into the environment (or ambience). Ideally, then, variables (or valences) can be reduced by a hybrid method, "combining the advantages of human thought-process structures and processes of biological development with the logical and analytic accuracy of computers" to arrive at a series of "aggregated one-dimensional numbers."[53]

Such numbers are in large part what the ecological footprint seeks to produce: These numbers lay such environmental aggregates "boldly and clearly on the table."[54] Pinkus continues, "Climate science, then, takes into account 'a human dimension' but attempts to reduce its variables as much as possible."[55] This reduction is of a particular quality; it is made by defining "a boundary between subject (humans) and object (ambience)."[56] In other words, to reliably measure an environment as a finite resource, humans must be made an externality, as that which is not a part of but apart from. The nature/culture distinction here flexes its muscles. It is not so much that we need to factor human complexities into our models but that humans cannot be abstracted or pulled from out of the environments they are a part of in order to understand those very same environments. The footprint marks not only impact but also interaction.

***

My tracing of the footprint is far from complete, choosing as I have only some key works throughout the trope's life and operation. I have emphasized the trope's emergence and its contemporary status within the scientific literature. This work demonstrates our irreducible entanglement with our environment. In its necessary insistence upon the quantitative and the economic, however, the ecological footprint suggests a certain way of being. Despite the move toward more ecological models designed to supplement otherwise purely economic models, it should be clear by now how the logic of *economy* dominates the footprint trope. For Bruno Latour, economy is itself a "mode of existence," which he describes at length in *Politics of Nature* and more recently in *An Inquiry in the Modes of Existence*. Latour argues that economics is an attempt to integrate the diverse elements of the world quickly and efficiently, and in terms of "producers," "consumers," and "goods."[57] This integration allows economy to "establish a hierarchy of solutions" "in order to discover the optimum in its allocation of resources."[58] As part of establishing this hierarchy, Callon, Lascoumes and Barthe argue in *Acting in an Uncertain World*, economics must "draw a strict dividing line between that which is taken into account and that which is not."[59] "The market," they write, "is efficient because it is able to *frame the problems* and *not get entangled* in all the overflows and side effects that it might generate."[60] As we have seen already, economics needs to simplify. However, simplification is not the only potential trouble that comes with economic models. Economics does not just simplify by drawing lines between what is counted and what is not; it also simplifies by truncating how those lines are drawn and by whom. "[I]n order to produce the optimum," Latour writes, "economics does not burden itself with any consultation."[61] Economy is a mode of existence, a way of being, that wants to abrogate all other modes unto itself. Economics attempts to short-circuit politics, another Latourian mode of existence, and with it negotiation, compromise, and the slow, continual work of composing a common world.

Economic appeals to what the earth is and what its limits are can cut both ways. Note Gary Johnson's 2011 remarks (recently recirculated) on climate change and what the government's role might be in ameliorating it. He grants that the earth is getting warmer *and* that this is "man caused." But, basically, he intones, *so what?* He has *accounted* for the impact and interaction; he just does not feel *responsible*. "In billions of years the sun is going to actually grow and encompass the Earth, right? So global warming is in our future."[62] That said, "If government gets involved [in fighting climate change] we are going to be spending trillions of dollars and have no effect whatsoever on the desired outcome."[63] Given the inevitability of radical global warming, we'd be better off making money building more coal-fired plants. Economics finds that optimum even if (or especially if) it is not what the body politic desires. Johnson's relationship to climate change is purely economic. What his troubling statements reveal is a cost-benefit analysis. It is as if

Johnson used the WWF footprint calculator and decided he could afford it and make a little money at the same time.

What makes this soundbite so outrageous is not its bizarre pessimism or subsequent unvarnished capitalism but its erasure of our living between now and the end of the world. The earth really is going to melt away when the sun impacts it. The sun's footprint knows no mercy. When it goes supernova, the earth will well and truly be used all up. This, we know, will probably happen. What is disturbing about his callous logic is the *meantime* we will all have to live in. Whatever the destiny of the earth—whatever the earth might be without us—we will still have to live here until then. The footprints we leave are not simply blemishes on a doomed planet but the making of a place we will have to inhabit. It is also troubling that the "we" assumed by Johnson is too small, too human. There is no sense of interdependence or care for the world that presents itself to Johnson as an economic resource. Any ethics that disparages relations as such and the traces they leave can be no ethics, which are fundamentally relational.

A more robust and complex ecology, which I develop in the next section, adds wrinkles and hiccups to the seemingly smooth and cold calculations of the footprint as deployed above. Such a complex ontology would necessarily complicate the idea that a simple shrinking of our impact is sufficient in imagining our possible relations with the whole wide world.

Returning Footprints

If there are ontological limitations in the way we currently employ the footprint trope, then how might we re-turn the trope to something more moving? The footprint synecdoche we are familiar with does good work, but it hems us in. But there is more to the footprint than the scientists above suggest. While the footprint calculator and much of the language on the WWF website suggests a purely quantitative performance of the trope, the moving image of the footprint in the sand captures an intimacy in which a qualitative aspect resides. It could be not only a way of modeling and measuring our impact quantitatively but also a way of motivating qualitatively different ways of being in the world. After all this I do not want to give up on the footprint just yet. It has an appeal I cannot deny. As I described it in the introduction, there is an intimacy and immediacy to the footprint trope: *You are here.* There is also a traceability to the trope: *Here is how you came to be here*. And there is a futurity and consequentiality: *Where will you go next*? Rather than disregard them entirely, how can we otherwise inhabit our own footprints?

Tim Ingold, an anthropologist working primarily with non-Western ontologies, provides one way to intensify our notion of the footprint. Ingold works to more or less track down and root out the nature/culture distinction and the limits it places on anthropology's ability to understand non-Western ways of being in the world. The nature/culture split undergirds much of the thinking around the footprint in activating the assumption that nature is apart from us as something that we act upon. Echoing other treatments of walking, such as Michel de Certeau's, Ingold writes, "I suggest that to understand how people do not just occupy but *inhabit* the environments in which they dwell, we might do better to revert from the assembly to that of the walk."[64] Ingold provides an attitude adjustment that does not necessarily preclude the current scientific deployment of the footprint trope. His is another turn, another version of the trope. In short, the distinction he makes between *transport* (wherein one travels *across* the surface of the earth) and wayfaring (wherein one moves *within* and *along* the earth in such a way as to generate its surfaces) allows for an articulation of the footprint trope defined less as a solely quantitative impact and more as an equally qualitative engagement.

For the last two decades, Ingold has written eloquently and persuasively of what he calls wayfaring or wayfinding[65]:

> Wayfaring, I believe, is the most fundamental mode by which living beings, both human and non-human, inhabit the earth. By habitation I do not mean taking one's place in a world that has already been prepared in advance for the populations that arrive to reside there. The inhabitant is rather one who participates from within in the very process of the world's continual coming into being and who, in laying a trail of life, contributes to its weave and texture.[66]

Wayfaring is about "place-*making*," and it is a way of understanding place outside the nature/culture and human/nonhuman distinctions.[67] The world is not already here for us as a resource that we then either use or abuse, impact or preserve. Footprints are not marks upon a complete world but are the very means by which a world is brought into existence.

There is no carrying capacity as such, which is not to belittle this concept but rather to point again toward the ways the earth is not ready-made but always becoming, being made. There is, then, an ethical orientation in wayfaring that I see as a necessary complement or complexification of the standard footprint trope. This ethical orientation is embedded in the very idea of "coming into being."[68] We make the world, in part, alongside others. And so *along* is an important preposition for Ingold. There is a subtle but important difference here. We cannot simply measure the quantity of our impact as our sole form of accountability; we are on the hook for so much more. Like the straw in the sea turtle's nose, we are on the hook with our movement, the trails our footprints compose and contribute to the world's texture, its being, which is neither knowable in advance nor settled for very long. "The inhabited world," Ingold writes, "is a reticulate meshwork of such trails, which is continually being woven as life goes on along them."[69] These trails are made of our footprints, which, rather than leaving marks upon some static substrate, make up the threads weaving that very surface. This is an ethical relation not at play in our current, commonsense employment of the footprint trope with its emphasis on minimizing our impact.[70]

Our current commonsense use of the footprint trope is much more in line with what Ingold labels *transport*, which he contrasts with wayfaring as a mode or way of being. "Transport…is essentially destination-oriented," Ingold writes.[71] "It is not so much a development *along* a way of life as a carrying *across*, from location to location, of people and goods in such a way as to leave their basic natures unaffected."[72] Ingold's term, transport, applies to the ecological footprint as developed by Rees and Wackernagel. Transport takes the world as given and sees its movement as taking place *across* or *on* the surface. With respect to the ecological footprint, there is a painful irony here. In many ways, the ecological footprint wants precisely to attend to the *along* of wayfaring: It wants to see people and natural resources as complexly enmeshed. However, in its predominantly quantitative model it winds up privileging *across*-ness. Again, the WWF website uses language like "natural resources withdrawn from the Earth."[73] This language, perhaps unintentionally, treats the earth as a distinct object to be mined (or preserved) from which we are apart, rather than an ongoing process of which we a part.

To return to Rees and Wackernagel, the language of carrying-capacity that lives in the ecological footprint suggests to me that we are *on* the earth, which carries us. That carrying-capacity is then rendered in terms of resources suggests an attitude of transport. The prepositions of *along* and *across* matter for the distinction between transport and wayfinding, which is fitting as prepositions attend to relations, and the manner of those relations. The prepositions very much serve my argument concerning an expanded treatment or appreciation of the footprint metaphor. "To go along … is to thread one's way *through* the world rather than routing from point to point *across* its surface. Indeed for the wayfarer the world, as such, has no surface."[74] What kind of intimacy is suggested by certain prepositions? What ways of being are packed into them? It is fair to say that intimate sociality inheres in some relations more than others?

When the footprint is a purely quantitative measure, it exists as a mode of transport. The reduction of our response-ability to accountability, and from there to forms of responsibility that privilege detachment, favors transport. Do I simply want to *account* for the impact of the interventions described in my prologue or do I need to think through them as the kinds of interactions to which I am able to *respond*? In transport we move *across* or *on* the landscape. It is the mode of transport that Gary Johnson performs in skipping *across* time and place to the end of the world; he has not imagined (or simply does not care about) living *along* from now until then. However, and as Ingold argues, "We cannot get from one place to another by leap-frogging the world."[75] We thus need footprints to mark not transport but wayfaring. It is important to note, though, that I am not arguing that the ecological footprint, and other such models and measurements, mark transport *de jure*, only *de facto*. The distinction between wayfaring and transport is neither hard and fast nor set in stone. It is an attitude that can be adjusted. "Transport … is distinguished not by the employment of mechanical means but by the dissolution of the intimate bond that, in wayfaring, couple locomotion and perception."[76] With a change in attitude, quantitative measures too can wayfare.

Wayfaring points toward our inescapable relatedness. There is no way out, no way of moving that does not leave marks. We can measure those marks as best we can, but those measurements will not settle the question. Furthermore, when we treat our footprints only as a measure of that which we pull out of the earth, we mistakenly place ourselves outside and upon the earth rather than inside and along the earth. Here is where a steadfast devotion to the nature/culture distinction (and the co-morbid human/nonhuman binary) is at work. Here I follow

Jane Bennett's argument in *Vibrant Matter* to begin complicating that distinction, and its manifestation in the ontology of the footprint:

> But one thing I have noticed is that as I shift from environmentalism to vital materialism, from a world of nature versus culture to a heterogeneous monism of vibrant bodies, I find the ground beneath my old ethical maxim, "tread lightly on the earth," to be less solid.[77]

Distinctions that draw strong ontological lines between supposedly unique human activity and the world do just that. They take us out of the world and our responsibility to it. "What matters is not how fast one moves" for instance, Ingold writes, reminding us that transport is marked by mode rather than means, "but what this movement should be in phase with, or attuned to, the movements of other phenomena of the inhabited world."[78] In his articulation of wayfaring Ingold gets very close to rhetoric when he cites Mary Carruthers: "'The rhetorical concept of *ductus*,' Mary Carruthers explains, 'emphasizes way-finding by organizing the structure of any composition as a journey through a linked series of stages, each of which has its own characteristic flow.'"[79] Imagined as *ductus*—the Latin word for to lead, and from which we get such words as seduction, abduction, and transduction—rhetoric is that which *leads us along*. The invocation of composition *and* rhetoric likewise introduces the concept of characteristic flows that suggests there is a range of ways to wayfare. I appreciate the nod to rhetoric here because that is where the footprint, fully considered, inexorably leads us.

Flows and footprints are journeys and compositions. These flows, like those of rivers, glaciers, migrating wildebeests, and the jet stream, make the earth. We can and should measure and model them all. But those models would benefit from recognizing such movement as wayfaring. Glaciers do not move *across* the surface of the earth; their paths move *along*, weaving what will be the surface. Glaciers leave, to risk a momentary flight of anthropomorphic fantasy, footprints in the form of valleys, rivers, and gravel. The sediments deposited by glaciers, to borrow from Kathleen Stewart, constitute the contingent ground in which we stand.[80] Wayfaring, in intensifying the footprint trope, ratchets-up our ethical obligations. We are not simply using or using-up the earth; we are weaving, in part, what the earth will be. That cannot be resolved purely by the accuracy of our models. Rather, it has to be negotiated among possible flows.

Epilogue: Turtles (and Everything Else) All the Way Down

I now turn, finally, to the title of this chapter: What does *better* do? What does it matter to speak of *better* footprints? How do these questions help me sort through and move through my ambivalence in the face of the two turtle videos in the prologue? First, *better* does not prohibit asking after the relative size of footprints. It does not dismiss or undermine scientific and/or economic models and measurements. Rather, it places those forms of accounting *along*-side other forms of response-ability. The question here is not only about accounting for what the earth has to offer—what it can provide—but in imagining, deliberating, and being responsible for what the earth might yet be. Second, *better* prompts the question of response-ability in ways *smaller* does not. That is, to speak of the *quality* of footprints immediately foregrounds the rhetorical stakes of footprints and our adjudication of them. *Better* ratchets up the evaluative stakes and rhetorical dimensions of decisions about the environment. When someone says *smaller*, one grabs the scales; when someone says *better*, one generates a list of pressing questions and claims. We ask ourselves *better for what*? Latour reminds us that economics works precisely to circumscribe complex and collective consultation.[81] I remain fully committed to the footprint: In all of its manifestations, it can and does attend to our emplacement in the world. We live here and nowhere else, but we need a more intense enactment of footprints; not simply as the impacts we make as we move *across*, but as our way of being *in* the world. *Smaller* suggests transport; *better* compels wayfaring.

Returning to the videos with which I started, I must admit that it is not any easier to sort them now than it was then. Indeed, I am still unsure of my feelings about the examples as well as my judgments. What wayfaring suggests to me, though, is that the qualitative approach to judgment I have been pursuing is always going to be hard. No shortcuts, which only look easier because they skip so much. We cannot leapfrog our own being in the world. There is only ever the long way, which we have to walk together. We cannot tread lightly, only differently, and, we hope, better. Judgment itself has to be a form of wayfaring. The instinct to measure and account is an important one. What are the discernible consequences of helping the baby turtle? What is at stake in removing the straw? Is it worth considering that the turtle with the straw in its nostril was in danger because of a prior, thoughtless engagement with humans and that the other was simply a result of an ecology red in tooth and claw? Perhaps this is precisely the different that makes a difference in this case. But the difference surely cannot become axiomatic: it is not teleological but contingent.[82]

Ingold argues that humans, far from living lives abstracted from the world, "play a small part, along with the innumerable other beings—human, animal, spiritual—that have inhabited [in this particular case] the forest at one time or another, in creating the environment in which people now live, and from which they draw their sense of being."[83] This is the work I think the footprint trope can do: the work of developing "techniques for engaging with fellow humans" and "for engaging with the animals and plants on which life depends."[84] Such techniques do not emancipate us from "an alien world of nature," but "draw the inhabitants of that world into an unbounded sphere of intimate sociality."[85] What are the turtles writing when they slowly pull themselves to shore? Thinking fully about our footprints, where the meaning of *better* is precisely what is at stake, is to think complexly about how the world is constantly composed. Intimate sociality rather than lessened impact ought to be the trope's aim.

Another turtle story here at the end; another set of footprints to trace. The setting is the Ostional Wildlife Refuge in Costa Rica. Humans (both visitors and locals) gather in large numbers on the beach as Olive Ridley sea turtles (also in large numbers) come in from the ocean to lay their eggs.[86] The visitors (frequently referred to as *tourists* or *ecotourists* in the various news items) begin snapping photographs and posing for selfies (Fig. 7.4). Children are placed, it is claimed, on the backs of turtle shells. A traffic jam ensues with turtles blocked from their usual nesting sites. According to some reports, anxious or otherwise frustrated turtles returned to the sea having not laid eggs. At the same time, locals, according to accepted, legal practice, are there to collect eggs to eat during the first three days of the nesting period (Fig. 7.5). "Under a law written especially for Ostional," writes John Burnett, "the government allows [not without the objections of some scientists] an egg harvesting cooperative to collect all they can during the first 36 hours of every arribada [large scale nesting] ... In return, the community must protect the olive ridley."[87] The three or so park rangers on duty are quickly overwhelmed by the sheer number of people present. Confusion abounds.

Fig. 7.4 Visitors at the Ostional Wildlife Refuge in Costa Rica

There is no sure way to sort these behaviors, these footprints, ahead of time. These two sets of human footprints on this Costa Rican beach surely are not equal, but their difference cannot simply be assessed in terms of quantity. This is not to say that such measurements have no place here; carrying capacity is not infinite, and knowing what we are using and what is available is crucial. Such a sorting, however, prevents us from making more fine-grained, kairotic distinctions, which are precisely what we need in adjudicating these moments, these movements. To measure solely in these quantitative terms is to repeat the mistake of transport: to move *across* rather than *along*. The footprints here could only be impacts upon a separate and stable substrate distinct and divorced from those who walk across it—who can only ever be threats at worst, users at best. Impeding turtles from laying eggs or collecting those eggs, when all is said and done, are perhaps equally impactful: We could certainly measure that. What we cannot quantify is the attitude at work in both cases. I suspect one could make a case that park visitors are moving *across* the beach—they are engaging in a form of transport where the beach is but a destination. The locals on the other hand are

Fig. 7.5 Locals collect eggs to eat

wayfaring. They are moving *along* the beach not as a form of transport but as a place they are inhabiting. This does not mean, by the way, that locals will always enact wayfaring. Anyone can occupy anywhere. Anyone can inhabit anywhere. Each time, in short, we will have to trace the modes of being at work and decide accordingly.[88] There is no place outside of this place where we can finally measure what we've done—no earth that is finally *the* earth. We all get the earth we make together—turtles all the way down, sand all the way down, ocean all the way down, footprints all the way down.

Notes

1. CostaRicanSeaTurtles, "Sea Turtle with Straw up its Nostril—"'NO' TO PLASTIC STRAWS," YouTube video, Published on Aug 10, 2015, https://youtu.be/4wH878t78bw.
2. G4ViralVideos, "Baby Turtles Being Born on the Beach," YouTube video, accessed November 14, 2015, https://youtu.be/WEIzTDu-jx4.
3. These raw numbers were accessed using Mark Davies' modified version of the Google Books ngram viewer interface.
4. "Reducing Energy Intensity and Emissions," Walmart, accessed January 26, 2016, https://corporate.walmart.com/.
5. "Sustainability," BP Global, accessed January 26, 2016, http://www.bp.com/.
6. "BP Sets Out Gulf of Mexico Costs," BP Global, accessed January 26, 2016, http://www.bp.com/.
7. Ibid.
8. "UN and Sustainability," United Nations, accessed January 26, 2016, http://www.un.org/.
9. "NATO Role in Energy Efficiency," North Atlantic Treaty Organization, accessed January 26, 2016, http://www.nato.int/.
10. "Special Operations Forces—A Small Footprint," North Atlantic Treaty Organization, accessed January 26, 2016, http://www.nato.int/.
11. Alex Swerdloff, "This British Brewery is Using Reindeer to Deliver Beer," Munchies, Published Dec 19, 2015, https://munchies.vice.com/.
12. "The Human Footprint," WWF, accessed September 1, 2016, http://www.worldwildlife.org/.
13. Strangely enough, my thoughts also turn to snow and to penguins, and to Werner Herzog. His documentary *Encounters at the End of the World* captures two heartbreaking scenes of disoriented penguins—Herzog speculates that the penguins might be "deranged"—walking away from their colony and the open ocean. They are waddling, assuredly, to their deaths. This happens in the documentary on two occasions: once from afar and later close up, when the crew encounters a penguin already 80 kilometers "from where it should be." The rule on Antarctica is to not interfere with the movement of penguins no matter what. "Do not disturb or hold up the penguin," Herzog is told. "Stand still. And let him go on his way." So the crew watches and films as long as they can while the penguin waddles away. They cannot intervene; and besides, someone explains to Herzog, the penguin, even once retrieved and returned to the colony, would likely repeat its run to "certain death." *Encounters at the End of the World*, directed by Werner Herzog (Discovery Films, 2007), DVD.

14. Kenneth Burke, *A Grammar of Motives* (Berkeley: University of California Press, 1969).
15. "The Human Footprint."
16. Burke, *Grammar*, 505, emphasis added.
17. Tim Ingold, *Being Alive* (New York: Routledge, 2011). 149.
18. Samantha Senda-Cook, "Materializing Tensions: How Maps and Trails Mediate Nature," *Environmental Communication* 7, no. 3 (2013): 357.
19. Work in areas like environmental justice has often led the way in bringing to environmentalism this critique or moving beyond approaches that see the environment as existing elsewhere as wilderness (i.e., conservation and preservation). See Robert Bullard, *Dumping in Dixie: Race, Class, and Environmental Quality* (Boulder: Westview Press, 2000); Ronald Sandler and Phaedra Pezzullo, *Environmental Justice and Environmentalism: The Social Justice Challenge to the Environmental Movement* (Cambridge: MIT Press, 2007). The focus on wilderness and environmentalism as the reduction of impact can create blindspots with respect urban areas for instance. With environmental justice, the water crisis in Flint Michigan, for instance, is an environmental issue.
20. William Cronon, "The Trouble with Wilderness; Or, Getting Back to the Wrong Nature," *Environmental History* 1, no. 1 (1996): 16.
21. Mathis Wackernagel and William E. Rees, *Our Ecological Footprint: Reducing Human Impact on the Earth* (Gabriola Island: New Society Publishers, 1996).
22. Laurence Wright, Simon Kemp, and Ian Williams, "'Carbon Footprinting': Towards a Universally Accepted Definition," *Carbon Management* 2, no. 1 (2011): 62.
23. William Safire, "Footprint," *The New York Times*, accessed November 14, 2015.
24. William Rees, "Ecological Footprints and Appropriated Carrying Capacity: What Urban Economics Leaves Out," *Environment and Urbanization* 4, no. 2 (1992): 121.
25. Ibid.
26. Ibid., 122.
27. "The Human Footprint."
28. "Reduce your impact," WWF, accessed October 1, 2016, http://wwf.panda.org/.
29. Rees, "Ecological Footprints," 123.
30. Ibid.
31. Ibid., 124.
32. Lisa Deutsch, Åsa Johnson, Max Troell, Patrick Rönnbäck, Carl Folke, and Nils Kautsky, "The 'Ecological Footprint': Communicating Human Dependence on Nature's Work," *Ecological Economics* 32 (2000): 354.

33. Paul Templet, "Externalities, Subsidies and the Ecological Footprint: An Empirical Analysis." *Ecological Economics* 32 (2000): 383.
34. G. Cornekis van Kooten and Erwin H. Bulte, "The Ecological Footprint: Useful Science or Politics?" *Ecological Economics* 32 (2000): 385.
35. For a lengthy commentary on the ecological footprint model, see *Ecological Economics* 32 (2000). For additional critiques, see Nathan Fiala, "Measuring Sustainability: Why the Ecological Footprint is Bad Economics and Bad Environmental Science," *Ecological Economics* 67 (2008): 519–525; Jeroen C.J.M. van den Bergh and Harmen Verbruggen, "Spatial Sustainability, Trade and Indicators: An Evaluation of the Ecological Footprint," *Ecological Economics* 29 (1999): 61–72.
36. van Kooten and Bulte, "The Ecological Footprint," 383.
37. Ibid., 388.
38. Fabio Grazi, Jeroen C.J.M. van den Bergh and Piet Rietveld, "Spatial Welfare Economics Versus Ecological Footprint: Modeling Agglomeration, Externalities and Trade," *Environmental and Resource Economics* 38 (2007): 137.
39. van Kooten and Bulte, "The Ecological Footprint," 385.
40. Wright et al., "Carbon Footprinting," 61.
41. Ibid.
42. Randall Teal, "The Process of Place: A Temporal View of Sustainability in the Built Environment," *Environmental Philosophy* 7, no.1 (2010): 63–78.
43. Wright et al., "Carbon Footprinting," 64.
44. Ibid., 69, emphasis added.
45. In this way Wright et al. echo John Locke, when the latter writes, "It is of great use to the sailor to know the length of his line, though he cannot with it fathom all the depths of the ocean. It is well he knows that it is long enough to reach the bottom, at such places as are necessary to direct his voyage, and caution him against running upon shoals that may ruin him. Our business here is not to know all things, but those which concern our conduct." John Locke, *An Essay Concerning Human Understanding* (New York: Penguin Classics, 1998), 58.
46. Wright et al., "Carbon Footprinting," 69.
47. Ibid., 68.
48. Ibid.
49. Teal, "The Process of Place," 4.
50. Ibid., 13.
51. Allan Stoekl, "Chartreuse," in *Prismatic Ecology: Ecotheory Beyond Green*, ed. Jeffrey Jerome Cohen (Minneapolis: University of Minnesota, 2013), 134.
52. Teal, "The Process of Place," 14.

53. Karen Pinkus, "Ambiguity, Ambience, Ambivalence, and the Environment." *Common Knowledge* 19, no. 1 (2012): 95.
54. Teal, "The Process of Place," 4.
55. Pinkus, "Ambiguity," 95.
56. Ibid., 94.
57. Bruno Latour, *Politics of Nature: How to Bring Science into Democracy* (Cambridge: Harvard University Press, 2004), 132. See also Bruno Latour, *An Inquiry into Modes of Existence* (Cambridge: Harvard University Press, 2013).
58. Ibid.
59. Michel Callon, Pierre Lascoumes, and Yannick Barthe, *Acting in an Uncertain World: An Essay on Technical Democracy* (Boston: MIT Press, 2011), 236.
60. Ibid., emphasis added.
61. Latour, *Politics of Nature*,134.
62. Jeremy Schulman, "Gary Johnson Wants to Ignore Climate Change Because the Sun Will Destroy the Earth One Day," Mother Jones, Published Sep 22, 2016, http://www.motherjones.com/.
63. Ibid.
64. Tim Ingold, *Lines* (New York: Routledge, 2007). 75.
65. In earlier versions of this argument (*The Perception of the Environment*) Ingold used the term *wayfinding*. More recently (*Lines* and *Being Alive*) he has used *wayfaring*. His definitions of both terms are quite similar.
66. Ingold, *Lines*, 81.
67. Ibid., 101.
68. Ibid., 81.
69. Ibid., 81 and 84.
70. "The Human Footprint."
71. Ingold, *Being*, 150.
72. Ibid.
73. "The Human Footprint."
74. Ingold, *Lines*, 79.
75. Ingold, *Being*, 152.
76. Ingold, *Lines*, 78.
77. Jane Bennett, *Vibrant Matter: A Political Ecology of Things* (Durham: Duke University Press, 2010), 121.
78. Ingold, *Lines*, 101.
79. Ibid., 95.
80. Kathleen Stewart, "Precarity's Forms," *Cultural Anthropology* 27, no. 3 (2012), 522.
81. Latour, *Politics of Nature*, 134.

82. Another turn toward a revised, intense ecological footprint comes from a scientist thinking through climate change and human responsibility and accountability. "From my perspective as an astronomer," Adam Frank writes, "human beings and the cultures we've created are just another expression of the planet…We're just something the Earth has done." I like the link between expression, the impression of footprints, and the creative, compositional activity of living. Expressions invite judgment and footprints invite measurement; this need not be mutually exclusive. Indeed, they must necessarily be entangled, but we need to think of ways beyond simple measurement and management and to other, nuanced forms of judgment. See Adam Frank, "Climate Change is Not Our Fault," *NPR*, Published Oct 6, 2015, http://www.npr.org/.
83. Tim Ingold, *The Perception of the Environment: Essays in Livelihood, Dwelling and Skill* (New York: Routledge, 2000), 140.
84. Ibid., 321.
85. Ibid.
86. Chris Kitching, "Sea Turtles' Chance to Beat Extinction Destroyed by a Selfie: Mob of Eco-tourists Stop Endangered Creatures from Laying Their Eggs by Invading Costa Rica Beach to POSE with Them," *Daily Mail*, Published 15 September, 2015, http://www.dailymail.co.uk/.
87. John Burnett, "Costa Rican Villagers Sell Turtle Eggs to Save Sea Turtles, but Feud with Scientists," *UT-LANIC*, Published Aug 16, 1997, http://lanic.utexas.edu/.
88. "In measuring carbon emissions, it's easy to confuse morality with science," speaks a *New Yorker* headline, in a piece by Michael Specter, who writes, "An excessive carbon footprint has become the equivalent of a wearing a scarlet letter." But this confusion of the morality and science might just be exactly what we need. See Michael Specter, "Bigfoot," *The New Yorker*, Published Feb 25, 2008, http://www.newyorker.com/.

CHAPTER 8

Making Worlds with Cyborg Fish

Caroline Gottschalk Druschke and Candice Rai

How can we present a proposal intended not to say what is, or what ought to be, but to provoke thought, a proposal that requires no other verification than the way in which it is able to "slow down" reasoning and create an opportunity to arouse a slightly different awareness of the problems and situations mobilizing us?
Isabelle Stengers.

riprap of things:/ Cobble of milky way./…all change in thoughts/ As well as things.
Gary Snyder.

This chapter invites a provocative awareness of the riprap and cobble that co-compose a watershed assemblage we call the cyborg fish: comprising connections, interruptions, and viscosities among, at minimum, a researcher, a river, an estuary, water temperature, lunar tide, public policy, fish with a biological imperative to migrate, two fish-blocking dams, and a human monkeywrencher committed to helping those fish bypass

C.G. Druschke (✉)
University of Wisconsin-Madison, Madison, WI, USA
e-mail: caroline.gottschalk.druschke@wisc.edu

C. Rai
University of Washington, Seattle, WA, USA
e-mail: crai@uw.edu

B. McGreavy et al. (eds.), *Tracing Rhetoric and Material Life*,
Palgrave Studies in Media and Environmental Communication,
https://doi.org/10.1007/978-3-319-65711-0_8

those dams at all costs. Gilles Deleuze and Felix Guattari theorize assemblage as a dynamic coming together of heterogeneous elements—semiotic and material, human and other-than-human, natural and built—into a multiplicity of relational ties.[1] These elements, held together and apart through various viscosities and velocities, establish particular orders, ways of being, strata, territories, and social imaginaries within places and times, all the while undergoing constant movement, reordering, mutation, and disintegration.

Our interest in assemblage is deeply tethered to our commitment to understanding the connections between rhetoric and ecology in the places of persuasion: pointing not only to the connections among arguments, people, technologies, and texts, for example; but also to the connections among species through distribution patterns, migrations, trophic relationships, energy fluxes, and the like; and how those connections are stitched together and held apart in various places and times. By expanding the concerns of rhetoric through an ecological perspective—and expanding the concerns of ecology through a rhetorical perspective—we point rhetorical studies beyond a systemic, networked orientation to discourse and text, and toward the diverse energetics that weave among and through the biotic and abiotic materials of the world. This is an orientation toward the possibilities of work "from the inside," an approach that, as Karen Barad explains, "doesn't presume to take a position outside of science but rather constructively and deconstructively engages with science from the inside (not uncritically but not as critique)."[2]

Drawing on immersive ethnographic fieldwork that documents and theorizes encounters with rivers, river herring, dams, symbols, emotions, and humans, this chapter employs an ecological sensibility to offer a speculative inquiry into what we refer to as diadromous methodology.[3] The term diadromous derives from the Greek roots *dia*—across or through—and *drómos*—run, path, or way, referencing something that moves across or through channels. Diadromous methodology, as we propose and employ the concept here, patterns itself after the diadromous, migratory alewives (*Alosa pseudoharengus*) at the center of this cyborg fish assemblage, who adapt physiologically and behaviorally to thrive in ocean and fresh water. The concept of diadromy when applied to methodology points us toward the threads, connections, and speeds of the assemblage; to capacities to migrate, adapt, and even transform; and to the role of something we might name rhetoric or ecology in holding these fleeting alliances together and apart. Diadromy pressures assemblage thinking by placing explicit emphasis on this methodology's ability to transform *us*, in part by employing Isabelle Stengers' call to "'slow down' reasoning and create an opportunity to arouse a slightly different awareness of the problems and situations mobilizing us."[4] Engaging in criticism that insists on taking a pause—to explore and even inhabit the interconnectedness of elements that bleed into one another—can open new lines of sight that are necessary for creating transgressive ways to mobilize anew.

This methodology—where we take "methodology" to always also mean "politics"—as it is employed here allows us to assemble with monkeywrenchers, rivers, dams, and fish in a coastal Rhode Island watershed to create playful, consequential, *diadromous* versions of "trans-species commoning."[5] It allows us to imagine how a diadromous, ecological practice of rhetoric from inside science and activism—in their earthly and ethereal manifestations—might move us into the creation of dynamic, albeit fleeting, trans-species alliances through our actions and our texts. In this instance, diadromous methodology points us toward our participation in the assemblage of the cyborg fish—a creature composed of fish-human-river-dam and inspired as much by biological knowledge of diadromous fish as by the creative vision of Donna Haraway.[6] In so doing, diadromous methodology and its cyborg fish offer an addition to what Woelfle-Erskine and Cole point to as "naturalcultural strategies for resisting logics of domination, hierarchy, and homogenization in particular situations."[7] Through diadromous methodology, we gathered with fish, dams, and other humans to co-theorize and co-participate in a cyborg fish assemblage. By offering that work here—through a description of our shared place; an introduction to its humans, fish, and dams; and an exploration of the controversies, collaborations, and sinews that run through that place and those actors in the assemblage—we propose a speculative vision for rhetorical studies that orients the field toward playful and consequential engagements with a more expansive array of co-creators.

Ecological and Methodological Approaches to Distributed Agency

A recent turn toward thinking about rhetoric ecologically comes with the conceptualization of rhetoric as ontological: the idea that rhetoric is of and about the conditions of being and not simply a way of knowing about those conditions.[8] Rhetoric, in this sense, is part and parcel of the dynamic entanglement of all the elements of the world, while also offering the tools, capacities, and orientations that allow us to attune to and act within this world. This move suggests that rhetoric exceeds human desire and intention, as its enmeshments within complex assemblages are tied to forces and vectors that motor with and without human intention. As such, rhetoric's scope includes things obviously rhetorical and things seemingly not: fish, emotion, public memory, temperature, discourse, dreams, bodies, tools, and on. Rhetoric becomes something like a discipline and practice that, as S. Scott Graham envisions, treats "the symbolic, the affective, and the physical as ontological coequals."[9] We suggest here that diadromous methodology, and the cyborg fish it renders, align with this framing.

Thomas Rickert defines rhetoric as "one of the modalities for *attunement* to the world," as "something world-transforming for individuals and groups immersed in vibrant, ecologically attuned environments."[10] Diane Davis points to "an originary (or preoriginary) rhetoricity—an affect*ability* or persuad*ability*—that is the condition for symbolic action," while Rickert similarly calls rhetoric an "emergent result of environmentally situated and interactive engagements, redolent of a world that affects us, that persuades us prior to symbolicity."[11] Rhetoric is "ambient," as Rickert argues, because of "its embodied and embedded or situated character, its dispersal across things that themselves have gradations of agency, and its dynamic emergence within an environment that occasions certain effects."[12]

While Graham rightly criticizes Rickert's Heideggerian insistence on authenticity and withdrawal and its tendency to configure rhetoric as "result" and "dynamic emergence" rather than a constitutive, ontological force unto itself, Rickert's notion of distributed agency remains useful for our fish-work here. In Rickert's view, the rhetorical concept of *kairos* is a form of attunement and dispersed agency—ontological, material, emplaced—a quality that manifests, ambiently, through the interactions of elements in the world, as an opportunity for action that is "dispersed into the material environs."[13] Jane Bennett also conceives agency as dispersed throughout the environs of place, insisting that we take seriously the agentive qualities and political dimensions of matter. Thus, she calls for a political theory of vital materialism where human and non-human elements interact in dynamic assemblages in which agency resides not with human beings and "human efforts" but rather "becomes distributed across an ontologically heterogeneous field."[14]

Human agency is, therefore, decentered and leveled out, existing on a horizontal plane alongside vibrant matter. Where human agency exists, it manifests as embedded and made possible within and against the whole. This is something like being caught in a sticky web where one lifts a leg to make a movement in this direction only to find it not only harder to move than anticipated but also that one is necessarily always moving the web and all elements therein in unpredictable ways. This unpredictable stickiness is akin to Woelfle-Erskine and Cole's insights on the stochastic responses of non-human species and evokes the need for a diadromous approach. Diadromous methodology recognizes that, rather than hindering movement, this stochastic stickiness can be productive; it emphasizes that the surprising interplay of the symbolic, the affective, and the physical actually provides the conditions for co-creativity and agency.

Rhetoric as eco-ontological names a manner in which we discover ourselves in the world—emplaced, embedded, entangled—and becomes the art of, or more like a capacity for, attunement to the world and its everyday activities. Because we, and the thing we call rhetoric, emerge in and from place and situation, rhetorical activity might be imagined, in Rickert's terms, as "kairotic, adaptive responses to evolving situations" where these situations are actually co-created through rhetoric itself.[15] In our analysis here, rhetoric—and ecology, for that matter—names the thing(s) that hold together and apart this watershed assemblage we come to refer to as the cyborg fish.

In our case, the elements of fish, river, human, tide, temperature, and dam—the materials of the cyborg fish—are stuck together in a stochastically sticky way, at times arranging themselves in related waves that offer *kairotic* opportunities for action, as we soon detail. The diadromous methodology proposed here—and the cyborg fish that diadromous methodology helps to discover and co-create—offers a way to study the stickiness of place through field immersion in place. Diadromous methodology, as we will describe, allowed us as researchers to approach a distributed, complex assemblage through long-term engagement with place, people, and fish: co-author Caroline Gottschalk Druschke's (CGD's) dwelling in a particular ecosystem; the transgressive participation it sparked in an assemblage we dub the cyborg fish; and the speculative consequences of that participation for rhetoric, environmental communication, and politics.

Spring Migration[16]

A love story.

On a stormy night in April 2014, CGD crossed two bridges over a New England bay to arrive at a film festival screening of the documentary film *DamNation*.[17] Commissioned and conceived by Patagonia's founder Yvon Chouinard and aquatic ecologist Matt Stoecker, *DamNation* tells a powerful story of dam construction and removal in the American West and its consequences for migratory fish species. It is a beautiful film: epic, powerful, hyperbolic, clear-cut. Native Chinook salmon and steelhead trout emerge as the clear protagonists, buttressed by the efforts of activists fighting on their behalf against misguided, capitalist, expansion-hungry, mid-century American politicians and engineers and the concrete behemoths they wrought. The sides are clear: pristine nature pitted against technology and capitalism run amuck, what Woelfle-Erskine and Cole point to as the ongoing impacts and logics of Manifest Destiny in the American West. There are right and wrong answers. And the right ones are explosive, energized, expansive. Cathartic.

But as CGD sat in that auditorium with hundreds of other people watching the film, cheering the electrifying demolition of Washington's Condit Dam, and later asking questions of co-producer and co-director Travis Rummel, who traveled in for the screening, there was an equally epic, though humblingly modest, story playing out all around us in the small rivers and mill streams of coastal New England. That very month, as it does up and down the US Atlantic coast, an ancient, evolutionary force pulled an estimated 70,000 alewives and bluebacks—collectively known as river herring—back into the 7.5 mile long Missituk River that birthed them, to fight flashy flows in hopes of migrating into their freshwater spawning grounds upstream.[18]

While it might seem that this biological, migratory force has nothing to do with rhetoric, we will, in fact, be making that very case. For now, let it suffice to say that if we understand rhetorical force as not simply the arguments contained within alphabetic texts or wielded by human agents, but as ecological and ambient, as exceeding human desire and intention, as all that might move the collective us, we can begin to see how this chemical, climatic, and biological force can become entangled in that which constitutes persuasion and its conditions: words, images, hormones, processes, ideologies, myths, desires, fish sex, and so on. As we see it, something we could call persuasion, and maybe rhetoric, is clearly at work in the complex assemblage of physiology, lunar high tides, rainfall, water temperature, memory, smell, chemistry, and more that combine to pull river herring—*kairotically*, right on cue—from their meanderings in the Atlantic Ocean back into the tiny ponds of their birth. This urging, signaling, and calling is a form of persuasion, and maybe argument, to these river herring that it is time to return home. And the audience obliges.

As diadromous fish, river herring depend as much on the ocean as on their upstream, freshwater habitat, where each female and male alewife work together to fertilize up to 100,000 tiny pink eggs that, if they survive the miniscule odds of not getting eaten or destroyed along the way, will drift on to the substrate below, hatch, mature, and finally migrate downstream to the open ocean in autumn. River herring are known to travel to the edges of the continental shelf in their three to five years at sea: surviving predation, acidification, disorientation, commercial trawling, and a stunning physiological transformation that allows them to thrive in the freshwater habitat of their births, then in saltwater, and back to their freshwater origins to spawn the next generation. So, as CGD sat watching *DamNation*'s iconic, powerful, *visible* salmon in April 2014, thousands of humble river herring inconspicuously made their way into the mouth of the Missituk River, past a bevy of hungry osprey and cormorants, around the river's first dogleg, and smack into the stone face of the Main Street mill dam. Thousands of years of evolution could not surmount ten feet of stone dam or a failing fish ladder—the concrete structure meant to serve as a stairway for migrating herring to bypass the dam.

While river herring's dorsal fins rarely break the surface, the buildup of fish in the brackish water below the dam is a sight to behold: a pulse of silver flashing beneath the clear, amber water. Herman Melville's Ishmael was describing the Atlantic Ocean when he mused, "the waves rolled by like scrolls of silver; and, by their soft, suffusing seethings, made what seemed a silvery silence, not a solitude," but he may as well have been describing the spring river herring run on the Missituk River.[19] Those lucky enough to know about the pulse of life beneath the unbroken water keep their eyes glued to the river each spring, riveted to the Sisyphean task at hand. The alewives fight heroically against the surging current, the poorly designed fish ladder, and the stone face of the dam, meditative in their consternation. They labor together, writhing in unison at a speed that appears a slow and perfect standstill, a speed they can persist at, as a federal fisheries biologist friend put it, "infinitely."

Their biological, infinite, even hopeful patience is a jarring affront to the avatar of efficient, early industrial capitalism they face: the nineteenth-century mill dam. In their coordinated movement, the annual congregation of river herring bodies serves as a provocative argument against the seemingly impenetrable dam.

Beginning in 2010, the river herring's purgatory at the dam was arrested by the most modest of saviors: a large, green, bent, ripped, zip tied, PVC pipe-handled net that scooped them up rather unceremoniously and launched them over the face of the dam. Not much worse for wear, the river herring continued their journey upstream until they were scooped again over a second failing fish ladder, and completed their journey up into the freshwater habitat in which they spawn. From 2010 to 2016, a growing number of net-wielding community volunteers joined forces with the fish at those two ladders—communicating through cell phones, standing on temporary structures, filming with GoPros, donning waders, navigating currents and tides—to hoist the fish ten at a time over the two dams and on to new life.

Human Activism on the Missituk River

One way to engage with the story of the Missituk River would be to focus on its loudest and boldest human characters. A significant body of work in rhetoric of science and environmental communication takes this tack, focusing primarily on the agency of human actors that advocate, argue, interject, and act in the realm of public debate, especially in environmental controversies.[20] Our story of the Missituk herring, too, could be told through its heroic, larger than life human actor: the man who bought those green nets and who single-handedly started the work of the Missituk herring lifters—we will call him Hayduke[21]—with a simple announcement to the state fisheries biologist:

> Tom, I'm going to start tossing some herring over the dam.

In 2010 Hayduke noticed the amber water was thick with herring because of a recent fish stocking effort in the state. He realized the increasing numbers of migrating river herring were not making it through the fish ladder (built in the late 1960s for now-extirpated Atlantic salmon), thanks to poor placement of the ladder's opening and higher than passable water velocity through the ladder. Hayduke immediately called the state fisheries biologist, Tom, to tell him there was a problem with the fish ladder that would have a long-term negative impact on the river herring restoration program. Tom already knew of the problem, but the state was limited in its ability to address the issue by the labor and cost required to pass river herring over the dam by hand or by tanker truck. Hayduke promptly announced that he planned to step in.

State representatives quickly placed Hayduke's effort within rhetorical ecologies of institutions, channeling his work into technological procedures, processes, and paperwork. Hayduke, for example, had to seek approval from the town as the legal owners of the dam and complete paperwork, apply for scientific collector permits, and carry liability insurance. But Hayduke pointed to the earliest days of the fish lift, before the paperwork and permits:

> The town turned a blind eye and said "Do what you've got to do." And the state didn't say anything. And I had no insurance, no herring permits, nothing… It was just that simple. Nobody called me. I went down on my own… And I was violating like every law. I took a yellow caution vest and I just wrote [Department of the Environment] on the back. And Frank and I went down in there and started throwing fish over… And then people would start to show up and I would say, "Do you want to help? Come on down!"

Hayduke worked against the current of the river and of town, state, and federal management to take matters into his own hands because, as he explained:

> Nobody else was doing it. I've always had a passion for migratory fish. I saw a problem. I knew it was a problem. And I decided to do something about it.

The work, for Hayduke, was an all-consuming passion:

> It's just like this is like my Disneyland. It's my Super Bowl and my World Series. It's, you know, it's everything.

Hayduke used to engage with formal river herring policy discussions through the New England Fisheries Management Council and the National Oceanographic and Atmospheric Administration (NOAA) but explained that he finally became fed up with what he saw as regulatory inaction on river herring protection. Hayduke's activism changed its focus to immediate (and sometimes unsanctioned), pragmatic, *kairotic* action, namely, getting the fish over the dam as soon as possible:

> I mean I could try to win some of those, but I don't really bother; I just move fish.

For those who joined Hayduke's movement, there was something seductive about the plight of the underdog river herring. The fish lifters stood up for river herring against what they saw as a mismanaged bureaucracy that thwarted efforts to get these fish past New England's 14,000 dams. Hayduke saw hope and salvation in these fish, magic and amazement, and many lifters did, too. A powerful sense of righteousness and wonder embedded itself within the culmination of elements in this place that constituted rhetorical force and *kairotic* opportunity.

For five years, the group labored to lift fish over the dam, until 2015, when a planned renovation of the fish ladder promised to resolve the situation. But a slow start to construction in late fall 2014 meant the renovation of the Main Street ladder was abandoned for winter. Back at work in spring 2015, the crew demolished the lower reaches of the Main Street fish ladder, blocked a portion of the dam, and rerouted the streambed and flow to accommodate their work. Construction inched along while anxious federal fisheries managers and town residents worried the ice would soon melt on the downstream salt pond, coinciding with a new or full moon and heavy rains that would signal the herring to make their way upstream. River herring run on seasonal, *kairotic* time; the timing of their spring migration is a moving target dependent on a variety of ambient factors, and that uncertainty added to the anxiety of the river herring's advocates—both public and governmental—as construction crept along.

Despite Hayduke's focus on just moving fish, the seemingly glacial pace of the work crew and the impending arrival of the river herring offered an exigence for Hayduke to vent his frustration at state and federal management. As spring progressed, and water temperature rose, the fish lifters were cc'ed on a series of anxious emails from Hayduke to NOAA, state, and local officials, which built to a March 2015 all-caps email demanding federal officials cease construction and reroute the flow of the river:

YOU NEED TO MAKE A DECISION AND YOUR DECISION NEEDS TO BE SOON.

Without a sufficient response, Hayduke offered his own solution two days later, emailing schematic drawings of a slight site alteration that would allow for netting, and imploring officials to stop work and execute plan B.

Kairos emerged from the interaction of email, anxiety, warming temperatures, and soon-to-be-migrating fish, as local, state, and federal officials decided to halt work and adopt Hayduke's design. Officials followed up with a series of emails forwarded through Hayduke to the lifters containing strict instructions about applying for and carrying state licenses to access the gated construction area and locked up nets. Where, in other years, dynamic groups of up to 50 people would meet at self-organized times to lift fish together, in spring 2015, netting became a policed and regulated activity, dispersed to individual participants only when they took the initiative to attend a training, secure state permits, and remember the combination for the newly locked nets. Even Hayduke's impressive skills at community organizing were not enough to consolidate the energy that had existed in previous years. The window of opportunity, it seemed, had passed. *Kairos* was swept downstream.

In late 2015 construction crews finished work on the Missituk fish ladders, and the state environmental agency distributed a celebratory press release about the herring's now "seamless" run to their historic spawning grounds. News coverage of the state's opening ceremony at the ladder referred to Hayduke as "a simple bystander." In light of the reconstructed ladders, "[The group's] efforts weren't needed." But when herring returned to the Missituk in spring 2016, it became clear that thousands of river herring, once again, gathered in the pool below the upstream ladder, unable to reach their spawning grounds further upstream. The fish lifters responded by gathering, once again, too: an arrangement of human bodies below the fish ladder, mingling with bodies of in-migrating alewives and bluebacks and assorted aquatic creatures, navigating permit genres, and emails, and technical equipment, and moving in unison with the bodies and life histories of these fish.

The partial failure of the technical solution at this fish ladder served as a reminder that countered prevailing public narratives in which a single action, like the Elwha River dam removal in Washington State, can seem to restore an entire river system. New England's estimated 14,000 dams, paired with myriad anthropogenic impacts, mean that one dam removal or improved fish ladder cannot "fix" a system, if fixing means something like returning it to a "natural" state. Instead, the present and foreseeable future of Missituk river herring life history involves a complex assemblage of alewife, blueback, dams, fish ladders, nets, fish lifters, researchers, river, rain, tide, arthropods, moon, gravel, permits, nutrients, and on and on: "lines of articulation or segmentarity, strata and territories" as one actor is segmented from the next, "but also lines of flight, movements of deterritorialization and destratification" across them as phenomena speed up and slow down through tides, days, weeks, seasons, people to "accelerate and rupture."[22] The insights of assemblage theory—which the Missituk river bears to us so starkly—is that while our forms of agency are always discovered in a place and time, and that there is always opportunity everywhere for agency and change, this change is limited, contingent, unpredictable, partial, here and now, not always and forever, fragile and incremental. We know not what it is that we can really do—such is a form of wonder and of hope that keeps us moving and reaching in the face of such radical unknowability and stickiness. Among other things, a diadromous methodology can help us archive the forms of reaching present in a place (such as fish lifting designed to solve a problem) and imagine the forms we might take up wherever we dwell.

So we can imagine environmental communication and rhetoric of science researchers being interested in the Missituk lifters; they offer an elegant community-based solution to the technical problem presented by the malfunctioning ladders, while posing poignant questions about legitimacy and authority over this seasonal, biological migration. The group is interesting, too, because while some members seem to engage in "legitimate" forms of participation with state and federal agencies (e.g., attending public fisheries forums), many seem to participate in the annual fish migration not so much to engage with the work of federal agencies, but to step outside of frustrating formal processes of public participation in environmental decision making. By self-organizing fish lifts, the group sidesteps technocratic models of environmental governance and policymaking, a lack of productive forums for citizen involvement, and the futility of the decide-announce-defend model of public participation processes.[23] Community members literally and figuratively work upstream and engage with ongoing conversations about stakeholder interest in agency-led environmental efforts and the possibilities for active community engagement.[24]

But we want to push readers outside of this anthropocentric, non-distributed frame. We suggest that conceiving the Missituk events through a diadromous methodology focused on the complex assemblage of the cyborg fish (that includes CGD) offers more potential for environmental communication and rhetoric of science, allowing us to understand this tale not as a human-centered display of activist agency, but as a creative, dynamic, consequential cyborg assemblage that works with, through, and beyond, altering the ecologies of the Missituk watershed, both material and symbolic.[25] In other words, we suggest that it may be more useful, more fundamentally and creatively productive, to look instead to how human joined with and was joined by fish, net, dam, and river to co-create the assemblage we point to as the cyborg fish. Diadromous methodology—enacted above as the tangle of forces bundled together by the thing we call rhetoric that co-creates the spring migration, and below as we work through the figure of the cyborg fish—begins to get at how we might reorient our rhetorical perspectives to slow down and capture the deep interconnectedness of elements in the assemblage (animal, mineral, symbolic, and more) that work together in and against a sort of consonance of place. We argue that the attempt to attune to such consonance, to perceive and tentatively and partially represent it in narrative form, is an affordance of this field-based diadromous approach.

Assembling the Cyborg Fish

The assemblage of fish lifter, river, dam, and fish on the Missituk River exemplifies and becomes accessible through theoretical and physical diadromy, in the figure we refer to as the cyborg fish. Haraway conceives the cyborg as animal and machine, organic and inorganic, alive and not alive; as she puts it, cyborgs are "creatures simultaneously animal and machine, who populate worlds ambiguously natural and crafted."[26] Importantly, Haraway seeks to muddy the borders between animal and human, human and machine, biotic and abiotic, you and me, us and them. Through the figure of the cyborg fish, we seek to fuse rhetoric with ecology, human with fish, nature with culture, metaphysics with ontology, and find inspiration for a diadromous methodology that is marked not so much by specific practices so much as an orientation to analyze and participate in such fusing and dissipating across things as manifested and emergent in place. We build from the way that, as Haraway argues, "cyborg imagery can suggest a way out of the maze of dualisms in which we have explained our bodies and our tools to ourselves."[27] As she insists, "A cyborg does not seek unitary identity… there is no drive in cyborgs to produce total theory, but there is an intimate experience of boundaries, their construction and destruction."[28] Cyborg fish demonstrate our orientation toward assemblage, pointing us toward what is and is not included in the cyborg fish at any given time, as the cyborg fish accelerates and slows, moves fluidly across species and policies, migrates from Missituk watershed, to researcher, to MacBook, to printed book chapter, to reader: maybe altering your own human-fish assemblage.

The Missituk herring-human-river-dam-ladder assemblage is a cyborg fish because its existence and the conditions of this existence are facilitated, constrained, and shaped by the workings together of many factors and actors that are stitched together, always tenuously, through the exchange of energies that are both *trophic*—which is to say ecological(-rhetorical), related to the exchange of energy as organisms associate with one another through acts of eating and decomposition—and *tropic*—which is to say rhetorical(-ecological), related to turning toward another body or force. This tropic and trophic orientation that we are advocating for is to perceive of one's objects of study not as objects at all, but more as rhetorical-ecological process and connective tissue. In this configuration, the unit of observation and analysis stresses this simultaneous, inseparable interface of what is often called *rhetoric* (symbolic systems, human cultural stuff, ideology, social matters, shared desires, designs, memories, and affects that tenuously enable the coordination of humans within assemblages) and what is often called *ecology* (natural and built environments and forces, matter, biological rhythms, bodily logics, knowledge, and ways of being that become entangled with and exist beyond human design and perception) in order to reframe the ontological equality of the physical, the symbolic, and the affective. Where, for Graham, if things "gather through electromagnetism and gravity, that gathering is not ontologically distinct from their gathering through representation or ideological hegemony," in our own rhetorical-ecological diadromous context, if things gather through gravity, salinity, chemical signature, nutrient budget, or hunger, that gathering is not ontologically distinct from their gathering through capitalist logics, or narrative, or metaphor.[29] And all of the above have interest for us here.

We bring rhetoric together with ecology, the exchange of symbolic and physical energies, the exchange of calories and affective responses, as actually part and parcel of the same thing at base, and our synthesis emphasizes this energetic as extending beyond the intentional. Where Emily Plec describes a related sort of internatural communication as including, "the exchange of intentional energy between humans and other animals as well as communication among animals and other forms of life," the trans-species, diadromous rhetoric that we are advancing here is not always or only intentional or emerging from motive.[30] Nor is it necessarily an "emergent *result* of environmentally situated and interactive engagements," as Rickert argues, though rhetoric strikes us as a useful name for the connective force at work on the Missituk, holding together the cyborg fish assemblage and naming the conditions that pull it apart.[31]

This conception of rhetoric as affinity or persuadability, as a form of attunement with symbolic, affective, physical environment or, better, *ecosystem*, seems to label the *tropic* turnings we see at work on the Missituk, as leaf turns toward sun, alewife turns toward headwaters, fish lifter turns toward fish.[32] And yet, we understand this theoretical perspective—this attention to the turning and attraction of *tropic* conceptions—as more useful, more revelatory, more productive in CGD's deep engagement on the Missituk, when paired with an attention to the *trophic* dimensions of energy's exchange—as alewife eats zooplankton, human eats alewife, human resists eating alewife, alewife resists being eaten by human. The cyborg fish, then, becomes manifest as assemblage through both *tropic* persuadability and *trophic* flux: through the communing of nature, culture, fish, human, and biology in processes more-than-human and more-than-biotic; and through Missituk activists who assemble and are assembled by mill dams, fish ladders, and mill ponds as integral components of the herring's life cycle. Here, the figure of the cyborg fish offers a way to understand the life and death of an activist project that is not really a death, one that offers the possibility for perpetual life if life becomes something more like energy, or movement, or capacity to act on and through and with others, or possibility, or hope, as Haraway rendered it:

> From another perspective, a cyborg world might be about lived social and bodily realities in which people are not afraid of their joint kinship with animals and machines, not afraid of permanently partial identities and contradictory standpoints. The political struggle is to see from both perspectives at once because each reveals both dominations and possibilities unimaginable from the other vantage point.[33]

Energy, in this perspective, continues to flow—from sun, to phytoplankton, to zooplankton, to alewife, to osprey, to maggot, and on. This conception of the cyborg—and the diadromous methodology that enables it—asks us to inhabit a political orientation that insists we "see from both perspectives at once," and more than both, which is also to say that we truly open ourselves up to being moved, changed, and persuaded ourselves. Taking on the perspective of plankton, fish, river, learning to see from these multiple perspectives, offers a sort of perspectivism that, as Peter Skafish describes, offers, "a way of thinking in which the first reference is not to the self but to the other person: every thought, subject, or collective depends on and comes from the other," where personhood extends beyond the human.[34] This perspectival politics is a transformative move that can open up creative space for the sort of speculative new awareness for which Stengers calls. As Skafish notes, "When our basic ontological tenets end up 'twisty,' ... there are political effects."[35] Diadromous methodology—and its positioning of the researcher within the assemblage—warrants these political effects as ones that are embedded in place, sticky, contingent, and limited, even as they remain vehicular and agentive. Diadromous methods can help us identify and co-invent deeply situational and *kairotic* forms of, and capacities for, reaching (toward change or problem solving) in place and time, and to identify and co-invent the supports and conditions (be they powerful arguments, material resources, or otherwise) required to help us reach toward whatever political effects seems to occasion our collective striving, even if we never fully reach them.

In our experiences on the Missituk River, deep engagement in the field and with its people and things, and the development of supporting knowledge about ecology, economics, political science, anthropology, and more, have been essential to the awareness of constant disturbance, to adopting a different perspective, to forming politics rooted in change, contingency, dynamism, and ecology, to learning to see anew, to being persuaded in manners not planned on or even, in some cases, desired, to being jolted fundamentally through encounter with others. To the recognition of the cyborg fish and to the transfer of energy that it performs and represents. And to the political consequences of that recognition. In short, the cyborg fish warrants the queer possibilities of diadromous methodologies.

The cyborg fish assemblage exists in a real, experienced, tangible place of persuasion—the Missituk watershed—where river herring bodies gather annually to fulfill their biological imperative against all odds;

where technology and capitalism conjoined to build a dam fueled by human desire for progress, and later by sentimentality when the dam outlived its functional use, embedding itself into alewife and blueback life history; and where various actors—including CGD—are compelled to gather, act, argue, and intervene together. Such a place catalyzes and constitutes the terrain of invention, a location both material and conceptual composed of enmeshments, in this case, of river herring biology and life history, mill dam, fish ladder, fish lifters, state and federal fisheries policies, circulating topoi, and all manner of other things that weave and are woven among these things that help them gather here to co-create playful and even transgressive naturalcultural assemblages.

Where Rickert points to rhetoric as "one of the modalities for *attunement* to the world,"[36] CGD's ambient fieldwork on the Missituk—composed of watching, waiting, wading, touching, reading, listening, talking, laboring, kayaking, teaching, lifting in coordination with and with growing awareness of tides, temperatures, arthropods, predators, storms, nutrients, offshore trawlers, sediment, sand, cobble, dam—offers ecology as another essential modality for the type of creative, world-changing, world-making work that can and does emerge from the assemblages within which we immerse ourselves as researchers. CGD's spiritual practice and study of ecology is apt here, as what the Cary Institute calls, "The scientific study of the processes influencing the distribution and abundance of organisms, the interactions among organisms, and the interactions between organisms and the transformation and flux of energy and matter."[37] This ecological orientation offers another perspective on how we might deepen our understanding of the vast enmeshments and collective, agentive forces that compose any assemblage. In the case of the Missituk River assemblage discussed here, immersion in the field prepared CGD to compose together with other humans and other-than-humans: creating relationships, vibrant spring herring runs, informational signage, experiences, cyborg fish, and attuning CGD to the possibilities of a more-than-natural diadromous rhetoric. In other words, ecology reminds us of *trophic* energy and materiality, offering a useful correction for *tropic*-heavy symbolic views of rhetoric. While it may not be correct to divide ecology and rhetoric in this way—in fact, we are certain it is *not* correct—we offer the distinction here as a useful heuristic. A thought experiment with legs or, in this case, fins. CGD's fieldwork on the Missituk and the trans-species labor it blossomed into offered a diadromous methodology of ecology alongside rhetoric as a modality for attunement with the capacity to make change. It offered rhetoric as an active force, a multidimensional energetics with the attractive capacity to hold assemblages together, and rip them apart.

Back on the Missituk itself, as time passes, more dreams, desires, actions, nutrients, chemicals, bacteria, and arguments entangle themselves in that place, providing more resources from which to solve its contentious and urgent problems—though what constitutes a problem or appropriate action is always, continuously, open for debate—as we (an open, trans-species we) mobilize in inventive assemblages. We can think of this rhetorical action as deeply situated monkeywrenching that attends to and loops in wide-scale environmental advocacy, but does so in a manner saturated with the tools and resources and details and biota and abiota of very specific places and times: the creation of a diadromous methodology. Indeed, it is our contention that capacities for response must be (are always already) situated: *kairotic*, embedded, emplaced, entangled, and attuned to the here and now.

Joining the river herring, who are beckoned to their natal rivers each spring by a mysterious assemblage of ambient biological forces, are fish lifters, who leave children and family at home to gather at the dam, donning waders or warm coats, nets and buckets in hand, in water and along the banks, and organize their bodies into the collective to work in the most mundane and (un)common ways. All the bodies and things, biologies and desires in deep connection, materiality tethered to persuasion as rhetoric-ecology: ripped flesh of fingers, flashy river flows, flying herring scales, freezing water and briny smell, slippery rocks, human with herring, dam and river, failing fish ladder, aging nets, arguments for how to move through or control, catch and release, the life of the river—all and more engaged in *phronetic* study, gaining the sort of wisdom concerned with action to which Aristotle long ago pointed. The fish lifters of the Missituk have dropped out of disembodied, immaterial discussions of fisheries policy in nondescript meeting rooms across the region and have stepped into the cold spring water to gain a sort of experiential knowledge, a practical or applied wisdom inaccessible through other means. More than anything else, this diverse group—fish, human, river, dam, net—engages with and through the natural, built, and symbolic ecologies of this place; each participant is lost at moments to the flow of the river, the argument, the action, the herring's plight. Flows of ecological and rhetorical energy—*trophic* and *tropic* interactions—become indistinguishable. Rhetoric is part of the ecology of this place and ecology becomes rhetorical. Here, again, the unit of analysis is rhetoric-ecology, the eye trained on the place of interface and confusion. Persuasion pulls the herring upstream, pulls the lifters to and in the river, orchestrates their collective movements down the line of lifters, seems to self-organize the coordinated movement of wader-clad bodies in the river, the hoots, the yells, the cheers, the silence. Pulls the lifters to take action: action that is mostly illegitimate, sometimes ill-informed, usually collaborative, often contentious, and always consequential; and action, too, that becomes entangled not merely with the plight of the fish, but also with the ambient and biological forces to which these fish are responding. And, in so doing, the lifters assemble with the river herring and the fish ladder and the river and the dam—and the researcher—to collectively monkeywrench a future not for but *with* these fish.

Creating Human-Fish Worlds: The Activism of Cyborg Fish

True to our ecological, speculative mode, we want to decenter the role of Hayduke and his fellow human activists engaged in important political work and, instead, draw attention to an ecological view of rhetoric as an attractive force that shapes and holds together the infinitely complex workings of the elements of the world and to a diadromous methodology that hopes to grant tentative and fraught access to these complex worldings. While migrating river herring might be thought of as beyond the scope of rhetoric, we are arguing in no uncertain terms that the force of river herring migration—as a physical and social force that exceeds human beings—is entangled with our understanding of rhetoric. That even though the annual fish migration is beyond human control or even past our capacity to know and predict the myriad factors that trigger the calling in of river herring from miles away at sea to the precise freshwater pools in which they were spawned, pairing rhetoric with ecology through a diadromous methodological approach allows us to see, learn from, participate in, *and be transformed by* this persuasive, trans-species force. Because rhetoric is immersed within and emerges in concordance with emplaced ecologies and never apart from them, rhetoric can act as a means for reshaping and mutating ecologies as one of many elements within the assemblage that can exert force.

In focusing on the persuasive forces of river herring, we join rich conversations among rhetoricians and communication researchers interested in the rhetorical qualities of biology, environment, and animals.[38] We engage with that work, but we note that what our work does differently, among other things, is to move us out of discussions of "natural" animals, into the figure of the cyborg, a move that necessarily expands our definitions of rhetoric and rhetoric's possibility. The case of the Missituk herring assemblage demonstrates actants engaged in a project together. It exemplifies *praxis*—a getting something done thanks to a particular dynamic assemblage of time, space, people, policies, and things. It shows people participating in forms of public engagement but recognizes that engagement is deeply woven into that wider assemblage. Human actors school with river herring—communing with them—as piscine actors school with humans. The figure of the cyborg fish helps us consider the persuasive capacities of river herring migration, actual and analog, and to recast activism and engagement beyond the human agent.

Like Zoe Todd's "fish plurality," which attends to the plentiful and competing, codified and ephemeral, "multiple ways of knowing or defining fish," the cyborg fish suggests that one aim of diadromous methodology is the attention to a deeply emplaced and intimate form of communing and acting within and in response to a constellation of forces, people, things, desires, affects, frictions, and arguments within an ecosystem.[39] As the diadromous cyborg fish is capable of knowing in multiple ways—swimming across and through strikingly different ambient environments, existing, sensing, and responding to new elements as they cross different aquatic terrains—so, too, must a diadromous view of rhetoric adopt an immersive orientation toward seeing, sensing, knowing in multiple, and perceiving the places and ways things, people, and elements fold into and create each other, and then folding and creating themselves anew.

In short, without understanding the river system of the Missituk, or the life history of the alewife, it is easy to focus on the Missituk Herring Association as a rhetorical case study that marks the gathering of anthropocentric, agentic rhetorical energy: through Hayduke's impassioned pleas to join him; through the community's *kairotic* interest in the spring migration; through the big personalities of the state and federal fisheries managers and of the even bigger personality of Hayduke himself. By adopting a view of rhetorical ecologies, we begin to see how strategic, relational, and material arguments in favor of fish migration travel across time and space, between and among the many players described above, and we might even take stock of the genres, permits, technologies, and images through which arguments about fish passage travel. But a biophysically grounded ecological orientation to the system—a diadromous methodological orientation—demands an expanded interpretation. Fish lifter, fish, and net may have assembled to create the cyborg fish, but even when the community activists and the construction crew recede into the distance, frustrated, on the one hand, work completed, on the other, the cyborg fish goes on, and carries with it the rhetorical energy that once gave life to the activist movement. The cyborg fish carries on: maybe biological fish and concrete ladder and wooden baffle but no human. The cyborg fish is perpetual energy, rendered in alternate form. Energy, like water, finds a way.

This is a view of rhetoric as velocity, flow, force, communion, and embeddedness tangled with materiality and nature, a force welded as much to biology as persuasion or the influence of crafted human argument. It is, in its sparest form, to say that understanding rhetorical ecology is to understand the forces and flows of one's place in the world. The urgency toward action and the shape that action might take emerges in the forms allowed by the amalgamation of these workings together. Human actors are only a sliver of the actors, and our—a trans-species our—ability not only to act but also to perceive how, why, and when to act is a capacity manifest from and imbricated within the places we inhabit in the world. Thus, the will to act, as well as the dispersal of action, are always flows that are tied to human and more-than-human factors.

The successful endgame, then, of this type of ecological engagement is generally not the dramatic overthrow of systems of power in a grand narrative peopled with heroes and villains, where good overcomes evil in an epic battle that results in a new world. Rather, we advocate for the creative exploration of methods and modes more subtle, nested, and complicit—where one is never sure how one's actions make a difference, where one is willing to give up one's notion of oneself, where making a difference might mean slight alterations in what others can see or do, where sometimes imperceptible micro-changes are, in fact, the thing to be celebrated and noticed as victory. Even that we are writing and you are reading this now indicates a certain success for the ecological engagement that Hayduke, the river herring, the fish lifters, the Missituk, CGD, and CR practiced together—in that the rhetorical velocity of those actions has resulted in the circulation and open possibilities of those stories and bodies and energies for us, and maybe for you. That their work might alter us—and you—slightly is something. It is not a grand narrative of change that sweeps us away, but it is a profoundly hopeful message of small changes in the face of radical complexity, contingency, and impossibility.

Just as diadromous river herring rely on both freshwater and saltwater to survive, rhetoricians interested in our places—and in acting in our places—might rely on diadromous approaches that account for rhetoric-ecology to attune to Gary Snyder's "riprap of things:/ Cobble of milky way" in order to creatively foster "all change, in thoughts,/ As well as things."[40] If we have any hope of consequence, of worldmaking in current earthly conditions, we need to engage with field methods and work to understand both rhetorical theory and biophysical materiality. We need to try to understand things like fish, human, dam, river, and tide, even if we fail at it. We need to embrace the ethical demand to know—both know about and intimately know—what we are talking about.

We urge environmental rhetoricians and communicators to play with diadromous approaches—in our case, qualitative, experimental, and ecological field methodologies that allow us to listen to, learn from, and create with the connections between humans, river herring, rivers, tides, and dams. We encourage researchers to learn and practice from both the outside and the inside of science; to run upstream and down through watersheds, theories, and power structures; to become more porous to the diverse and changing ambient environments in which we work and which we co-create, in their symbolic, affective, and biophysical forms. At its heart, this move requires an expansion of the matters of concern—both conceptual and material—that we attune to in learning about the world and acting within it. This deep engagement with people and things—the interruption it provokes and concordant attention to divergent, diffractive, surprising, transgressive threads across and among beings and things—offers the possibility for work in rhetorical ecologies, environmental communication, and rhetoric of science to contribute to informed analyses and actions that can support broader normative goals of sustainability, environmental and social justice, and what we take to be the root goal of rhetoric: the negotiation of a more equitable and ethical life in common.

Like the diadromous fish that can miraculously adapt from freshwater to salt water and somehow return home to their natal rivers from thousands of miles away, rhetoricians can strive to be similarly flexible but emplaced. To consider what we could gain from homing to our natal rivers, digging in, and listening. Sensing. Attuning. Communing with others. *And acting*. Borrowing the rhetorical energy of the fish, the river, the dam, the riffle, getting out of its way, hitching a ride, and engaging this power as it pulls us downstream and then up. That may not be a satisfying answer, but only a pregnant question: *What if*?

NOTES

1. Gilles Deleuze and Felix Guattari, *A Thousand Plateaus*, trans. Brian Massumi (Minneapolis: University of Minnesota Press, 1987).
2. Karen Barad, "Nature's Queer Performativity," *Qui Parle: Critical Humanities and Social Sciences* 19, no. 2 (2011): 51.
3. Pronounced \dī-ˈa-drə-məs\.
4. Isabelle Stengers, "The Cosmopolitical Proposal," in *Making Things Public: Atmospheres of Democracy*, eds. Bruno Latour and Peter Weibel (Cambridge: MIT Press, 2005), 994.
5. Cleo Woelfle-Erskine and July Cole, "Transfiguring the Anthropocene: Stochastic Reimaginings of Human-Beaver Worlds," *TSQ: Transgender Studies Quarterly* 2, no. 2 (2015): 311.
6. Donna J. Haraway, "A Cyborg Manifesto: Science, Technology, and Socialist-Feminism in the Late Twentieth Century," in *Simians, Cyborgs, and Women: The Reinvention of Nature* (New York: Routledge, 1991), 149–181.
7. Woelfle-Erskine and Cole, "Transfiguring the Anthropocene," 300.
8. Ecological approaches to rhetoric include Anis Bawarshi, "The Ecology of Genre," in *Ecocomposition: Theoretical and Pedagogical Approaches*, eds. Christian R. Weisser and Sidney I. Dobrin (Albany: State University of New York Press, 2001); Jenny Edbauer, "Unframing Models of Public Distribution: From Rhetorical Situation to Rhetorical Ecologies," *Rhetoric Society Quarterly* 35, no. 4 (2005). For examples of key ontological texts, see Thomas Rickert, *Ambient Rhetoric: The Attunements of Rhetorical Being* (Pittsburgh: University of Pittsburgh Press, 2013); S. Scott Graham, "Object-Oriented Ontology's Binary Duplication and the Promise of Thing-Oriented Ontologies," in *Rhetoric, Through Everyday Things*, eds. Scot Barnett and Casey Boyle (Birmingham: University of Alabama Press, 2016).
9. Graham, "Binary Duplication," 121.
10. Rickert, *Ambient Rhetoric*, xviii–xv.
11. Diane Davis, *Inessential Solidarity: Rhetoric and Foreigner Relations* (Pittsburgh: University of Pittsburgh Press, 2010), 2; Rickert, *Ambient Rhetoric*, 34.
12. Rickert, *Ambient Rhetoric*, 36.
13. Ibid., 95.
14. Jane Bennett, *Vibrant Matter: A Political Ecology of Things* (Durham: Duke University Press, 2010), 23.
15. Rickert, *Ambient Rhetoric*, 29.
16. Caroline Gottschalk Druschke conceived and conducted all fieldwork for this chapter from April 2011 to present, including participant observation, email exchanges, pedagogy, and 52 semi-structured interviews.

From 2012 to 2016, Druschke worked in collaboration with Kristen C. Hychka, Peter Schooling, Evan Dunphy, and with community fish lifters, migratory fish experts, and local, state, and federal officials involved in freshwater restoration and marine fisheries. Much of the narrative voice is Druschke's. Candice Rai contributed theoretical and writerly support.

17. This paper builds on Druschke's broader fieldwork, an Institutional Review Board-approved research study that requires the use of pseudonyms for many people and places throughout.
18. Missituk is not the river's actual name. Missituk translates as great tidal river (missi = great; tuk = river pushed by tide or wind) in the family of Algonquin languages that provide so many place and river names throughout New England.
19. Herman Melville, *Moby Dick* (New York: Bantam Books, 1964), 219.
20. See Phaedra C. Pezzullo, *Toxic Tourism: Rhetorics of Pollution, Travel, and Environmental Justice* (Birmingham: University of Alabama Press, 2009); Carl G. Herndl and Danny A. Bauer, "Speaking Matters: Liberation Theology, Rhetorical Performance, and Social Action," *College Composition and Communication* 54, no. 4 (2003); Danielle Endres, "Science and Public Participation: An Analysis of Public Scientific Argument in the Yucca Mountain Controversy," *Environmental Communication: A Journal of Nature and Culture* 3, no. 4 (2009).
21. Fans of Edward Abbey may understand this pseudonym and its parallels. With a dogged passion for migratory fish, a lack of inhibition in the face of state and federal fisheries regulators, a sharp mind for engineering fish passage solutions, boundless energy despite sometimes troublesome health, and a night job that allowed him daytime hours for his activism, Hayduke became in many ways the voice of river herring on New England's southern coast.
22. Deleuze and Guattari, *A Thousand Plateaus*, 3–4.
23. For more on the difficulties of public engagement in environmental policymaking, see Stephen P. Depoe, John W. Delicath, and Marie-France Aepli Elsenbeer, *Communication and Public Participation in Environmental Decision Making* (Albany: State University of New York, 2004).
24. Harry Collins and Robert Evans describe "upstream" science studies, while Gregg Walker, Susan Senecah, and Steven Daniels offer environmental communication frameworks for considering community engagement. See Harry M. Collins and Robert Evans, "The Third Wave of Science Studies: Studies of Expertise and Experience," *Social Studies of Science* 32, no. 2 (2002); Gregg B. Walker, Susan L. Senecah, and Steven E. Daniels, "From the Forest to the River: Citizens' Views of Stakeholder Engagement," *Human Ecology Review* 13, no. 2 (2006).

25. See Caroline Gottschalk Druschke, "Watershed as Common-place: Communicating for Conservation at the Watershed Scale," *Environmental Communication: A Journal of Nature and Culture* 7, no. 1 (2013).
26. Haraway, "A Cyborg Manifesto," 149.
27. Ibid., 181.
28. Ibid.
29. Graham, "Binary Duplication," 121.
30. Emily Plec, *Perspectives on Human-animal Communication: Internatural Communication* (New York: Routledge, 2013).
31. Rickert, *Ambient Rhetoric*, 34, emphasis ours.
32. Recent examples of rhetoric scholars interested in *tropic* and *trophic* dimensions of rhetoric include Caroline Gottschalk Druschke, "What are We Talking About When We Talk About Ecology," (roundtable presentation, Rhetoric Society of America Conference, Atlanta, May 29, 2016); John Muckelbauer, "Implicit Paradigms of Rhetoric: Aristotelian, Cultural, and Heliotropic," in *Rhetoric, Through Everyday Things*, eds. Scot Barnett and Casey Boyle (Birmingham: University of Alabama Press, 2016); Diane Keeling and Jennifer Prairie, "Trophic and Tropic Dynamics: An Ecological Perspective of Tropes," Chap. 2 this volume.
33. Haraway, "A Cyborg Manifesto," 154.
34. Peter Skafish, "The Metaphysics of the Extra-Moderns: On the Decolonization of Thought—A Conversation with Eduardo Viveiros de Castro," *Common Knowledge* 22, no. 3 (2016), 404.
35. Ibid., 410.
36. Rickert, *Ambient Rhetoric*, xviii.
37. "Definition of Ecology," Cary Institute of Ecosystem Studies, accessed May 30, 2016, http://www.caryinstitute.org/.
38. Contemporary rhetorical animal studies kicked off with George Kennedy, followed thereafter by a small but steady swell of interest represented by, among others, Debra Hawhee and Diane Davis. See George Kennedy, "A Hoot in the Dark: The Evolution of General Rhetoric," *Philosophy & Rhetoric* 25, no. 1 (1992); Debra Hawhee, "Toward a Bestial Rhetoric," *Philosophy and Rhetoric* 44, no. 1 (2011); Diane Davis, "Creaturely Rhetorics," *Philosophy and Rhetoric* 44, no. 1 (2011).
39. Zoe Todd, "Fish Pluralities: Human-animal Relations and Sites of Engagement in Paulatuuq, Arctic Canada," *Études/Inuit/Studies* 38, no. 1–2 (2014): 217.
40. Gary Snyder, *Riprap and Cold Mountain Poems* (Berkeley: Counterpoint Press, 2009).

CHAPTER 9

Embodying Resistance: A Rhetorical Ecology of the Full Cycle Supper

Samantha Senda-Cook and George F. McHendry, Jr.

Introduction

Just after dusk, huddled around a table near a greenhouse we wait in a disorderly line for a pint of locavorous beer and locally made ice cream. We have just completed the Full Cycle Supper, an event during which we biked around local urban farm plots and ate various courses of food made with local ingredients. As we wait, volunteers disappear into an overgrown area and prepare a bonfire. Nearby a person begins to strum an acoustic guitar. Christmas lights hastily strung up over the table provide just enough light to make out the scene. Surrounded by trees, weeds, and rows of vegetables it was easy to forget we were in the heart of Omaha. Just hours earlier we were alternating between riding Omaha's hills and enjoying pesto-deviled eggs, mere feet from the chickens who laid them. The sweat has subsided from our brows and a general attitude of revelry has settled in. During the ride, we ate local, learned

S. Senda-Cook (✉) · G.F. McHendry, Jr.
Department of Communication Studies, Creighton University, Omaha, NE, USA
e-mail: SamanthaSenda-Cook@creighton.edu

G.F. McHendry, Jr.
e-mail: GMcHendry@creighton.edu

B. McGreavy et al. (eds.), *Tracing Rhetoric and Material Life*,
Palgrave Studies in Media and Environmental Communication,
https://doi.org/10.1007/978-3-319-65711-0_9

about local food sources and practices, and biked *en masse* on streets usually dominated by automobile traffic. The goal of the supper is to teach people how to eat locally and engage in multi-modal transportation; but this event also revealed how rhetorical practices of resistance exist ecologically, forming a web of social support and materially reshaping the local communities of people who engage in them.

Mundane acts of resistance can articulate powerful rhetorical practices for living with the environmental problems pervasive in contemporary life.[1] As such, rather than focus on extraordinary rhetorical acts that shape environmental policy and global action, concentrating on banal materiality can encourage rhetorical critics to recognize the varying ways that resistance becomes part of a community. Everyday practices make and unmake our world by manifesting and sometimes challenging abstract cultural norms and ideals.[2] For example, practices of transportation and food selection, preparation, and consumption have received increased scrutiny recently; contemporary choices regarding food and transportation are sites of political embodiment. Although we agree that systematic changes are necessary to combat environmental problems, we invest hope and see power in the everyday action of individuals to transform their communities.

On the micro-scale, ordinary acts of resistance can push against dominant food systems by supporting local food growers and practicing multimodal transportation. To investigate how these practices become part of the everyday, we approach rhetoric from an ecological angle, conceiving of this event as a complex system of intertwining human and non-human existence and action. Jenny Edbauer explains, "Rhetorical ecologies are co-ordinating processes, moving across the same social field and within shared structures of feeling."[3] Nathaniel Rivers and Maarten Derksen note that the concept of ecologies allows scholars to move beyond the zero-sum approach to agency. In taking an ecological approach to deception, they undermine the idea that "one individual takes away agency from another, but only by concealing her own."[4] Likewise, we use an ecological approach to emphasize how power *circulates* among communities, particularly at one event where attendees learn about how they can integrate resistive practices into their own lives. An ecological focus such as this traces how social capital operates within a community when members come together to prepare for and respond to environmental threats.[5] The value of an ecological approach is that it attends to "the dynamic, multilevel, evolving nature" of this resistance.[6] While the concept of an ecosystem illuminates the interconnections between human, non-human, and non-living entities, it does not entirely communicate the habitual nature of these connections. We use *ecology* metaphorically to: (1) describe the web of environmental micro-practices that are contained within local food and multi-modal transportation systems, which we call an *ecosystem*; and (2) explain our methodological approach to this research—in situ rhetorical ecology. We also turn to Pierre Bourdieu and Michel de Certeau to help us unpack the banality and repeated nature of choosing what foods to eat and how to get around. The Full Cycle Supper offers a bounded act that granted access to what would otherwise be disparate and individualized activities. In short, the Full Cycle Supper revealed an emergent and interconnected community of diverse material entities (e.g., people, bikes, plants, and streets), working together to enact positive environmental practices rather than studying individual environmentalists persisting in their idiosyncratic ways.

The actions we observed at the supper, we argue, rhetorically construct practices of *ecological habitual resistance*, stemming from the concept of *habitus*.[7] Specifically, through material tactics, ecological engagement, and cultural capital exchange, volunteers and organizers of the Full Cycle Supper embody resistant political positions. In this paper, we begin by discussing the interrelated concepts of habitus, tactics, and cultural capital.[8] Then we concretize these ideas through a case study of an event called the Full Cycle Supper. This case study has implications for theories of resistance because, first, it demonstrates how groups that are not historically oppressed respond to systems that marginalize their practices of environmental concern. Next, it illustrates how a habitus can create structures of resistance, not only dominance. Lastly, it provides insight into how places operate tactically and strategically.

Ecological Habitual Resistance: A Conceptual Framework

Although many protests and events give attendees an opportunity to exercise political will and glimpse ways to embody resistance, few provide a map of developing *ecological habitual resistance*. We define ecological habitual resistance as rhetorical practices that consistently push against marginalizing structures (such as the damaging environmental practices associated with industrial farming and an automobile society, which is a car-dominated culture[9]) while also advancing alternative practices that allow people to act in accordance with a variety of political concerns. We examine the ways in which rhetorical acts of resistance become concomitant with everyday life. For example, in an automobile society, biking can become an act of habitual resistance when people incorporate it to meet their ordinary transportation needs. On the one hand, they are battling against the dominance of motorized transportation, and on the other, they position biking as an alternative means of getting around. In essence, material practices of resistance such as this can be embedded in the ecologies of everyday life even when that resistance seems to produce little in the way of systemic change. Material acts of resistance such as these function rhetorically and, when analyzed from an ecological perspective, reveal not only how individuals advance productive forms of resistance—that is, forms of resistance that produce alternative ways of being—but also how those forms can take root within entire communities.

Bourdieu's concept of the habitus provides a theoretical foundation for habitual resistance by establishing how small, ordinary practices construct everyday cultures and govern decisions, thoughts, and actions on "aconscious, pre-intentional" levels.[10] Although Bourdieu often invoked the habitus to refer to strategic participation with the grain of culture, we deploy it in this chapter to suggest that it can also be used to structure resistive practices. Utilizing an ecological approach to study a habitus of resistance makes salient the way that social change evolves because the concept of ecology uncovers how small changes can develop into larger ones through its sensitivity to initial conditions and feedback.[11] These practices can function as what de Certeau calls tactics in that they provide opportunities to redirect the flows of power.[12] Finally, as the habituated practices become accepted and respected, they reinforce themselves through systems of accumulating and trading cultural capital.

Habitus

Habitus is a set of social structures that guide and dictate the roles into which an individual will evolve, and it is the conditions under which we come to see, practice, and act in the world. Bourdieu explains, "as an acquired system of generative schemes, the habitus makes possible the free production of all thoughts, perceptions, and actions inherent in the particular condition of its production—and only those."[13] Zizi Papacharissi and Jan Fernback argue, "Bourdieu's habitus is a type of organic system that encompasses more than just beliefs, but also an entire propensity for how a person thinks and behaves—action is a key component of habitus. The habitus is based on previous thoughts and experience, without intention."[14] Thinking of the habitus as an ecosystem recognizes the interactivity of people, ideas, practices, and material objects that develop into patterns across space and time. Bourdieu theorizes a determinist philosophy where social forces generate the habits, practices, and behaviors that function as "reasonable" or "common-sense" in the everyday.[15] In addition to "being made explicit in institutions, laws, and educational curricula," Janet Atwill also suggests, "The structures of the habitus can be embedded in material objects and embodied performances."[16]

The practices of a habitus are material themselves, manifesting through embodied actions and emplaced discourses.[17] Bige Saatcioglu and Julie L. Ozanne argue that an "embodied sensibility" governs the material ways that the habitus unfolds.[18] Ways of being, such as choosing what food to eat, what clothes to wear, and how to get from one place to another, structure norms and expectations in our society. Steven Marsh's evocation of habitus makes salient its connection to bodies and places when he writes, "habitus is inscribed by habit and inhabitation."[19] In other words, the habitus impacts ingrained actions and is tied to particular places, ecologies themselves, making some practices make sense in some places and not others. Stacey Sowards explains that the construction of a habitus includes sets of dispositions that one learns growing up in a specific place and that compel particular actions and ways of being.[20] Sowards makes an important break from Bourdieu's conception of a totalizing, singular habitus and suggests that, like sub-cultures, multiple habituses can exist simultaneously and operate on distinct levels.

Like ecologies, co-existing and interacting with one another, habituses develop simultaneously and create varying situational cultures. Although disconnected from place, Saatcioglu and Ozanne argue that *emotional* habitus and *moral* habitus can both create the same structural components that define habitus, and they depend on other ways of embodying identity in the world. Papacharissi and Fernback articulate that a habitus can exist for a specific group and produces "a homogeneous character that creates a taken-for-granted nature among the group's practices and beliefs."[21] In this way, habitus can function as a cultural space in addition to being tethered to a specific place. It also creates an opportunity to undermine the centralized power of a dominant *habitus*. Papacharissi and Fernback argue that some habituses, even dominant ones within a group-specific habitus, can be "brutally critiqued and marginalized."[22] Therefore, even though the structuring capacity of habitus discourages behaving in ways that go against the grain—positioning them as inconvenient, dangerous, or simply making them unrecognizable—alternative habituses can develop and operate as sources of resistive agency.

For Sowards, a habitus can produce opportunities for rhetorical agency. She explains,

> Rhetorical agency is not just the process of negotiating one's individual and social habitus but, rather, the agent embraces enabling mechanisms, such as collective and collaborative efforts for social organizing and elements that facilitate the constitution of identity, from past and present dispositions, while resisting rhetorically constructed social conventions that limit or foreclose rhetorical options.[23]

While a habitus refers to the interacting material and conceptual structures that interpret and suggest some choices, behaviors, and thoughts rather than others, tactics are the embodied manifestations that reflect, construct, and (potentially) unmake a habitus.

Tactics of Resistance

To compose a habitus of resistance, community members must engage in practices that might operate individually but ultimately coalesce into a recognized form of resistance. De Certeau is useful here because he offers theories of resistance as articulated through everyday, material practices. Grounding his arguments in Michel Foucault's theories of power, de Certeau extends the ecological idea that power flows and circulates rather than moves uni-directionally.[24] He contends that power is enacted through everyday discourses and small-scale behaviors. Power also circulates and reproduces through practices that both establish norms and challenge them. While power of oppression is certainly enacted through subtle measures of discipline, as Foucault emphasizes, "With Michel de Certeau one can always perceive an optimistic élan, a generosity of intelligence, and a trust given to others in such a way that no situation appears to him a priori fixed or hopeless."[25] In his conceptualization, people are knowledgeable without knowing, engaging in practices that make sense without considering their actions, foundations, or consequences. If people can find microresistances, then they can find, "microfreedoms, mobilize unsuspected resources hidden among ordinary people, and in that way displace the veritable borders of the hold that social and political powers have over the anonymous crowd."[26] From looking for evidence of resistance in the everyday practices of disempowered people, de Certeau puts forwards theories that help articulate from where and under what circumstances people can and do resist.

De Certeau's conceptualization of strategies and tactics recognizes that both are ways to obtain and rework power. De Certeau argues that strategies and tactics enact different loci of power. He explains that a "strategy assumes a place that can be circumscribed as *proper*."[27] Strategies operate in static spaces to stabilize. By contrast, tactics "cannot count on a 'proper' (a spatial or institutional localization)."[28] Tactics are mobile, recomposing spaces rather than being contained by them. For example, roads themselves are evidence of a centralized power source that encourages particular forms of travel (i.e., personal, motorized vehicles that have the capacity to travel at high speeds). Whereas roads strategically maintain recent norms of getting around, individuals can use tactics to challenge those norms by employing alternative means of transportation or engaging in friendly acts of protest such as Park(ing) Day.[29]

Said another way, strategies come from institutions—centralized locations of power—and tactics originate from individuals and organizations that lack an epicenter. "A tactic is determined by the *absence of power* just as a strategy is organized by the postulation of power."[30] The word absence here is problematic in a Foucauldian sense of power, that is, where people negotiate power rather than hold it. We situate what de Certeau calls an "*absence of power*" as a lived imbalance in the flow of power between various groups in the social milieu. William E. Duvall elaborates on tactics, defining them as, "the inventiveness of ordinary people, not merely intellectuals and experts, as they develop tactics of resistance and open spaces for heterogeneity and difference ... in the face of totalizing power and dominating discourses."[31] Because strategies emanate from centralized sources, both physical and abstract, power is subject to checks of public knowledge. Tactics typically fly under the radar, performing resistances on small, decentralized scales. Anna Dezeuze writes, "This is the kind of resistance that relies neither on a future utopian vision nor on a nostalgic primitivism. It is situated in the specific rather than the general, and in an accumulated mass of collective activities rather than in individual agency."[32] Strategies and tactics can be both immaterial and material. They take the form of language and embodied ways of being and both result in material conditions in terms of monetary allocation and economic (dis)advantages. With these potential consequences, people who engage in tactics depend on the recognition and cooperation of like-minded groups.

Cultural and Social Capital

Whereas tactics of resistance come together to form habitual resistance, the structures themselves rely on habitus-dependent ways of reinforcing the resistant practices and norms that comprise these unique cultural spaces. Ecologically speaking, the small inputs contributed by community members will necessarily incur feedback. If that feedback comes primarily from dominant habituses, then it most likely does not encourage resistive actions. However, if the feedback legitimizes and reinforces tactics of resistance, then slowly the ecosystem itself can begin to change, responding to the uncoordinated tactics instead of the centralized strategies. Bourdieu's concept of cultural capital helps explain how positive feedback not only cultivates tactics of resistance but also rewards members by allowing them to gain status within resistive collectives.[33] Habituses, as ways of living and practicing, establish relationships and competencies that create and sustain cultural relationships. Like in an ecosystem, this give-and-take—the trading of benefits and hardships—governs the relationships between human and non-human entities. Cultural capital privileges specific subject positions over others and intensifies those advantages by linking material relations with social practices. Zhongwei Song explains, "In general, cultural capital is the totality of one's knowledge, skills, experience, competencies, and worldview that eventually determines how great a social and financial advantage or status its owner could have in a given society."[34] Bourdieu "articulates three forms of cultural capital: embodied (e.g., manners, tastes, skills, and dispositions), objectified (e.g., cultural possessions), and institutionalized (e.g., degrees and certificates)."[35] Cultural capital recognizes the learned skills and traits that, while not economic themselves, will yield economic advantages that accompany status in particular habituses.[36] Families teach these skills and traits to their children to reinforce social and economic hierarchies that privilege some at the expense of others, but recall that Bourdieu developed a theory of how dominant structures become entrenched.[37] As we established above, other kinds of habituses can exist, ones couched in resistance and dedicated to redirecting the flow of power. However, without resources, such as stronger social connections and respect among peers, resistance is difficult to sustain because of the corrosive forces of dominant practices. The benefits of ecological habitual resistance are not typically economic; economic capital remains a strategy of the dominant (i.e., capitalism functions as a dominant force restricting habitual resistance). Therefore, we turn to other kinds of capital that enable resistance.[38]

Of importance is the notion of social capital, which is based on collective identity. Bourdieu explains that social capital rests on the relationships one develops with different members of the same group.[39] Yu Pei Chang explains, "Social capital generally refers to the ability to approach a group or a community in which interpersonal relationships based on trust, norms and feelings exist; for the game generation, social capital indicates interpersonal networks of individual gamers, which comprise families, school colleagues and partners in game play."[40] Cultivating such relationships is one of the benefits of participating in ecological habitual resistance. Furthermore, by chatting and riding a bike with other like-minded people, participants at an event like the Full Cycle Supper have an opportunity not only to foster social connections but also to understand how they fit into the ecosystem. Finally, Lyon contends, social capital "provides access to others' resources. Individuals who belong to influential social groups have 'connections' to more resources than those with fewer and less influential relational ties."[41]

Bourdieu not only meant capital in the sense of wealth, however. Cultural capital refers to the cultural resources one can muster and deploy in various social situations. Bourdieu gave the example of Maliwnowski's "'tribal banker' who accumulates food only to lavish it on others and so build up a capital of obligations and debts that will be repaid in the form of homage, respect, loyalty, and new accumulation of material goods."[42] Another example would be the urban farmer who shares crops and growing advice in exchange for land, logistical aid, and future financial backing. These future favors owed are cultural capital that the urban farmer can call in, but are real and accumulated by deftly moving through the politics of urban food culture.

Describing a global marketplace, Tuba Üstüner and Douglas B. Holt theorize a *deterriorialized cultural capital* that runs contrary to Bourdieu's articulation of cultural capital as a sedimented product of local life.[43] They explain, "In contemporary consumer cultures, there is no epicenter from which cultural capital flows; virtually any category of consumption or locale is susceptible to appropriation via cultural capital."[44] In response to this global phenomenon, resistance movements have embraced place-based actions, producing place-based, resistant cultural capital. The deterriorialization of resistance can be particularly vexing because apparent contradictions (advocating for local food while buying global goods, for example) can appear to be logically contradictory despite being part of the fabric of a group's or an individual's resistive practices. An example of this paradox at the Full Cycle Supper can be found in the bitter essence of the hops infused in the keg of beer at the tour's last stop. Nebraska's climate is (mostly) inhospitable for the mass production of hops. While we drank *local* beer, that beer was a product of local, national, and global ingredients. Local consumption is, under contemporary capitalism, a permanent performative contradiction.

Cultural capital is learned and social capital is cultivated; community members use both to foster resistance.[45] Gokcen Coskuner-Balli and Craig J. Thompson explain, "Social groups who occupy lower positions in the socioeconomic hierarchy could create their own systems of status and ascribe positional value to subcultural capital that would typically not have currency in elite social fields."[46] Among resistant subcultures, cultural capital, social capital, and economic capital do not necessarily have positive relationships. In fact, possessing economic capital can undermine one's cultural or social capital in some subcultures that distrust economic-focused pursuits. In these and other subcultures, cultural capital connects to the bodies and places that play host to practices.

Like habitus, cultural capital is embodied and emplaced.[47] Craig J. Thompson and Tuba Üstüner write, "Embodiment is central to the acquisition of cultural capital."[48] They give the example of a "muscular physique or a suntan" as evidence of the time invested in specific practices that produce embodied cultural capital. Cultural capital goes beyond what Bourdieu calls "the embodied state," though.[49] Extending this idea, Alexander Lyon argues, "Cultural capital is displayed through preferred behaviors, mannerisms, pronunciation patterns, and routinized practices learned through interactions with others."[50] Additionally, a person's location matters as much as the actions in which they engage. Practices that communicate cultural capital in some circumstances are suspect or idiosyncratic in other places. Cultural capital depends on knowledge of both the physical place and the dominant culture and subcultures of an area. Therefore, an ecological knowledge of the practices helps us, as critics, uncover the nuances of these practices. Although Bourdieu advanced these theories to postulate how practices maintain dominant social relations, we suggest activists use them to alter the flow of power and offer alternative modalities. In reading cultural and social capital together, we unveil their complexities and interconnections, in turn making a rhetorical ecology approach sensible. Drawing situations together with complicated and ambient contexts offers a unique insight into how community members articulate ecological habitual resistance.[51]

Habitual Resistance

When taken together, these concepts—habitus, tactics, and cultural and social capital—offer a lens through which to analyze material and ecological rhetorical practices. For rhetorical critics, they grant access to discourses revealed through spaces and bodies, and provide a means of understanding how these material forms of communication can fit into larger ecologies of cultural reproduction and resistance. They also offer a means and rationale for accessing everyday systems of exchange and provide explanatory power and potentially improved decision-making about how and when to intervene as rhetorical critics. The construction of culture is material because it rests on the embodiments of cultural norms enacted by ordinary people. In focusing on these material forms of everyday rhetoric, we posit a theory of a *habitual resistance*, which depends on tactics and can ultimately produce the cultural and social capital that binds together tactical efforts and renders them intelligible to other group members.

Critical theory conceives of the habitus as an array of strategic acts that document dominant structures of power. Recall that de Certeau argues, "Strategies ... conceal beneath objective calculations their connection with the power that sustains them from within the stronghold of its own 'proper' place or institution."[52] Such structures produce systems of logic that reward people for embodying practices correctly—that is, in a valued and recognizable fashion. Emer O'Toole suggests, "Social conditionings in specific circumstances have produced the habitus, and it is this history which imbues the habitus with cultural capital and determines its functioning within the logic of a cultural field."[53] However, habitual resistance is an accumulation of tactics for living against the grain, at once a pushing against dominant powers and generating practices of microresistance (that can exercise power in their own way). Through language and embodied actions, those who wish to resist dominant structures, can cultivate habitual resistance—that is, a structure in which practices of resistance become recognizable and sensible. As Duvall contends, "our daily practices may be salvaged from the ever-present danger of falling into unconscious patterns and routines and become inventive forms of resistance and subversion in which we renew ourselves as the other and through which possibilities emerge."[54] In other words, our daily modes of living enact resistance.

One such tactic could be the use of dominant places to establish ecological habitual resistance. Bike riders who group together for Critical Mass function as an example of this concept.[55] The point of Critical Mass is to ride as a group, obeying traffic laws to show that bikes are a legitimate form of transportation and can function with cars on streets. Thus, deploying the use of streets for their own ends, these riders take their place in traffic but simultaneously work on their own project. In another, more local, example, in one neighborhood in Omaha, the city owns the tennis courts, but because the neighborhood association wants them to be multi-use, the association has paved them in concrete and installed chain link fencing as nets. These more durable materials, arranged in the conventional side-by-side tennis court format allow tennis to continue while people also use them to ride bikes and play soccer. In this way, the neighborhood remade a space to suit their needs while allowing dominant interpretations of space to continue, which creates a space for accepting ecological habitual resistance.

Cultural and social capital allow habitual resistance to exist because they reinforce and support practices that go against the grain, providing a respite from the constant pressures to conform. In this context, capital can function both ecologically and socially and thus can organize collectivities of resistance. When serving ecological functions, it brings together tactical groups that rely on one another by incurring and returning forms of support. In doing so these organizations build their status in the community and become more (or less) reputable. Additionally, cultural capital functions socially, bringing together individuals who may or may not be part of larger organizations of resistance. However, by volunteering and participating in resistive activities, they exchange cultural capital and cultivate status within communities as benefactors or dedicated advocates. Accumulating cultural and social capital allow groups and people to continue to resist even in the face of overwhelming problems and entrenched systems of being. Our case study demonstrates how habitual resistance can develop and be supported through tactics and exchanges of cultural and social capital.

Full Cycle Supper: A Case Study

We posit that ecological habitual resistance exists in the everyday material, tactical ways in which bodies maneuver against the grain of dominant food practices and enact their own alternatives. The Full Cycle Supper was a "four-mile bicycle tour of...neighborhood farm plots" and featured "farm-fresh courses at each site prepared by local chefs featuring produce grown at each site."[56] One participant described the events as a progressive dinner moving to different sites for each course. Each event raises funds for a local urban farm, Big Muddy Urban Farm, and the Community Bike Project Omaha (CBPO), which repairs and restores bikes, educates community members, and has free bike programs for children and adults. Roughly forty people attend each supper, including thirty paying participants and ten volunteers helping to usher riders from stop to stop, serve food, and coordinate logistics. George has attended two events as a paying ticket-holder and Samantha has helped organize all five. Micro-practices of resistance are lived, daily, and dynamic, making them hard to track. However, we examine The Full Cycle Supper, which celebrates and encourages the use of local food and multi-modal transportation, as an event that demonstrates these daily practices. Rhetorical ecologies reverberate through the interactions of humans, non-humans, and non-living entities. Engaging ecologies critically often involves in situ research. We accessed the ecosystem of the Full Cycle Supper through participatory critical rhetoric.[57] Participatory critical rhetoric is a collection of methodological approaches for doing rhetorical criticism in the field. We engaged in participant observation and unstructured interviews to explore the supper's ecology. In doing so, we participated in and with the ecosystem of local urban farming and transportation. This contact is crucial, because, as Edbauer explains, "The contact between two people on a busy city street is never simply a matter of those two bodies; rather, the two bodies carry with them the traces of effects from whole fields of culture and social histories. This is what it means to say that the social field is networked, connected, rather than a matter of place, sites, and home."[58] As participants, we made ourselves into bodies that came into contact with one another at the Full Cycle Supper. While the suppers exist outside of everyday lived habits, they function as a touchstone for advocates of multi-modal transportation and eating locally. The Full Cycle Supper creates habitual resistance by engaging in material tactics, locating participants within the ecosystem, and creating a space for exchanging cultural and social capital.

Material Tactics

The Full Cycle Supper engages in material tactics to challenge dominant transportation and food consumption practices. The material actions of participants and settings of the dinner contrasted sharply with dominant food and transportation practices in Omaha and the US. For example, the dinner took place entirely outside, locating itself at the farm sites that grew much of the food consumed during the event. The material proximity of growing food to consumed food challenges even contemporary farm-to-table restaurants and demonstrates the ability of places to resist alongside human participants. Additionally, during the event, movement was a central feature, requiring participants' bodies to perform actions to propel them to the next site. Most centrally, though, participants in the supper move tactically on bikes through Omaha during the event. Although the ride was on mostly side streets, our pack of hungry bikers occasionally slowed down traffic, existing as a Critical Mass that asserted its presence on roads. Some motorists seemed bewildered and unable to cope with the deluge of bikes—all with the right-of-way in theory though seldom in practice. Biking was a tactical exercise during the event by subverting the right-of-way politics of motorists and making streets legible to the movements of bikes in a visible demonstration of movement, viability, and solidarity. While making these actions legible may seem to be a movement from tactics to strategies, such appearances are mirages; legibility does not fix the efforts to one location or centralize power as enacting strategies does. Legibility allows community members to recognize one another and build an affinity, which we discuss in the following sections. An example of strategies along these lines would be when cities alter roads to accommodate bike riders and pedestrians with wider sidewalks or devoted bike lanes. In this case, by contrast, participants rode their bikes through streets designed only for automobile traffic.

At one particular corner, after eating basil pesto deviled eggs from chickens raised by Big Muddy, a group of fifteen bike riders waited for a car to proceed. The driver waited for the bike riders to proceed. The basic rules of right-of-way came undone and upset the dominant power relationship cars assume with urban transit in Omaha. Yet, this example underscores the eventual nature of the Full Cycle Supper—the supper as spectacle that engaged in tactical resistance. However, material practices at the event connected to more durational tactical acts. The thoroughfares required to move people around the city became the means by which community members performed local food and bike culture. Mile by mile, the streets became the material by which volunteers disseminated local food as a way of life and participants (re)made the city as legible to bikes.

The event's engagement of tactical resistance is an especially complex rhetorical strategy, given the mixture of ideologies present at the supper. On the one hand, many of the people at the supper do not practice the types of habitual resistance the supper engages in. Dominating streets in this way was unusual (and uncomfortable) for some. On the other hand, the material function of the meal, the biking, and the speeches and performances of the organizations impelled participants' bodies into resistive acts. Participants interpellated one another through their behaviors at the supper to gauge their place in the ecosystem, the event itself hailed them to act differently in the world (even if for some that material transformation only lasted a few hours).[59] By virtue of their attendance and followership, participants behaved tactically. This mix of ideological commitments is suggestive of the tactical terrain under which the Full Cycle Supper (and by extension CBPO and Big Muddy) operates. Organizers and volunteers occupy a habitual resistance by putting the city, and its dominant places, to work for other ways of living and doing and demonstrate these practices for participants.

The presence of participants who were outsiders created another instance of tactical resistance. Though everybody at the supper had some interest in the event, rhetorically, organizers constructed the event as a showcase for living locally, and supporting local food and multi-modal transportation. Material practices, such as identifying the specific garden bed the food came from, emphasizes the possibility of eating differently. Biking from stop to stop (the only means of transportation possible at the supper) hails bodies able to ride to do alternative transportation. These elements of the supper introduce the sponsoring organizations as parts of a network to resist dominant ideologies.[60] These hails called diverse sets of participants to both engage in temporary acts of resistance and to engage in rhetorical remapping of what is possible in their city, to reconsider their habits and to take on forms of ecological habitual resistance. Speeches at each stop implored participants to join various ways of living, support the hosting organizations, and engage in resistive material practices. Before and after the speeches at each location, participants had a chance to talk to one another, which provided an opportunity to sort out ideological commitments.

Ecological Engagement

The Full Cycle Supper displayed a diversity of messages, each with their place in the ecosystem, and the coordinated bodily practices became a webbing that brought participants from different worlds together. As we moved from table to table throughout the supper's various stops, a familiar ritual took place—a kind of membership check—as someone would inevitably ask, "Are you with the bike project or Big Muddy?" This basic rhetorical exercise, identification, did not ask participants if they were friend or foe, but rather something analogous to asking with which party at a wedding are they acquainted. The organizers (and probably most participants) assumed that participants heard about the event through one organization or the other, had a foot in one world of habitual resistance or the other. While everyone was there for similar political reasons, not everyone was part of the same specific ecosystem. Participants were engaged in elaborate routines of sense making, trying to define and map the paths that brought each another to the supper. Together, they comprised a complex, evolving, and never self-identical ecosystem where a variety of tactical forces were at play.

Over dinner, for example, the vegetarian main course arrived before the plates for those who consumed meat. This caused a moment of dissonance that opened space for dialogue and identification. A small group of vegetarian participants and organizers gathered in surprise, noting that usually vegetarian dishes are served last in what seems like an afterthought. In part because Big Muddy grows the vegetables, the chef incorporated the vegetarian dishes as a feature of the meal and placed them up front in the presentation of the food. The chef's emphasis on vegetarian plates sparked a discussion among those at the table who were vegetarians about their food choices and created moments of identification and spontaneous conversation. At each of the Full Cycle Suppers, organizers have been surprised when more people want vegetarian entrees than those who said they did when they bought tickets. Recognizing a place of habitual resistance, some of the participants may feel safe to experiment with a different culinary practice.

The diversity among participants at the event was chaotic at times because the event drew from audiences that shared overlapping political interests but very different ways of living and moving in the world. In other words, supporters of the CBPO and Big Muddy came together without necessarily being familiar with the political commitments of both organizations. This political complexity was clear when the volunteers from CBPO and Big Muddy, who we contend take on habitual resistance, are juxtaposed with participants who may not inhabit such a habitus. A function of the event, then, was to foster habitual resistance by trying to turn practices of alternative transportation and local food into routine habits. Many participants drove their cars to the Full Cycle Supper (while stowing their bikes on their vehicles). Transportation to the event (as opposed to during the event) makes salient the incongruity between the habitual resistance of the volunteers and the casual resistance of participants. And, yet, the ecosystem of the event ensured that disparate groups co-mingled, learned from one another, and came to depend on the different ways of performing local food and multi-modal transportation.

There were different levels of engagement in the core practices of eating local food and exercising multi-modal transportation; organizers, volunteers, and participants all played different roles and enacted different rhetorical postures while interfacing with the material practices of the supper. Some at the event were invested in multi-modal transportation and/or local food already; for them, this event was a celebration of their commitment to a sustained ecological habitual resistance. Samantha falls into this category, as a volunteer and board member for CBPO, she comprises her identity in part through multi-modal transportation practices and buying food from Big Muddy, which are part and parcel of the Full Cycle Supper's resistive aspirations. For others, the event was about seeking and gathering information or trying out these resistive practices. The supper provided a showcase for Big Muddy's community supported agriculture (CSA) program and for CBPO to discuss its work in the community. In both cases, participants who live at the cusp of resistive practices were enticed to do more. George fits in this category; he decided, for example, to move from one CSA program to Big Muddy's after the event. For participants like George, the common opening question—"Are you with the bike project or Big Muddy?"—left them trying to explain their interest in other, less obvious, ways. Nevertheless, it brought together people who were living habitual resistance with those who were curious or committed to differing degrees. This bringing together importantly happened over specific material practices with the capacity to take on aspects of ecological habitual resistance.

The event dramatized the embedded, routine, material habitual resistance of the organizers, volunteers, and some participants; it presented a way of working tactically, appearing to do one thing while really doing many things. Such efforts culminate in taking advantage of gaps in strategic formations and allowing participants a view of what their role in the ecosystem could be. For example, what appears to be an abandoned lot to most people, looks like a potential site for growing food to the members of Big Muddy. Their farm plots exist on urban land, often the result of charitable lease deals costing them one dollar a year. In planning urban farming sprawl, Big Muddy farmers have to see empty lots as potential space for tactical resistance; they have to seek out landowners and negotiate deals. The informal deals and trades reconfigure the commodity value of the land (Big Muddy's produce yield is worth more than one dollar a year), and these deals set up intricate webs of cultural capital exchange.

Cultural and Capital Exchange

While Big Muddy leases plots of land for one dollar a year, CBPO also pays one dollar a year to rent an old apartment building from the same group of people who own the plots leased to Big Muddy. Both Big Muddy and CBPO pay for any capital improvements (e.g., raised beds) or repairs (e.g., a broken window), but the group that owns these properties stays involved, usually helping with labor and fundraising. Facilitating land for Big Muddy and a building for the CBPO makes this group of people visible as community donors and that visibility has important repercussions. The growth of Big Muddy is high stakes; the rise of urban farming and non-profit bike education in Gifford Park (the Omaha neighborhood where these groups reside) works to make something different happen in a neighborhood where crime used to be a much bigger concern. Instead of letting buildings rot because of city inaction or letting lots become impromptu trash dumps, the CBPO and Big Muddy see an opportunity to repurpose the space, not just for their own ends but also with the goal of a better community in mind. Thus, CBPO, Big Muddy, and their benefactors are actually advancing the causes of the community simultaneously and creating the material space for a functional ecological habitual resistance. Three intricate elements here form a relationship of mutual interdependence that can be unpacked using the concepts of cultural and social capital. Big Muddy and CBPO (1) rely on their benefactors (2) who provide low-cost land, and the community (3) relies on Big Muddy and CBPO for local food and access to alternative transportation, knowledge, and labor. Big Muddy, CBPO, and their donors accrue cultural capital through their work in the community, which makes their resistance to dominant systems viable.

While Bourdieu's examples of cultural capital mainly express the ways the rich may use capital over others, or one may hold capital over another as a series of cultural debts, cultural capital can also be expressed as the mutual entanglements a community feels to work against unjust and oppressive situations. In other words, cultural capital can work toward resistance, which recirculates power, not merely reversing it but pulling its threads when the opportunities arise. Bourdieu acknowledged that cultural capital can include "moral obligations and emotional attachments created and maintained by [a] generous gift."[61] The sharing of food, transportation, knowledge, and skills all become forms of leverage that create bonds, communities, relationships, and obligations in mutual networks of belonging that are determined by and in opposition to material economic relations. As the community, the farm and bike shop, and their supporters become intertwined they can call on each other for favors that reify their social position in the community. These basic capital exchanges explain how cultural capital functions as a material practice of ecological habitual resistance.

For example, during the first Full Cycle Supper one of our stops was at the Gifford Park Community Garden. This garden is and is not part of Big Muddy. The garden is its own separate community space with individually owned plots; however, Big Muddy provides knowledge and labor to maintain the general parts of the garden, such as the chicken coop and compost piles. At the request of one of Big Muddy's benefactors, who is a community organizer, the garden was the only stop on the supper where organizers did not serve food. The benefactor wanted to showcase the garden. Moreover, the stop at the garden occurred out of sequence from the night's plan to honor a request from the benefactor that participants observe an annual community potluck, which celebrated the garden and the end of the summer. The request to include the garden and the granting of that request is cultural capital in action. Participants in the supper stood and watched as community members gave speeches that shared news and information about their food, and said goodbye to an important community member. Organizers positioned participants to bear witness to this community activity, to see and feel the import of a community stitched together by a large and dense garden space. During another Full Cycle Supper, the same benefactor was present at Big Muddy's main site for the start of the event. In extemporaneous remarks, one of the event's organizers praised the benefactor at length for their generosity and commitment to the community. Participants were encouraged to give them a round of applause. The recognition of the benefactor at both events is a form of cultural capital—a mutual exchange of land, support, and resources for recognition and status. The benefactor's embodied (i.e., disposition) and objectified (i.e., possessions) cultural capital derives from the possessing and holding of property throughout the Gifford Park community and the willingness to donate some of that property to community-centered organizations. The benefactor's status as property owner and community member fosters a series of relationships that produce the recognition of their work that occurred at both suppers.

Embodied cultural capital is also an integral part of the Full Cycle Supper for participants. This form of cultural capital flows from skills, tastes, and dispositions. The supper functions as a means of demonstrating and persuading participants to learn to live differently, and possibly more sustainably, with an eye to environmental conservation and limits. During one of the suppers, organizers took participants to a plot of land where sweet potatoes had been grown. The field, now clear, was ready for a winter cover crop. In an extemporaneous speech, one of the event's organizers explained the purpose of a cover crop (to protect and enrich the land over the winter) and described the crop that they would plant (legumes). Then organizers asked participants to sow the earth. Handful by handful, person by person, participants walked the plot of land and threw seeds from a bucket. In this way, participants embodied a small amount of cultural capital as they experienced the specific expertise that accompanies cover crops, soil enrichment, and land management. Cultural capital enables habitual resistance by affirming resistive efforts through non-monetary systems of support and praise. Through their embodied actions, participants and organizers illuminated the structures of habitual resistance that create different manifestations of everyday opportunities for resistance, as well as the subtle systems of recognition that reward these actions.

Utilizing material tactics, ecological engagement, and exchange of cultural capital adds to and reinforces existing structures of an ecological habitual resistance. Deploying material tactics and repurposing material spaces (as an exchange of cultural capital) demonstrate the power participants have to create a world (albeit a local and contained one) where their practices make sense and are valued, which reveals the rhetorical significance of such practices. The participants' ritual of ascertaining information about one another allowed them to gauge the degree to which they each committed to the existing ecological habitual resistance and thus to gain a sense of the power the group contained.

Discussion

In this essay, we considered the material practices of resistance taken by local activists that carve out room for food and transportation alternatives. At a theoretical level, we focused on how existing literature on resistance and materiality can provide a frame for understanding the daily, lived experiences of those who work to define their lives in opposition to dominant modes of transportation and global food ways. We argued that a habitual resistance is a structure defined by practices of tactics and resistive expressions of cultural capital. To explicate these concepts we have grounded our theory in the Full Cycle Supper, both a celebration of local food and multi-modal transportation and a deftly constructed rhetorical event that produced a different set of politics. Approaching this project from a rhetorical ecology perspective has given us the opportunity to reveal the interconnected nature of not only the participants but also the practices in which they engage. It has demonstrated an underlying ecosystem in which organizers and participants moved together to challenge dominant cultural practices. The Full Cycle Supper, as a locus for sparking and supporting habitual resistance, demonstrates how people who are not historically marginalized can engage in resistive acts and feel at odds with dominant society. Additionally, it demonstrates how a habitus can serve both resistant and dominant functions. Finally, it shows how places themselves, and not just behaviors, can function tactically or strategically, emphasizing the role that the farms and streets played in this ecosystem.

First, the Full Cycle Supper is an interesting example of resistance because activists for multi-model transportation and local food can be, but are not necessarily, part of historically marginalized groups who can feel slighted in contemporary society. Because the practices in which they engage are only recently gaining recognition as possibilities, especially in a city like Omaha, it can feel isolating to advocate for these hyper-local alternatives. In other words, without knowledge of an ecosystem like the one we examined here, environmentalists may feel alone. One strategy for undermining these resistant efforts is to frame them as elitist. However, the individuals that run these organizations eschew the cultural hallmarks of elite organic food and cycling cultures in everyday life. For the Full Cycle Supper, those who bought tickets appeared comfortably middle- to upper-class based on their clothing and gear, their ways of speaking, and our brief conversations that revealed occupations and education. But, engaging in practices of eating locally grown food and biking through a city, practices that are still largely marginalized in Omaha, should not automatically characterize a person as elitist. The critique that such practices are elitist is a clever way to position the institutionalized agriculture industry as an underdog. The event was also expensive, a limiting feature, especially given that both organizations that sponsor the event work to benefit economically marginalized parts of Omaha as a central part of their missions. For example, Big Muddy donates to Table Grace, a local restaurant that has a "pay what you can" pricing scheme and also accepts Electronic Benefits Transfer (EBT) and will give EBT tokens if people do not use their entire amount so that they can spend it another week. The Community Bike Project Omaha (CBPO) sells bikes cheaply and gives away bikes to people who cannot afford other means of transportation. Through sales and Youth and Adult Earn-a-Bike programs, CBPO reported that it provided over 500 bikes in 2016 to community members of all ages. The event also clearly privileges bodies that can ride bikes. The ride is physical; Omaha has many hills that can be difficult for novices or those with disabilities to ride. While the organizations promote local food and multi-modal transportation, the practices of the dinner prefigure a degree of ability that may exclude some of the target audience for their mission. In short, while some elements of local food consumption are elitist and pro-status quo, in practice here that is not the entire truth. The event is, in part, a compromise that practices its ecological habitual resistance before an audience that may not share those practices, but in doing so garners monetary capital to continue to build an ecosystem of resistance.

By contrast, participating in a system of transnational capitalism through food and transportation has become a default way of living in contemporary US cities. Grocery stores feature goods that transect local boundaries, laws, climates, and nations. Global produce sacked in bags made from refined petroleum and hauled from the grocery store to home in an automobile is omnipresent as a reasonable way of living in contemporary middle-class America. This way of shopping and eating did not appear suddenly, but is carefully taught and rehearsed via social structures—"the habitus is what enables the institution to achieve full realization."[62] Notice also that the practices of the habitus are in material relations. Shopping, buying, eating, cooking, and moving throughout the city are material practices.

Second, Bourdieu deployed the concept of habitus to explain normative social structures. However, habitus can also be useful as a term for understanding the structural forces that produce enclaves of resistive practices. Joshua Atkinson explains, "resistance exists as a continuum from widespread resistance (emancipatory) to total integration (hegemonic)."[63] To challenge a dominant habitus, a resistant one must be more emancipatory than hegemonic, but it must also be recognizable. Events like the Full Cycle Supper support habitual resistance not only through the practices we delineated above but also by rendering multi-modal transportation and local food practices as sensible, or at least understandable, which in turn contributes to structures that support such practices. We examined the ways local urban farmers and urban cyclists express resistance to "common-sense," such as resistance to multi-national corporate food stores and reliance on automobiles for transportation. Discontent with these contemporary modes of living produce ecological habitual resistance that becomes an altogether different habitus.

Finally, at times, participants explicitly discussed the globally networked transportation and food systems made normal through our contemporary society, highlighting how places themselves can be examples of strategies and tactics in the struggle for social change. Over dessert, a few participants discussed the concept of a neighborhood market. The area around Big Muddy's main garden plot has a number of small shops that carry a limited grocery selection. The area also has some larger chain grocery stores, but their selection and quality are limited. In the summer, Gifford Park, the neighborhood in which Big Muddy and the CBPO operate, hosts a market on Fridays. Big Muddy sells produce at this market and works to get local produce to the community. The Gifford Park Neighborhood Market is an enclave for local, community, and sustainable goods in a neighborhood that exhibits diversity of race and economic class. Just a mile down the road, the Walmart corporation recently opened their version of a neighborhood market. One of the farmers for Big Muddy spoke at length about Walmart's audacity to call their store a neighborhood market. This struggle between the two neighborhood markets is the essence of strategies and tactics.

De Certeau wrote, "The 'proper' is a victory of space over time. On the contrary, because it does not have a place, a tactic depends on time—it is always on the watch for opportunities that must be seized 'on the wing.'"[64] By contrast, "A tactic insinuates itself into the other's place."[65]

Walmart has a proper place in global capital, while Big Muddy grows in borrowed lots, using volunteer and underpaid labor. Walmart's neighborhood market is legible as a grocer—permanent and fixed. Gifford Park's market has a regular space (an otherwise vacant lot), but its structure is impermanent and itinerant. During the conversation over dessert at the Full Cycle Supper, the contrast in material practices between Walmart's strategic incursion into this neighborhood and participants' anti-Walmart rhetoric became sensible. The farmer's tactical resistance to Walmart came through the construction of alternative neighborhood markets and foodways. Getting people out of stores like Walmart (though only in limited ways) and into local food are tactical inversions of the strategic function of global capitalism, which underscores the tactical nature of the Full Cycle Supper.

Each of these examples underscores the ways in which resistance can become a way of living as expressed through material rhetorics of everyday life, particularly through an ecosystem like the one that we critiqued. This fundraiser proffered a chance for people already engaged in resistive practices to feel hope and see the potential for change. Simultaneously, it introduced people to forms of ecological habitual resistance and/or enticed some people to more fully commit to such resistance. Although the event itself was out of the ordinary, it functioned as a touchstone—a way to come together and engage in these daily acts. The idea of riding a bike to go out to dinner or pick up groceries is a foreign concept to most people. But the Full Cycle Supper created a space to showcase how those actions can bring hope in the face of uncontrollable global forces. The supper made legible a habitual resistance even for people who do not normally participate in such practices.

Notes

1. Barbara Kingsolver, Camille Kingsolver, and Steven L. Hopp, *Animal, Vegetable, Miracle: A Year of Food Life* (New York: HarperCollins Publishers, 2007); Michael Pollan, *Cooked: A Natural History of Transformation* (New York: The Penguin Press, 2013).
2. Greg Dickinson, "Joe's Rhetoric: Finding Authenticity at Starbucks," *Rhetoric Society Quarterly* 32, no. 4 (2002): 5–27; Samantha Senda-Cook, "Rugged Practices: Embodying Authenticity in Outdoor Recreation," *Quarterly Journal of Speech* 98, no. 2 (2012): 129–152.

3. Jenny Edbauer, "Unframing Models of Public Distribution: From Rhetorical Situation to Rhetorical Ecologies," *Rhetoric Society Quarterly* 35, no. 4 (2005): 20.
4. Nathaniel A. Rivers and Maarten Derksen, "Ecologies of Deception in Psychology and Rhetoric," *Quarterly Journal of Speech* 101, no. 4 (2015): 637.
5. Nikolay Mihaylov and Douglas D. Perkins, "Community Place Attachment and its Role in Social Capital Development," in *Place Attachment: Advances in Theory, Methods and Applications*, eds. Lynne C. Manzo and Patrick Devine-Wright (London: Routledge, 2014), 59–74.
6. Garrett M. Broad, *More than just Food: Food Justice and Community Change* (Oakland: University of California Press, 2016), 29.
7. Pierre Bourdieu, *The Logic of Practice*, trans. Richard Nice (Stanford: Stanford University Press, 1980).
8. Ibid.
9. Garrett M. Broad, "Ritual Communication and Use Value: The South Central Farm and the Political Economy of Place," *Communication, Culture, & Critique* 6 (2013): 20–40; Ross Singer, "Visualizing Agrarian Myth and Place-Based Resistance in South Central Los Angeles," *Environmental Communication* 5, no. 3 (2011): 344–349. Zach Furness, *One Less Car: Bicycling and the Politics of Automobility* (Philadelphia: Temple University Press, 2010).
10. Dana Anderson, "Questioning the Motives of Habituated Action: Burke and Bourdieu on Practice," *Philosophy and Rhetoric* 37, no. 3 (2004): 265.
11. Fritjof Capra, *The Web of Life: A New Scientific Understanding of Living Systems* (New York: Anchor Books, 1996); Fritjof Capra, "Speaking Nature's Language: Principles for Sustainability," in *Ecological Literacy*, eds. Michael K. Stone and Zenobia Barlow (Sierra Club Books, 2005), 18–29.
12. Michel de Certeau, *The Practice of Everyday Life*, trans. Steven Rendall (Berkeley: University of California Press, 1984).
13. Bourdieu, *Logic of Practice*, 55.
14. Zizi Papacharissi and Jan Fernback, "The Aesthetic Power of the Fab 5: Discursive Themes of Homonormativity in *Queer Eye for the Straight Guy*," *Journal of Communication Inquiry* 32, no. 4 (2008): 350.
15. Bourdieu, *Logic of Practice*, 55.
16. Janet Atwill, "Bodies and Art," *Rhetoric Society Quarterly* 36, no. 2 (2006): 168.
17. Michael Middleton, Aaron Hess, Danielle Endres, and Samantha Senda-Cook, *Participatory Critical Rhetoric: Theoretical and Methodological Foundations for Studying Rhetoric* In Situ (Lanham: Lexington Books, 2015).

18. Bige Saatcioglu and Julie L. Ozanne, “Moral Habitus and Status Negotiation in a Marginalized Working-Class Neighborhood,” *Journal of Consumer Research* 40 (2013): 693.
19. Steven Marsh, “The Haptic in Hindsight: Neighborhood Cinema-Going in Post-War Spain,” *Studies in Hispanic Cinemas* 2, no. 2 (2005): 115.
20. Stacey Sowards, “Rhetorical Agency as *Hacienda Caras* and Differential Consciousness through Lens of Gender, Race, Ethnicity, and Class: An Examination of Dolores Huerta’s Rhetoric,” *Communication Theory* 20 (2010): 223–247.
21. Papacharissi and Fernback, “Aesthetic Power,” 350.
22. Ibid., 360.
23. Sowards, “Rhetorical Agency,” 227.
24. Michel Foucault, *The History of Sexuality*, vol. 1. (New York: Vintage Books, 1990).
25. Luce Giard, introduction to *The Practice of Everyday Life: Volume 2: Living & Cooking*, ed. Luce Giard, trans. Timothy J. Tomasik (Minneapolis: University of Minnesota Press, 1998), xxi.
26. Ibid., xxi.
27. De Certeau, *Practice of Everyday Life*, xix.
28. Ibid.
29. Danielle Endres, Samantha Senda-Cook, and Brian Cozen, “Not Just a Place to Park Your Car: PARK(ing) as Spatial Argument,” *Argumentation and Advocacy*, 50 (2014): 121–140.
30. De Certeau, *Practice of Everyday Life*, 38, emphasis in original.
31. William E. Duvall, “Are We Postinterruption? Postmodernism, Resistance, and Practice,” *The European Legacy: Toward New Paradigms* 2, no. 4 (1997): 693.
32. Anna Dezeuze, “Assemblage, *Bricolage*, and the Practice of Everyday Life,” *Art Journal* 67, no.1 (2008): 37.
33. Bourdieu, *Logic of Practice*.
34. Zhongwei Song, “*The Art of War* in Retranslating Sun Tzu: Using Cultural Capital to Outmatch the Competition,” *Translation and Interpreting Studies* 7, no. 2 (2012): 179.
35. Saatcioglu and Ozanne, “Moral Habitus,” 692.
36. Joshua Atkinson, “Contemporary Crusaders and Timeless Elders: Building Cultural Capital through Alternative Media Texts,” *Popular Communication* 5, no. 3 (2007): 171–190.
37. Saatcioglu and Ozane, “Moral Habitus.”
38. See Chang, “The Game Generation and its Leisure Capital: A Study in the Taiwan Social Context,” *Westminster Paper* 9, no. 1 (2012): 137; Carol Pui-Ha Chow, “Image Capital, Field, and the Economies of Signs and Space,” *The Communication Review* 16 (2013); Coskuner-Balli and

Thompson; Lyon; Clive James Nwonka, "Diversity Pie: Rethinking Social Exclusion and Diversity Policy in the British Film Industry," *Journal of Media Practice* 16, no. 1 (2015); Saatcioglu and Ozane, "Moral Habitus"; Song, "*The Art of War*".

39. Pierre Bourdieu, "The Forms of Capital," in *Handbook of Theory: Research for the Sociology of Education*, ed. John G. Richardson (New York: Greenwood, 1986).
40. Chang, "The Game Generation."
41. Lyon, 374.
42. Bourdieu, *The Logic of Practice*, 125.
43. Tuba Üstüner and Douglas B. Holt, "Toward a Theory of Status Consumption in Less Industrialize Countries," *Journal of Consumer Research* 37 (2010): S261.
44. Ibid., S263.
45. Atkinson, "Contemporary Crusaders."
46. Gokcen Coskuner-Balli and Craig J. Thompson, "The Status Costs of Subordinate Cultural Capital: At-Home Fathers' Collective Pursuit of Cultural Legitimacy through Capitalizing Consumption Practices," *Journal of Consumer Research* 40 (2013): 20.
47. Middleton et al., *Participatory Critical Rhetoric.*
48. Craig J. Thompson and Tuba Üstüner, "Women Skating on the Edge: Marketplace Performances as Ideological Edgework," *Journal of Consumer Research* 42 (2015): 248.
49. Bourdieu, *The Logic of Practice*, 243.
50. Alexander Lyon, "The Mis/Recognition of Enron Executives' Competence as Cultural and Social Capital," *Communication Studies* 59, no. 4 (2008): 373.
51. Edbauer, "Unframing Models"; Rivers and Derksen, "Ecologies of Deception."
52. De Certeau, *Practice of Everday Life*, xx.
53. Emer O'Toole, "Cultural Capital in Intercultural Theatre: A Study Pan Pan Theatre Company's *A Playboy of the Western World*," *Target* 25, no. 3 (2013): 412.
54. Duvall, "Are We Postinterruption?" 694.
55. Danielle Endres and Samantha Senda-Cook, "Location Matters: The Rhetoric of Place in Protest," *Quarterly Journal of Speech* 97, no. 3 (2011); Joseph Heath and Andrew Potter, *Nation of Rebels: Why Counterculture Became Consumer Culture* (New York: Harper Business, 2004).
56. Full Cycle Supper, Facebook, 2014, https://www.facebook.com/events/738708669574576/.
57. Middleton et al., *Participatory Critical Rhetoric.*
58. Edbauer, "Unframing Models," 10.

59. Louis Althusser, *Essays on Ideology* (London: Verso, 1984).
60. We want to note here that the supper itself can in many ways be read as exclusionary. One must have a bike in working order, money to purchase a ticket, and a body that is capable of riding a few miles. While we read the supper as an opportunity for doing resistance, the structures of the event do amplify a variety of economic and embodied privileges—including our own privileges.
61. Bourdieu, *The Logic of Practice*, 126.
62. Ibid., 57.
63. Atkinson, "Contemporary Crusaders," 183.
64. De Certeau, *Practice of Everyday Life*, xix.
65. Ibid., xix.

PART IV

Toward Justice and Care

CHAPTER 10

The Most Nuclear-Bombed Place: Ecological Implications of the US Nuclear Testing Program

Danielle Endres

The American West is the most nuclear-bombed place in the world. Although only two nuclear bombs have been used in warfare—devastating Hiroshima and Nagasaki, Japan, in August 1945—there have been over 2000 nuclear weapons detonated in tests conducted by eight countries over seventy-one years across the globe, with the largest concentration of tests located in the American West.[1] While the US conducted 1054 nuclear tests in the Marshall Islands, Alaska, Nevada, Colorado, Mississippi, and New Mexico over a fifty-year period as part of the US Nuclear Weapons Testing Program, the vast majority (928) of those explosions occurred between 1951 and 1992 at the Nevada Test Site, primarily at Frenchman Flat and Yucca Flat.[2] Although none of these tests matched the death and destruction inflicted by Fat Man and Little Boy on the citizens of Japan, the cumulative effects of nuclear testing in the American West, and globally, have exacted other forms of death

D. Endres (✉)
University of Utah, Salt Lake City, UT, USA
e-mail: danielle.endres@utah.edu

B. McGreavy et al. (eds.), *Tracing Rhetoric and Material Life*,
Palgrave Studies in Media and Environmental Communication,
https://doi.org/10.1007/978-3-319-65711-0_10

and destruction, not just to people and cities but also to the earth's ecosystem.

Nuclear testing—one consequence of the Pandora's box opened by the development of nuclear weapons—exemplifies the Anthropocene. The Anthropocene is a new planetary era, as evidenced in geological record, that expresses the impact human technology has had on global ecological systems. Some researchers have argued that the first nuclear bomb tested on July 16, 1945 in Alamogordo, New Mexico was the start of the Anthropocene.[3] Global radioactive fallout from over 2000 nuclear bomb tests—not to mention the stockpiles of undetonated nuclear weapons, the nuclear waste at nuclear weapons production facilities such as Hanford, and other traces from the global nuclear weapons production process—have left a geographic mark that will be detectable for millennia. This imprint reflects the increase in radioactive elements such as cesium-137 (no known natural sources) and plutonium-239 and 240 (only exist in trace amounts naturally). Although this stain on the earth's ecosystem is spread across the globe, the people, flora, and fauna of the American West have been disproportionately affected by nuclear weapons testing.[4]

The Nevada Test Site (NTS) was originally occupied by nomadic bands of Western Shoshone and Southern Paiute people, who still claim treaty-based ancestral rights to the land.[5] Many Shoshone and Paiute people had moved to nearby cities or were forcibly relocated to reservations in the early twentieth century before the land was first appropriated by the federal government to form the Las Vegas Bombing and Gunnery Range in 1940, and later the NTS in 1951.[6] Yet, according to a consortium of Western Shoshone and Southern Paiute tribes and organizations, "For many centuries, the [NTS] area has been a central place in the lives of American Indian tribes, continuously used by these tribes from antiquity to contemporary times."[7] Further, Western Shoshone and Southern Paiute governing councils, organizations, and people have actively resisted the federal government's jurisdiction over, and use of, the NTS.[8]

The NTS is approximately 1375 square miles of desert basin and range terrain that is part of what is called a "transitional zone between the Great Basin and Mojave deserts."[9] Although the NTS is a rich desert ecology inhabited by a robust diversity of human and more-than-human life, the US federal government justified locating the nation's test site there because "few areas of the continental United States are more ruggedly severe and as inhospitable to humans" and "although no locale can be said to be ideal or optimal for nuclear weapons testing, the Nevada Test Site was perhaps the best continental site available for avoiding collateral damage and radiation exposure to plants, animals, and, most importantly, human beings off site."[10] This logic assumes that nuclear weapons testing had to happen *somewhere*. That somewhere, wherever it was, would ideally minimize the amount of harm from nuclear weapon tests, but would not be immune from harm.

As the most nuclear-bombed place in the world, the NTS is a site of violence; it is a place where 928 nuclear bombs exploded. This legacy is visually apparent in aerial photos of a pockmarked landscape full of craters, one of which is enormous at a depth of 320 feet (for comparison, this would be the equivalent of a 20 or 25 stories high-rise building). This legacy can also be seen in the destructive material consequences of nuclear weapons on the local (and global) ecology.[11] The nuclear bombs exploded at the NTS not only contributed to illness and, in some cases, death to human bodies (including both NTS workers and downwinders) but also to the many more-than-human bodies inhabiting the NTS, especially those who happened to be at one of the many ground zero sites at the moment of nuclear detonation.[12] As such, the NTS is what environmental justice scholars call a national sacrifice zone, meaning it is a place that is set aside to be sacrificed to serve the broader interests of the nation, in this case national security during the Cold War.[13] The often-touted legacy of the Cold War, of which nuclear testing was a part, is that it prevented World War III. However, as downwinder Mary Dickson has explained, the Cold War can also be seen as an "undeclared war" on the people, the flora and fauna, the land, and the ecology of the American West.[14]

The violence at the Nevada Test Site can be explained through nuclear colonialism, a complex phenomenon within which indigenous people across the globe are disproportionately affected by the negative consequences of the nuclear production process from cradle to grave.[15] Scholars have examined nuclear colonialism in the context of the American West through studies of uranium mining, nuclear weapons production, nuclear weapons testing, and nuclear waste storage on or near Native American lands.[16] In this chapter I expand the concept of nuclear colonialism to account for its relationship with the more-than-human world. I argue that the Nevada Test Site is a colonized place wherein the entire ecological community—humans, animals, plants, soil, and water—is disproportionately affected by its interaction with nuclear weapons tests performed in the service of human national security. In extending the concept of nuclear colonialism to the more-than-human world, this chapter points to the material damage to animate and sensing beings that make up that more-than-human land community at the Nevada Test Site (and beyond, considering that radiation released from nuclear testing was not contained by the boundaries of the NTS). Furthermore, to avoid viewing the more-than-human land community as only an inert victim of nuclear colonialism, this extension also acknowledges the active role that more-than-human beings play within nuclear colonialism as a complex phenomenon of relations, and rhetorical practices, between humans, more-than-humans, and the nuclear production process. Extending nuclear colonialism to include more-than-human beings approaches the concept from a more ecological standpoint. By ecological (and by extension ecology), I am referring to the interrelationship and co-existence of beings and matter within the earth.[17] A more ecological standpoint on nuclear colonialism is concerned with not only cataloguing the environmental and human health implications of nuclear technologies but also with thinking about nuclear colonialism as a material-discursive phenomenon that highlights the interconnectivity and intersubjectivity between humans, more-than-human beings, land communities, and nuclear technologies.

Nuclear colonialism is, in part, rhetorical. I seek to highlight the role of rhetoric in the ongoing constitution of nuclear colonialism. While traditional theories of rhetoric assume that it is a human faculty, thinking about nuclear colonialism in relation to the more-than-human world relies on emerging theories of material rhetoric that not only acknowledge the material consequences of rhetoric, but also understand how places, more-than-human beings, and things are capable of rhetoric.[18] I rely on Natasha Seegert's definition of rhetoric as "the relational force of signals interacting with the world," which expands rhetoric to the more-than-human world by including "beings who surround us but who are frequently silenced."[19] This definition of rhetoric makes room for both discursive and material signals in a "world already speaking through affective networks of connection."[20] Critics can access these rhetorics by focusing on rhetorical performances and practices, as well as on the consequences, or force, of rhetorical dissemination. As this chapter will elaborate, nuclear colonialism entails a range of rhetorics—both in the sense of patterns of discourse with material consequences and forms of non-discursive material rhetoric—that justify, perpetuate, and challenge the practices of colonialism that underlie the nuclear production process. Instead of separating material and discursive forms of rhetoric, I view rhetoric as encompassing inextricably related material and discursive elements, as indicated by the imperfect term material-discursive. In other words, nuclear colonialism is a material-discursive phenomenon comprised by a variety of material and discursive, human and more-than-human rhetorics.

In the remainder of the chapter, I begin by developing my argument for expanding nuclear colonialism to the more-than-human world. Then I examine how this expanded notion of nuclear colonialism can be seen through an analysis of the Nevada Test Site as a place with rich rhetorical relations. In particular, I examine how characterizations of the NTS as a wasteland articulate with nuclear colonialism and open opportunities for material-discursive rhetorical resistance. The chapter concludes by considering implications for taking an ecological approach to nuclear colonialism.

Nuclear Colonialism

Nuclear colonialism is a material-discursive phenomenon that describes the power dynamics and interrelationships between indigenous peoples and the global nuclear production complex. Ward Churchill and Winona LaDuke first used the term "radioactive colonization" in their description of the disproportionate harms experienced by Native Americans and other indigenous groups as a result of the cradle to grave cycle of nuclear production.[21] The term, and its variant nuclear colonialism, has proliferated among indigenous activists and scholars.[22] According to the Indigenous Environmental Network:

> The nuclear industry has waged an undeclared war against our Indigenous peoples and Pacific Islanders that has poisoned our communities worldwide. For more than 50 years, the legacy of the nuclear chain, from exploration to the dumping of radioactive waste has been proven, through documentation, to be genocide and ethnocide and a deadly enemy of Indigenous peoples...United States federal law and nuclear policy has not protected Indigenous peoples, and in fact has been created to allow the nuclear industry to continue operations at the expense of our land, territory, health and traditional ways of life...This disproportionate toxic burden—called environmental racism—has culminated in the current

> attempts to dump much of the nation's nuclear waste in the homelands of the Indigenous peoples of the Great Basin region of the United States.[23]

Governments pursuing nuclear technologies justify sacrificing the sovereignty, land, and health of indigenous people in the interests of national security or the national interest. Nuclear colonialism in the US is embedded within a larger system of settler colonialism, wherein a population colonizes by taking over the land-base of an indigenous population and subjecting that population to a domestic-dependent relationship with the colonizer.[24] Nuclear colonialism relies on the intersections of settler colonialism and nuclearism.[25] It is a form of environmental injustice, but a unique form that takes into account Native American people's sovereignty and ongoing colonization.[26] Nuclear colonialism is also a form of violence, in its intersecting material-discursive sense. Nuclearism and colonialism in the US include instances of extreme physical violence (Indian wars and nuclear bombs), more subtle forms of incremental slow violence (slow-developing illnesses, such as diabetes and cancer), and psychological violence (colonization of the mind, assimilation, fear of nuclear threats).[27] These forms of violence are experienced through bodies, minds, and places. For example, the 928 nuclear bombs detonated at the NTS not only contributed to illness and, in some cases, death to people's bodies (from NTS workers to downwinders)[28] but also took a toll on indigenous peoples' sense of identity, culture, and place.[29] Nuclear colonialism is a material-discursive phenomenon in which the lived experiences, violence, and injustices of nuclear colonialism are inherently both material and discursive.

Extending Nuclear Colonialism to the Rhetorical

My previous research intervenes in the literature on nuclear colonialism by highlighting the specifically *rhetorical* elements of nuclear colonialism, describing how nuclear colonialism could not happen without extensive and complex systems of rhetorical justification.[30] Much of the research on nuclear colonialism focuses on demonstrating that the nuclear production complex has had a disproportionate burden on indigenous people—that is, this research provides geographical, empirical, and archival evidence which establishes that there is a verifiable pattern of disproportionate harm to indigenous populations from the nuclear production process. Yet there has been less emphasis in this research on the role of rhetoric in perpetuating, justifying, and resisting nuclear colonialism. These rhetorical practices are constrained and enabled by what Foucault described as a *discursive formation*, which is a pattern of discourse across a variety of discrete texts or statements that reflects regimes of power/knowledge.[31] My previous work focuses on articulating the rhetorical strategies and tactics within this discursive formation and responds to what I saw as a limitation in analyses of nuclear colonialism that lack sustained focus on the rhetorical and discursive aspects of nuclear colonialism.[32] In doing so, it shines a light on the rhetorical elements of nuclear colonialism while acknowledging that nuclear colonialism is not only a rhetorical phenomenon. This previous research relies on a more traditional definition of rhetoric—as a symbolic human practice—than the one I presented above. Further, although I noted above that nuclear colonialism is a material-discursive phenomenon, my previous research maintains a distinction between discursive and material elements by focusing primarily on illuminating one over the other— the rhetorical over the material. In the next two sections, I propose two interrelated interventions into current conceptualizations of nuclear colonialism that seek to better account for nuclear colonialism as an integrated material-discursive phenomenon within which human and nonhuman, material and discursive rhetorics circulate. First, I expand nuclear colonialism beyond the human by examining how the more-than-human land community is affected by, implicated in, and engaged in resistance to nuclear colonialism. Second, to account for the rhetorics of the more-than-human land community in the nuclear colonialism material-discursive formation, I continue in a line of scholars who seek to more fully account for the (rhetorical) agency of the material.[33]

Extending Nuclear Colonialism to the More-than-Human Land Community

Nuclear colonialism is usually conceptualized in terms of the human; it is perpetuated by human institutions, such as the federal government or members of the military industrial complex, and negatively experienced by particular groups of indigenous humans, Western Shoshone and Southern Paiute people in the case of nuclear testing at the NTS. Yet it is possible and desirable to extend the concept from its focus on the human to a focus on how an entire ecological community participates in nuclear colonialism, seeing more-than-human beings as possible agents, victims, resistors, and contributors. Doing so shifts attention to the many ways that nuclear colonialism, as a material-discursive phenomenon, is embedded with, produced by, and entails integrated repercussions on the *more-than-human land community* of animals (including humans), soils, plants, and water.[34] In the case of nuclear weapons testing at the NTS, expanding nuclear colonialism from a human to a more-than-human framework calls our attention to how this most nuclear-bombed place reveals a web of material-discursive ecological relations.

A more-than-human land community is an ecological concept for describing dynamic, animate, interactive places, ranging in scale from the local to the global.[35] The term describes earthen locations and all of the beings and processes that interanimate with them. Although many readers will recognize the term *more-than-human land community* as an amalgamation of Aldo Leopold and David Abram's thinking, and other readers will contemplate its resonance with new materialism's focus on non-human agency, I seek to highlight the term's indebtedness to and articulation with indigenous knowledges.[36,37] Native American scholars such as Vine Deloria, Jr. (Dakota), Jace Weaver (Cherokee), Greg Cajete (Tewa), and Megan Bang (Ojibwe) describe how both traditional and contemporary beliefs and practices across a variety of distinct Native American and pan-Native American cultures are characterized by ecological relationships with the land that acknowledge the agency and animus of non-human participants.[38] Realist animism and spiritual ecology recognize the possibility of participatory relationships between human beings and other beings.[39] For Abram, the term more-than-human world explains the animate and sensual qualities of everything that makes up a place: animals (including humans), lands, rocks, water, and air. The more-than-human world is capable of engaging in intersubjective relationships.[40] Cajete argues that this sort of intimate relationship between beings is at the core of many Native American spiritualities; it is "the essence of their survival and identity as people," recognizing that this plays out differently across the more than 500 Native American nations in the US.[41] These indigenous knowledges are expressed as "kinship with the land, its climate, soil, water, mountains, lakes, forests, streams, plants, and animals."[42] The notion of kinship with the land opens the possibility of thinking of land as a community. Leopold's land ethic "enlarges the boundaries of the community to include soils, waters, plants, and animals, or collectively: the land."[43] Nils Peterson, Markus Peterson, and

Tarla Rai Peterson use the term land community to suggest a place or means through which intersubjective relationships between human and more-than-human beings can occur. They write, "Leopold's land community is a society of interdependent human and extrahuman citizens that participate in decision-making and exercise other rights of citizenship."[44] I argue that a more-than-human land community draws on indigenous ecological knowledge to understand place as a dynamic community of intersubjective, sensing, and agential beings.

A more-than-human land community has the capability to engage in rhetoric, even though these rhetorical practices may be inaccessible, undetectable, or foreign to some human beings. This capability might be termed material rhetoric. As the editors and authors in this volume demonstrate, material rhetoric takes on a variety of meanings in the field. In this case, the animate nature of the more-than-human land community cues into defining material rhetoric as the non-discursive signals disseminated by the more-than-human world that Seegert terms *animate rhetoric*.[45] In his articulation of a materialist transhuman dialogic theory of communication, Richard Rogers asks, "We are very willing to talk about how discourse affects nature, but what about how nature affects discourse, and therefore, *us*?"[46] This reflects Rogers' critique of some early constitutive and materialist theories of rhetoric for ignoring the "natural material conditions within which we all exist—our physical bodies, geography, and climate," the agency of the more-than-human world, and the bidirectionality of communication.[47] The more-than-human land community concept assumes the possibility of communication between human and more-than-human beings. Yet, given the hierarchical dominance of human society and the difficulty many find in perceiving the animate more-than-human world, it is often up to human beings to "amplify and translate the voices of non-speaking human and extrahuman subjects."[48] In this case, this chapter is focused not only on how nuclear testing is a form of material-rhetorical violence that affects the entire more-than-human land community, but also on asking the question of what the legacy of the Nevada Test Site might be trying to tell us.

Before moving on, it is important to address how an extension of nuclear colonialism to the more-than-human land community could be perceived as problematic in relation to environmental justice and colonization. First, it could be argued that a move to de-center the human in nuclear colonialism represents a move toward the sort of mainstream environmentalism that environmental justice scholars have criticized for being too focused on nature to the exclusion of social justice for humans.[49] In a tension that has existed since the beginning of the environmental justice movement, environmentalism has been accused of putting the needs of the natural world before those of people and environmental justice has been accused of putting the needs of people before the natural world.[50] Extending nuclear colonialism to include the more-than-human land community, then, might be seen as putting the needs of animals, plants, and other beings in the more-than-human land community over the needs of indigenous people. Yet, despite how this tension has played out in a variety of specific struggles, thinking ecologically about nuclear colonialism avoids an exclusive focus on either human or more-than-human, and, therefore, avoids the binary between the two. Indeed, some strains of environmental justice scholarship and activism see it as an inherently ecological approach.[51] For example, the Principles of Environmental Justice adopted in 1991 at the First National People of Color Environmental Leadership Summit invoke both ecological and indigenous knowledges that attempt to reconcile the needs of humans and the more-than-human world. The Principles offer injunctions to affirm "the sacredness of Mother Earth, ecological unity, and the interdependence of all species" and to responsibly use land "in the interest of a sustainable planet for humans and other living things."[52] These aspects of the founding principles of environmental justice not only reflect the involvement of indigenous people in their crafting, but also promote justice as a concept that can be applied to the entire more-than-human land community (as Leopold also attempted to do). Similarly, the concept of a more-than-human land community values forms of realist animism and spiritual ecology in a way that resists reductions to either human or more-than-human. Kyle Powys Whyte (Potowatami) suggests that to adequately recognize indigenous experience, environmental injustice must be conceptualized not just as disproportionate harm to indigenous people and their lands, but also as the erasure of indigenous ecological beliefs. He writes, "Settler colonialism can be interpreted as a form of environmental injustice that wrongfully interferes with and erases the socioecological contexts required for indigenous populations to experience the world as a place infused with responsibilities to humans, nonhumans and ecosystems."[53] Following Whyte, environmental justice should acknowledge the interconnection and interrelationships between human and more-than-human beings, and the importance of that interrelation to indigenous people. It is in this spirit that I extend nuclear colonialism to the more-than-human land community.

Second, one might also argue that this extension dilutes colonialism and minimizes the unique experiences of human colonized subjects; accordingly, Eve Tuck and K. Wayne Yang argue that colonization is not a metaphor that can or should be extended broadly to explain all forms of oppression or social injustice.[54] Yet, it is possible to recognize settler colonialism as a system that links but does not equate the oppression experienced by indigenous people with that of more-than-human land communities. Indeed, Tuck and Yang argue that settler colonialism entails the management of "people, land, flora, and fauna" in the service of colonial power.[55] In this way, settler colonialism already includes consideration of the more-than-human land community. My approach does not see indigenous people as less-than-human nor does it promote the concerns of the more-than-human above those of indigenous people, but rather investigates the ways in which nuclear colonialism involves complex interrelations between human and more-than-human beings. Some of these complexities include the way in which indigenous peoples are disproportionately harmed by the nuclear production process *and* indigenous people are limited in their ability to experience the land as animate *and* other beings in the more-than-human land community are disproportionately harmed by the nuclear production complex. Nuclear colonialism, then, focuses on how the entire more-than-human land community (including human and more-than-human beings) experiences a place like the NTS in relation to colonial power and nuclearism.

A More Material Discursive Formation

Although previous research on nuclear colonialism has tended to highlight either its discursive elements or its material elements, this chapter seeks to analyze nuclear colonialism as a material-discursive phenomenon. Following Karen Barad, "*Phenomena*, according to my agential realist account, are neither individual entities nor mental impressions, but entangled material agencies."[56] Nuclear colonialism, then, is not a static thing, but a constellation of inseparable discourses and materialities expressed through multiple beings and agencies. It is neither solely discursive nor solely material, neither solely human nor solely more-than-human, and neither solely controlling nor solely resistive. This suggests that nuclear colonialism is not a one-way exercise of power by an isolated and independent agent, but is a set of interactive material-discursive practices that do things in the world. Rhetoric, when conceived as material-discursive interaction of signals, is a component of the phenomenon of nuclear colonialism, as it provides one medium of interrelation and expression of agency between humans and the more-than-human land community.[57]

One way to analyze nuclear colonialism is through the lens of a discursive formation; yet, as I will argue, the concept needs to be expanded to better highlight the interrelation between discourse and materiality and the rhetorical agency of the more-than-human land community. Foucault views discourse as a complex differentiated practice of representation that reflects the circulations and dispersions of power/knowledge in a particular historical moment. A discursive formation is a set of statements across multiple dispersed texts that reflect an order, despite their seeming difference. He wrote,

> Whenever one can describe, between a number of statements, such a system of dispersion, whenever, between objects, types of statement, concepts, or thematic choices one can define a regularity (an order, correlations, positions and functionings, transformations), we will say, for the sake of convenience, that we are dealing with a *discursive formation*. (Italics in original)[58]

Viewing nuclear colonialism as a discursive formation, then, allows for an analysis of the set of statements across multiple texts that contribute to an ordered understanding of nuclear weapons testing within nuclearism and colonialism. A discursive formation regulates the possibilities of what can be articulated within its boundaries. While there are boundaries, there are also numerous possibilities, including resistive possibilities (or what Foucault calls contradictions), within the formation. Foucault suggests that to uncover a discursive formation, the critic will "define the system of formation of the different strategies that are deployed in it; in other words, if one can show how they all derive (in spite of their sometimes extreme diversity, and in spite of their dispersion in time) from the same set of relations."[59]

While the discursive in discursive formation may be somewhat misleading in that it could imply the opposite of materiality, I seek to push

Foucault's concept of discursive formation such that this formation is both material and symbolic, and open to the more-than-human land community. As such, we might think of material-discursive formations. Although this is not a faithful reading of Foucault's concept, it is an extension that deepens the possibilities of the discursive formation as a heuristic for ecological phenomena. Karen Barad's critique of Foucault's (lack of) treatment of materiality in his theory of discourse argues: "the notions of materiality and discursivity must be reworked in a way that acknowledges their mutual entailment."[60] Barad's solution comes in the form of agential realism, which recognizes the "intra-activity" between discourse and materiality and the agency of more-than-human beings and things.[61] Following Barad, "The primary ontological units [within agential realism] are not 'things' but phenomena—dynamic topological reconfigurings/entanglements/relationalities/(re)articulations. And the primary semantic units are not 'words' but material-discursive practices through which boundaries are constituted."[62] I argue that a material-discursive formation provides a valuable analytic tool for uncovering the ways in which nuclear colonialism is justified and maintained through a variety of ordering statements, and simultaneously resisted through the productive possibilities within the formation. This approach also allows for consideration of what Stacey Alaimo calls "unpredictable material agencies," or the ways in which the more-than-human land community expresses its agency through rhetoric.[63]

In the previous two sections, I argued for two conceptual expansions of current research on nuclear colonialism. First, I moved to decenter the human in nuclear colonialism by drawing on indigenous ecological knowledge and other theorists to examine how the nuclear production complex affects and interrelates with an animate more-than-human land community capable of rhetoric. Human agents are certainly a part, but not the whole, of nuclear colonialism, which is more accurately described as a material-discursive phenomenon with intersecting more-than human agencies (and simultaneous modes of control and resistance). Second, drawing from Foucault and Barad, I use an expanded notion of a material-discursive formation as an analytic tool for understanding the material-discursive interrelations and the rhetorical practices within nuclear colonialism. In the next section, I turn to a brief analysis of nuclear testing at the NTS to illustrate what is gained through understanding the role of the more-than-human land community in nuclear colonialism.

Bombs Away: Violence at the Nevada Test Site

Touring the Nevada Test Site (in 2007) remains one of the more visceral and memorable field experiences I have had. After several years of research on the Atomic West, seeing the scarred cratered landscape of the NTS in person somehow triggered the enormity of the violence that occurred there in a way that I had not felt before. Being there I could see the mark that nuclear testing—this thing that seemed to me so ridiculous and destructive that I almost couldn't believe it really happened—left on this place. I remember standing at the edge of Sedan Crater—1280 feet in diameter and 320 feet in depth—being told by the tour guide that the crater was so big it could be seen from earth-orbit in space. The depth is the equivalent of the 11th tallest high rise building in Salt Lake City. I still feel discomfort and awe when I think about standing on the edge of this huge crater that was part of a project exploring the peaceful uses of nuclear weapons. I had known that the "peaceful use of nuclear weapons" is an oxymoron, but this massive hole in the ground showed me just how violent, jarring, and forceful nuclear explosions are to the earth. What was this 320-foot crater communicating to me? We often talk about the radiation released by these bombs, an invisible form of slow and accretive violence, but these marks on the earth communicated a form of violence to the land, the place, that I was unable to comprehend until I stood there, seeing and listening to what the nuclear bomb crater revealed. Like the permanent stain that the nuclear production complex has made on the geologic record of our earth, this crater similarly marks the long-term effects that a nuclear explosion can have. It also made a mark on me through the mental image that I may never be able to erase from my mind. As I continued to tour the NTS, I made an effort to look out for wildlife and plant-life as I sought to bear witness to the sort of radioactive apocalypse I had imagined in this place. Yet, something surprising happened as I saw seemingly healthy rabbits hop across the road. In spite of the visible craters and other markers of the over nine hundred nuclear explosions, I noticed that this place had not become the radiated and lifeless wasteland I had expected it to be. There were plants everywhere, and animals, all seemingly thriving in this place, revealing the resilience of this place. What were these more-than-human beings communicating to me? How does their presence mark the (un)healthiness of the NTS as a more-than-human land community? What became clear to me on my tour is that this place that has been bombed so many times is not just an innate background or scene for understanding nuclear testing, but a form of rhetoric that participates within nuclear colonialism.

This brief excerpt reveals the seed of the idea for what is now this chapter. Being at the NTS not only confirmed some of what I had imagined about the place but also surprised me. It made me more attuned to the land, the animal and plant presence, and the enormity of the violence that had occurred there. My thinking about nuclear colonialism changed in two important ways, facilitating: (1) a focus on how the more-than-human land community was also affected; and (2) an attention to the forms of resistance, survival, resilience, and rhetoric displayed by the more-than-human land community. In what follows, I turn my attention to an analysis of the NTS as a place where nuclear colonialism is entangled with the entire more-than-human land community, including but not limited to Shoshone and Paiute people, who have been particularly impacted by the decades of nuclear explosions in this place.

As a starting point, I give voice to implications of nuclear testing for the more-than-human land community of the NTS by reviewing some of the documented ecological effects of nuclear testing, not as a scene for my analysis, but as a way to expand our consideration of nuclear colonialism to the more-than-human land community. Recall that 928 nuclear bombs were detonated in the Nevada Test Site between 1951 and 1992, when the Comprehensive Test Ban Treaty (CTBT) came into effect. These above ground atmospheric and underground tests:

> served a variety of national security purposes. These include design testing for the verification of new weapons concepts, proof-testing of existing weapons, effects testing to determine the impact of nuclear weapons on man-made objects and structures, plants and animals, and the physical environment, and experimental testing in the search for possible peaceful uses.[64]

A description of Operation Big Shot (April 22, 1952) provides a snapshot of how nuclear testing affected the entire more-than-human land community at the NTS:

> Effects from the blast [Operation Big Shot] varied, depending on the distance from ground zero. The flash blinded sheep tethered above ground at 900 and 2,000 yards away. Heat from the blast started vegetation fires out to 2,300 yards, leaving numerous yucca plants and Joshua trees smoldering, and gave lethal burns to sheared sheep tethered above ground at 900

> yards. In foxholes, sheep at 900 yards received third degree burns and at 2,000 yards, in the open, first degree burns. Some "trinitite," sand turned to green glass first encountered at the Trinity test, formed at ground zero.[65]

A survey of the human health, environmental, and social justice legacies of the US nuclear testing program reveals that nuclear testing has had a palpable impact on the more-than-human land community of particular locations, such as the Marshall Islands or the NTS.

There has been a recent upsurge of research on the vast ecological implications of the military industrial complex. Sociologist Kenneth Gould contends that "Militarization is the single most ecologically destructive human endeavor."[66] Sociologists Brett Clarke and Andrew Jorgensen specify some of the forms of destruction, "including scorched earth practices, the diversion of rivers, the destruction of plants and animals, the burning of oil wells, and the use of chemical and biological weapons…The legacy of war includes toxic landscapes…posing persistent environmental and social threats."[67] Within the larger military industrial complex, the nuclear weapons production complex, from cradle to grave, has had particularly profound ecological effects. In *Nuclear Wastelands*, Arjun Makhijani, Howard Hu, and Katherine Yih, provide a rigorous account of the ongoing ecological hazards associated with the US nuclear weapons complex.[68] The Department of Energy estimates that the environmental legacy of Cold War nuclear weapons production includes: 1.7 trillion gallons of contaminated groundwater; 40 million cubic meters of contaminated soil and debris; and 100 million gallons of high-level radioactive waste; remediation that is expected to cost over 200 billion dollars and take over 70 years.[69] The NTS is particularly impacted from the 928 nuclear bomb explosions that happened there over a 40-year timeframe. According to Makhijani, Hu, and Yih, "Because of the many underground and near-surface explosions conducted here, the Nevada Test Site is highly contaminated," including an estimated: 2.8 million curies of strontium-90, 4.5 million curies of cesium-137, and 124,000 curies of plutonium underground; groundwater contamination that the DOE has not yet fully characterized; and 420,000 cubic meters of radioactive waste buried at the site. Beyond the site, an estimated 6 million curies of cesium-137 and 4 million curies of strontium-90 remain from atmospheric testing that spread across the nation.[70] The NTS' environmental monitoring project, while often framed in terms of human use and threats to human health, unequivocally shows that the more-than-human land community suffered a blow from the effects of nuclear testing.

Recall that one argument for the start of the Anthropocene is the new mark left in the geological record by the introduction of nuclear weapons (and concomitant testing of those weapons) that released unprecedented amounts of cesium-137 and plutonium-239 and 240 into the earth's ecosystem. Further, the ecological impacts described in this chapter are another type of mark of the consequence of nuclear weapons production, in the form of physical changes to land, flora, and fauna, including the pockmarked landscape, the 320-foot deep Sedan Crater, and the creation of trinitite at ground zero sites.[71]

The NTS does not reflect the entirety of nuclear colonialism, but offers an avenue for examining the complex interrelations and rhetorics between nuclear weapons, the more-than-human land community, humans, colonialism, and nuclearism in this most nuclear-bombed place. My analysis works in the interplay within the material-discursive formation that constrains and enables possibilities for engagement and the forms of rhetoric that emerge within this place, including both human and more-than-human rhetorics. As such, my object of analysis is not a discrete text, but a range of government documents, recollections from a half-day visit to the NTS, two visits to the Atomic Testing Museum in Las Vegas, and participation in a protest event at the NTS. I examine a tension between control and resistance in nuclear colonialism by revealing how nuclear colonialism constrains possibilities and harms the more-than-human land community and enables other possibilities through resistances within the formation. I begin with an analysis of how human rhetorics have constructed the NTS as "wasteland" and then turn to an analysis the rhetorics of the more-than-human land community, all of which interact within nuclear colonialism and a material-discursive phenomenon.

The Wasteland

The US federal government justified locating the nation's nuclear test site in the desert landscape of Nevada because of perceptions of a severe, inhospitable, and remote location that would minimize the amount of harm from nuclear bomb tests. Decision makers' widespread perception of this region as a sparsely populated desert wasteland reflects some European–American assumptions about land and aesthetics, which stand in contrast to indigenous Native American assumptions about and perceptions of the very same land as sacred, ecologically rich, and life-sustaining.[72]

The first European explorers who encountered the American West saw it as a desert wasteland. Early European–American settlers saw deserts as an obstacle to overcome or avoid.[73] According to historian Henry Nash Smith, to European–American people, whose cultural roots were agricultural, the desert represented a barrier to recognized patterns of living and economic prosperity.[74] Early explorers portrayed the desert as wasteland because of this perceived lack of cultivation potential and economic prosperity. John Charles Freemont characterized the Great Basin desert region in 1844: "It is called a desert and from what I saw of it, sterility may be its prominent characteristic."[75] Lieutenant Wheeler, who lead an expedition through Nevada in 1871 described what is now the NTS as "one of the most desolate regions upon the face of the earth," that was "known in common parlance among the settlers of the mining and mountain towns of Nevada as 'Death Valley'."[76] Moving from the late nineteenth century to the Atomic Age in the second half of the twentieth century, a new layer is added to evolving cultural perceptions of the American West. Starting in the 1940s, the "Atomic West" became a primary center of the nuclear weapons production complex. During the 1940s, 1950s, and 1960s, the federal government created a disproportionate number of militarized and atomic sites in the American West, including Trinity, Los Alamos, the Nevada Test Site, and Rocky Flats. Historians have documented the development of the American West in relation to common perceptions of the region as a low-population wasteland (although interestingly Robert Oppenheimer and other scientists in the Manhattan project loved and valued the desert region of Los Alamos, New Mexico).[77] Reflecting on this history, naturalist Ann Zwinger notes, "We Americans have tended to regard our deserts as wastelands, and nowhere has this been more literally true than the Great Basin."[78]

Descriptions of the NTS conform to this broader conception of the American West as a desert wasteland. According to a government document on atmospheric nuclear weapons testing:

> The Nevada Test Site consists of approximately 1,375 square miles of remote desert and mountain terrain owned and controlled by the Department of Energy and located in the southern part of the Great Basin

> northwest of Las Vegas...Water—or the lack thereof—is the dominating climatic characteristic. The lower elevations have hot, dry summers and mild winters and average six inches or less of annual precipitation. Higher elevations receive somewhat increased precipitation and have lower temperatures. Temperature extremes on the site range from below zero to 110 degrees Fahrenheit.[79]

An Atomic Testing Museum display describing how the NTS was chosen, states: "The test site possessed favorable conditions for year round testing and mountainous barriers that would prevent close observation." These favorable conditions included the desert environment and fewer humans living there than more populated areas. *Origins of the Nevada Test Site* describes the NTS as a "remote desert and mountain terrain" with a "harsh climate" and Frenchman Flat, the site of countless nuclear bomb tests, as a "remote desert valley."[80] These portrayals of harsh desert terrain are linked with arguments about how few people have lived in the region. According to *Origins of the Nevada Test Site*, "The site and the immediate surrounding area have always been sparsely populated," and "Even with a climate that has varied considerably over the last dozen millennia, the area that is now the Nevada Test Site has never been particularly conducive to human habitation and exploitation."[81] These characteristics of the region made it an attractive site for militarization, first as the Las Vegas Bombing and Gunnery Range and then as the Nevada Test Site. "In 1940, however, the precise characteristics that had made the region so unattractive—the desolation, lack of water, and general uninhabitableness—brought it to the attention of the federal government."[82] In the 1950s, an article in the magazine *Armed Forces Talk*, described the NTS as a "Damn good place to dump used razor blades."[83] This conception of the NTS as a wasteland is an important element in justifications for nuclear colonialism. Because of the perception of this place as a wasteland, the NTS was easily constructed as a national sacrifice zone that supported the interests of US national security.[84] Viewing the rhetorics highlighted here as part of the nuclear colonialism material-discursive formation indicates how these statements about the American West are influenced by relations between the material and the discursive, such that this evolving meaning of the American West, or the Atomic West, cannot be reduced to a simple causal relationship between the physical qualities of the place and the way humans talk about the place. Rather, this rhetoric participates within a complex interrelationship of place, humans, and more-than-human beings expressed in discursive and non-discursive ways.

Once a region is perceived, named, and valued (or not valued, in this case) as a wasteland, it is easy to think of it as a place for testing nuclear weapons. The consequences of weapons testing and other toxic and violent activities then contribute to the creation of a more literal wasteland through the rhetorical force of destruction, death, and ecological damage to the more-than-human land community. American Studies scholar John Beck notes, "Much of the Southwest is an achieved apocalypse, a space laden with invisible toxic evils."[85] Thus, the widespread human perception of the region as a wasteland has rhetorical force; it has material consequences for a land-use policy that sacrifices the more-than-human land community in the interests of national security.

Yet, these human descriptions of the Nevada Test Site do not occur in a vacuum that is separate from the more-than-human land community. The physical ecological features that make the American West appear "ruggedly severe" and "inhospitable to humans" play a role in the material-discursive formation of the region as wasteland. Indeed, we could not conceive of the region as wasteland without the entanglement between the material and discursive. Rhetoric, as signals shared between beings, encompasses the expressions of the more-than-human land community, how they are taken in by human beings and then expressed through terms like inhospitable, wasteland, and harsh environment, and how these different forms of expression come together as a part of the phenomenon of nuclear colonialism. As such, the construction of wasteland as a strategy of control within nuclear colonialism is itself reflective of the interrelationship between the human and more-than-human, and between materiality and discourse. In the next section, I examine how the more-than-human land community might be responding to the wasteland designation and other aspects of nuclear colonialism through expressions of damage, resistance, and resilience in the face of nuclear testing.

Rhetorics of Resistance and Resilience

From an ecological perspective, the Great Basin resists the wasteland moniker through physical expressions of diverse animal and plant life, hot springs, aquifers, and diverse basin and range geology. Historian Donald Worster challenges the common human rhetorical framing of the American West as a desert wasteland that is devoid of life. He wrote, "Almost nowhere was the American desert simply a stretch of empty sand; everywhere there was life. Americans had only to take the trouble to look."[86] In fact, according to a study by NatureServe in 2002, Nevada—made up of primarily desert terrain—ranks sixth in the US for its high level of endemic biological diversity.[87] Moreover, "Nevada is 11th among all states in total species diversity, 6th among all states in number of unique (endemic) species, including 64 recently discovered endemic species of springsnails, 8th among all states in butterfly diversity, and 9th among all states in mammal diversity."[88] Although water is not abundant in this region, springs, streams, and large aquifers span the desert terrain, providing water for the multiplicity of flora and fauna. When considered from the perspective of a thriving more-than-human land community, Nevada, and more broadly the American West, is a desert but is far from an "inhospitable" wasteland. In my own encounters with the Nevada Test Site and surrounding areas, I was often initially surprised, given my upbringing in a very different ecological region, to see just how many animal and plant species occupy this land.

The NTS in particular is home to a complex mosaic of more-than-human life that is representative of both the Mojave and Great Basin Desert ecosystems. According to the Department of Energy, there are 1500 animal species, including 924 species of insects, and 750 plant species that have been documented at the NTS.[89] According to the Department of Energy:

> Despite the harsh climate, the Nevada Test Site is home to a surprising array of plants and animals. The site is a transitional zone between the Great Basin and the Mojave deserts. Species from both deserts, including those native to one but not the other, are found in the area. Kit fox and the sidewinder rattlesnake, common only in the Mojave Desert, live in the southern reaches of the site, and mule deer and the striped whipsnake, favoring a Great Basin desert environment, reside in the northern parts. Other animals found on the site include coyotes, golden eagles, wild horses, mountain lions, and an occasional bighorn sheep and antelope. The range in elevation also helps provide diversity in flora and fauna. Mojave Desert plants such as the creosote bush dominate the lower elevations. Plants of the Great Basin Desert prevail above 5,000 feet, with open pinon-juniper and sagebrush woodland appearing at the 6,000-foot level. Between the two elevation extremes, sagebrush is the most common plant.

> Springs, the only perennial water sources on the site, sustain the wildlife population and are widely, if not abundantly, scattered across the area.[90]

While it may look like a wasteland because of a lack of charismatic mega flora and fauna, the ecology of the NTS is incredibly diverse. Los Alamos Biologist and Anthropologist Frederick Worman, who conducted ecological studies at the NTS, noted in his 1965 *Anatomy of the Nevada Test Site*: "Unfortunately, too many people think of the desert as an immense wasteland of drifting sands, populated by centipedes, scorpions, awesome spiders and reptiles—all dangerous and deadly."[91] In contrast, he argues, "To view the blooming desert [at the NTS] during a wet period is to see it at its most magnificent moment. One looks out upon a myriad of flowers designed to move the amateur taxonomist to complete ecstasy."[92] The more-than-human inhabitants have a strong presence in this place. This presence is an important form of rhetorical resistance to the wasteland rhetoric, and nuclear colonialism more generally. The more-than-human land community quite literally enacts a rhetoric of life that stands in contrast to the ways that European Americans have framed the very same place as a wasteland that is worthy of sacrifice in the interest of national security.

There are other more-than-human rhetorics that resist nuclear colonialism's controlling forces and highlight the interanimation between more-than-human beings within this material-discursive formation. Physical and geological markers of nuclear testing—such as the increase in plutonium-239 and 240 in the geologic record and the pockmarked landscape of the NTS—also act as rhetorical resistances to nuclear colonialism. Previous research indicates that national security and secrecy are essential components in attempts to maintain nuclear colonialism as a form of control and to justify the national sacrifice zones of the nuclear production process.[93] In his analysis of the Hanford Reservation, a former plutonium production site, William Kinsella argues that waste and other legacies from the nuclear weapons production process stand as material texts that undermine forms of discursive containment invented to maintain secrecy about the human and ecological health implications of nuclear weapons.[94] Similarly, the physical ecological legacies of nuclear weapons testing—itself a reality that many Americans do not realize actually happened in our own backyard—are a form of rhetoric that challenges and resists efforts to maintain secrecy and highlight the national security necessities of nuclear weapons. Recall that the invention of nuclear weapons is a factor, if not the defining factor, in the Anthropocene because of the permanent geologic record of increased amounts of plutonium-239, plutonium-240, and cesium-137 in the world. Likewise, the many craters within the NTS are a semi-permanent mark left by nuclear weapon tests as reminders of ground zero sites. These physical attributes cannot be (easily) erased, and present a permanent record of nuclear weapons production that can be seen by anyone willing to look. Nuclear weapons tests, then, leave an indelible trace of their own impact on the more-than-human land community of the NTS and beyond, and bear material witness to the global impact of nuclear weapons.

The more-than-human land community is an active agent in the ongoing phenomenon of nuclear colonialism, not only through its resistance to the wasteland and containment but also through its essential role in the nuclear production process. In other words, the testing of nuclear weapons is inextricably entangled with how the more-than-human land community of the NTS contributed to and reacted to nuclear bomb detonations. Descriptions of nuclear testing reveal that the more-than-human land community did not always act as expected during nuclear bomb detonations at the NTS, and that it was folly to assume that humans could predict exactly how the community would react. For example, according to A. Costadina Titus, "The geology of the area, which was to become so important in later years when testing moved underground, was given little consideration during the original selection of the site…But, like the weather, these geological formulations have not always behaved as expected."[95] Worman offers another example in his description of how native plants no longer appear in ground zero sites, but that Russian thistle "invaded the ground zeros en masse in the first growing season after the detonation of nuclear weapons."[96] The rapid growth of Russian thistle in ground zero sites is an example of how nuclear testing had profound implications on the ecology of the site, such that native plants were replaced by new plants in a material rhetoric of both violence and adaptation. The unexpected nature of nuclear testing and more-than-human land communities is also demonstrated by the effects of radioactive releases, such that the scientists and engineers conducting nuclear tests "could not have anticipated how these materials would travel in the environment, how the chemistry would change over time."[97] In these unexpected actions of the more-than-human land community, we see both the violence of nuclear testing and the resilience of the complex ecological system of flora, fauna, climate, and geology at the NTS.

Similarly, despite the seemingly destructive nature of nuclear testing, there are indications of the resilience of the more-than-human land community within the NTS. Even with over 900 nuclear bomb detonations, the NTS is not a literal wasteland where no beings can live. Indeed, I was surprised to see rabbits hopping around and the preponderance of plants that seemed to be thriving at the NTS. Ironically, it could be argued that perceived wasteland regions are actually some of the most protected ecological regions. For example, the US Fish and Wildlife Service recently opened the Rocky Flats National Wildlife Refuge in Colorado. The refuge was formerly the site of a nuclear weapons facility turned EPA superfund site. Although inconceivable to some skeptics, the refuge teems with a rich resident and migratory more-than-human land community.[98] However the remediation of this more-than-human land community is judged, and whether this site is seen as a form of green washing, the wildlife refuge demonstrates the resilience of the more-than-human land community in the face of the indelible ecological impact from the invention of nuclear weapons. Within a perceived wasteland lies the possibility of a more-than-human rhetoric that resists human domination and degradation.

Since the moratorium on nuclear testing in 1992, the Nevada Test Site has engaged in ecological restoration, environmental remediation, and species protection programs that, combined with the restrictions on human access to the area, are helping more-than-human beings recover from the decades of nuclear testing.[99] For example, the most recent environmental assessment of the NTS indicates:

> Though NNSS [Nevada National Security Site]-related radionuclides are detected in some plants and animals, the levels pose negligible risk to humans and biota. The potential dose to a person hunting and consuming these animals is well below dose limits to members of the public (see Section 9.1.1.2). Also, radionuclide concentrations were below levels considered harmful to the health of the plants and animals; the dose resulting from observed concentrations were less than 4 percent of dose limits set to protect populations of plants and animals (see Section 9.2).[100]

These data suggest that the level of radioactive contamination to the more-than-human land community is less than might be expected due to the lingering effects of the nuclear testing that occurred between 1951 and 1992. Of course, this does not mean there is not a negative ecological impact—indeed the very same report indicates that humans are advised not to consume flora and fauna from the NTS—but it does demonstrate forms of recovery and resilience in the more-than-human land community. The endangered desert tortoise, for example, is benefiting from the restricted human activities on the Nevada Test Site, and a 2008 report listed that the biggest threat to desert tortoises on the NTS was not exposure to lingering radiation but being hit by employee cars or injured by construction on the site.[101] In a milieu that still denies downwinders, atomic veterans, and atomic workers compensation for cancers and death associated with nuclear testing and fails to recognize the sovereign rights of Shoshone and Paiute people, the plant and animal diversity at the Nevada Test Site can be seen as an ecologically positive consequence within nuclear colonialism. It can be seen as a form of resilience that reveals the complexity of the ecological implications of nuclear testing and the possibility for both damage to and protection of the more-than-human land community.

Conclusion

Nuclear colonialism is ordinarily used to describe the relationships between the nuclear production complex and indigenous people. In this chapter, I extended the concept to consider the relationship between nuclear colonialism and the entire more-than-human land community (including humans). By viewing nuclear colonialism as a phenomenon that can be examined through the lens of a material-discursive formation, this chapter highlights how the more-than-human land community is not just a passive victim of violent nuclear bomb explosions but also an active participant within the larger phenomenon. Through an analysis of human and more-than-human rhetorics within the nuclear colonialism formation, I have shown how limiting analysis to human rhetorical practices, such as naming the American West as a wasteland suitable for nuclear testing, offers only a partial view of the complexity of actors and rhetorics within nuclear colonialism. An analysis of the more-than-human land community's responses to nuclear testing reveals the complex ecology of legacies of nuclear testing at the NTS. In so doing, this chapter also engages conceptual interrelationships between rhetoric, ecology, and materiality that suggest the importance of moving beyond: human-centered notions of rhetoric toward recognition of the more-than-human land community's rhetorical performances and practices; notions of seeing the materiality of rhetoric in the consequences of discourse toward seeing rhetoric as material-discursive; and viewing the environment as background for rhetorical action toward an ecological view that focuses on interconnections experienced by the variety of beings that inhabit our world.

This chapter is an admittedly partial foray into thinking through the relations between rhetoric, materiality, and ecology in the context of theorizing nuclear colonialism as a more-than-human phenomenon. As such, there are several implications and opportunities for opening further avenues of research. One possibility is to further consider how human-perceived and rhetorically designated "wasteland" places may actually allow for ecological restoration, and healthier more-than-human land communities. Similar to arguments that wilderness designations can lead to protection of "wild" spaces through reduced contact with humans, an ironic consequence of the nuclear production process has been the creation of protected zones with limited human engagement where wildlife populations thrive. Another possibility is to think about nuclear weapons themselves as rhetorical agents within nuclear colonialism. My analysis primarily focused on the rhetorical interplay between humans and more-than-human land communities as active agents within nuclear colonialism. Yet, following Barad and others, we might also consider how the nuclear weapon itself is rhetorical. Future inquiry might examine how nuclear weapons exceed their human creators and work in unexpected ways within nuclear colonialism.

In closing, I return to the question of what the legacy of nuclear weapons testing might be trying to tell us. Nuclear testing at the NTS is not just a form of violence and environmental injustice experienced by humans, but is also experienced by the more-than-human land community. That community is telling us, through permanent geologic markers, massive ground zero craters, destruction and resilience of flora and fauna, that it is an active participant in nuclear colonialism.

Notes

1. Kevin Schaul, "Eight Countries. 2,055 Nuclear Tests. 71 Years—Mapped," *Washington Post*, January 6, 2016, https://www.washingtonpost.com/.
2. Twenty-four of the 928 tests were conducted collaboratively by the United States and the United Kingdom. See US Department of Energy, *United States Nuclear Tests: July 1945 through September 1992* [DOE/NV--209-REV 15] (Las Vegas: US Department of Energy, December 2000).
3. David Biello, "Nuclear Blasts May Prove Best Marker of Humanity's Geologic Record," *Scientific American*, February 10, 2015; Jan Zalasiewicz et al., "When Did the Anthropocene Begin? A Mid-Twentieth Century Boundary Level is Stratigraphically Optimal," *Quaternary International*, 383 (October 5, 2015).
4. The disproportionate harm expands beyond nuclear weapons testing to include other stages in the nuclear production process—such as uranium mining, weapons assembly, and nuclear waste—that have occurred primarily in the American West. Valerie L. Kuletz, *The Tainted Desert: Environmental and Social Ruin in the American West* (New York: Routledge, 1998).
5. Although the site was renamed the Nevada National Security Site in 2000, I use the name Nevada Test Site in this chapter because it was the name of the site when nuclear weapons testing occurred.
6. Richard W. Stoffle et al., *Native American Cultural Resource Studies at Yucca Mountain, Nevada* (Ann Arbor: University of Michigan, Institute for Social Research, 1990).
7. Consolidated Group of Tribes and Organizations, "Appendix C: American Indian Assessment of Resources and Alternatives Presented in the SWEIS," in *Final Site Wide Environmental Impact Statement for the Continued Operation of the Department of Energy/National Nuclear Security Administration Nevada National Security Site and Off-Site Locations in the State of Nevada*, vol. 2 (Appendixes A Through I) [DOE/EIS-0426] (Las Vegas: US Department of Energy/National Nuclear Security Administration Nevada Site Office, February 2013), C-7.
8. Corbin Harney, *The Way It Is: One Water, One Air, One Mother Earth* (Nevada City: Blue Dolphin Publishing, Inc., 1995); Rebecca Solnit, *Savage Dreams: A Journey into the Landscape Wars of the American West* (Berkeley: University of California Press, 2000).
9. Terrence R. Fehner and F.G. Gosling, *Atmospheric Nuclear Weapons Testing 1951–1963* [DOE/MA-003] (US Department of Energy, September 2006), 9.
10. Fehner and Gosling, *Weapons Testing*, 10.

11. Makhijani et al., *Nuclear Wastelands.*
12. Ibid.
13. Robert D. Bullard and Glenn S. Johnson, "Environmentalism and Public Policy: Environmental Justice: Grassroots Activism and its Impact on Public Policy Decision Making." *Journal of Social Issues* 56, no.3 (2000): 555–578.
14. Mary Dickson, a journalist, playwright, and downwinder, introduced me to the concept that nuclear testing was an undeclared war on our own people, meaning the many downwinders who got cancer and other illnesses from their exposure to radioactive fallout from nuclear testing.
15. Ward Churchill,* "Radioactive Colonization: A Hidden Holocaust in Native North America," in *Struggle for the Land: Indigenous Resistance to Genocide, Ecocide, and Expropriation in Contemporary North America* (Monroe, Maine: Common Courage Press, 1993), 261–328; Nelta Edwards, "Nuclear Colonialism and the Social Construction of Landscape in Alaska," *Environmental Justice* 4, no. 2 (2011): 109–114; Donald A. Grinde and Bruce E. Johansen, *Ecocide of Native America: Environmental Destruction of Indian Lands and Peoples* (Santa Fe, NM: Clear Light Publishers, 1995); Kuletz, *The Tainted Desert*; Winona LaDuke, *All Our Relations: Native Struggles for Land and Life*, (Boston: South End Press, 1999); Nuclear Information and Resource Service, "Nuclear Waste + Native Lands = Environmental Racism [Press Release]," October 24, 2000, http://www.nirs.org/; Grace Thorpe, "Our Homes Are Not Dumps: Creating Nuclear Free Zones" *Natural Resources Journal* 36, no. 4 (1996): 955–964.
    *I am aware of the charges of plagiarism and academic misconduct against Ward Churchill that resulted in his dismissal from CU Boulder. After closely reading the report from CU Boulder, I found that Churchill's work on radioactive colonization is not indicted in the report. It does, however, appear that Churchill and LaDuke published the same information in a variety of journals. Although Churchill was the first to introduce the term "radioactive colonization," much subsequent independent research has substantiated the phenomenon. See Report of the Investigative Committee of the Standing Committee on Research Misconduct at the University of Colorado at Boulder (May 2006).
16. Danielle Endres, "The Rhetoric of Nuclear Colonialism: Rhetorical Exclusion of American Indian Arguments in the Yucca Mountain Nuclear Waste Siting Decision," *Communication and Critical/Cultural Studies* 6 (2009); LaDuke, *All Our Relations*; Kuletz, *The Tainted Desert*; Sarah Alisabeth Fox, *Downwind: A People's History of the Nuclear West* (Lincoln: Bison Books, 2014).

17. This is indebted to many thinkers, including both indigenous knowledge and new materialism. See Gregory Cajete, "'Look to the Mountain' Reflections on Indigenous Ecology," in *A People's Ecology: Explorations in Sustainable Life* (Santa Fe: Clear Light, 1999); Timothy Morton, *Posthumanities : Hyperobjects : Philosophy and Ecology after the End of the World* (Minneapolis: University of Minnesota Press, 2013); Stacy Alaimo, *Bodily Natures: Science, Environment, and the Material Self* (Indiana University Press, 2010).
18. For example, Thomas Rickert, *Ambient Rhetoric: The Attunements of Rhetorical Being* (Pittsburgh: University of Pittsburgh Press, 2013); Richard A. Rogers, "Overcoming the Objectification of Nature in Constitutive Theories: Toward a Transhuman, Materialist Theory of Communication," *Western Journal of Communication* 62, no. 3 (1998); Natasha Seegert, "Play of Sniffication: Coyotes Sing in the Margins," *Philosophy and Rhetoric* 47, no. 2 (2014); Natasha Seegert, "Rewilding Rhetoric with Animate Others," *Review of Communication* 16, no. 1 (2016).
19. Seegert, "Play of Sniffication," 160.
20. Seegert, "Rewilding Rhetoric," 79.
21. Churchill, "Radioactive Colonization"; LaDuke, *All Our Relations*; Winona LaDuke and Ward Churchill, "Native America: The Political Economy of Radioactive Colonialism," *The Journal of Ethnic Studies* 13, no. 3 (1985): 107.
22. For example, see Endres, "The Rhetoric of Nuclear Colonialism"; Nelta Edwards, "Nuclear Colonialism and the Social Construction of Landscape in Alaska," *Environmental Justice* 4, no. 2 (June 2011), doi:10.1089/env.2010.0023; Grinde and Johansen, *Ecocide of Native America*; Valerie Kuletz, *The Tainted Desert*; Nuclear Information and Resource Service, "Nuclear Waste + Native Lands"; Thorpe, "Our Homes Are Not Dumps."
23. "Indigenous Anti-Nuclear Statement: Yucca Mountain and Private Fuel Storage at Skull Valley," Indigenous Environmental Network, http://www.ienearth.org/indigenous-anti-nuclear-statement-yucca-mountain-and-private-fuel-storage-at-skull-valley/.
24. Unlike other forms of colonialism, in settler colonialism the colonizer takes over and stays on the land of original inhabitants. Patrick Wolfe, "Settler Colonialism and the Elimination of the Native," *Journal of Genocide Research* 8, no. 4 (2006).
25. A more thorough discussion of these systems can be found in Endres, "The Rhetoric of Nuclear Colonialism."
26. Tuck and Yang, "Decolonization Is Not a Metaphor"; Jodi A. Byrd, *The Transit of Empire: Indigenous Critiques of Colonialism* (Minneapolis:

University of Minnesota Press, 2011); Jace Weaver, "Indigenousness and Indigeneity," in *A Companion to Postcolonial Studies*, eds. Henry Schwarz and Sangeeta Ray (Hoboken: John Wiley & Sons, 2008).

27. Bryan McKinley Jones Brayboy, "Toward a Tribal Critical Race Theory in Education," *The Urban Review* 37, no. 5 (2006); Rob Nixon, *Slow Violence and the Environmentalism of the Poor* (Cambridge: Harvard University Press, 2011).
28. Carole Gallagher, *American Ground Zero: The Secret Nuclear War*, (New York: Random House, 1994); Richard L. Miller, *Under the Cloud: The Decades of Nuclear Testing* (The Woodlands: Two-Sixty Press, 1991).
29. Consolidated Group of Tribes and Organizations, "Appendix C: American Indian Assessment"; Harney, *The Way It Is*; Kuletz, *The Tainted Desert*.
30. Danielle Endres, "From Wasteland to Waste Site: The Role of Discourse in Nuclear Power's Environmental Injustices," *Local Environment* 14, no. 10 (2009); Endres, "The Rhetoric of Nuclear Colonialism"; Danielle Endres, "Sacred Land or National Sacrifice Zone: The Role of Values in the Yucca Mountain Participation Process," *Environmental Communication: A Journal of Nature and Culture* 6, no. 3 (2012).
31. Michel Foucault, *The Archaeology of Knowledge and The Discourse on Language* (New York: Pantheon Books, 1972).
32. Michel de Certeau, *The Practice of Everyday Life* (Berkeley: University of California Press, 1984).
33. This draws from the scholarly conversation around new materialism. See Alaimo, *Bodily Natures*; Karen Barad, *Meeting the Universe Halfway: Quantum Physics and the Entanglement of Matter and Meaning* (Durham: Duke University Press, 2007); Karen Barad, "Posthumanist Performativity: Toward an Understanding of How Matter Comes to Matter," *Signs: Journal of Women in Culture and Society* 28, no. 3 (2003); Jane Bennett, *Vibrant Matter: A Political Ecology of Things* (Durham: Duke University Press, 2009); Donna Jeanne Haraway, *Simians, Cyborgs, and Women: The Reinvention of Nature* (London: Free Association Books, 1991); Bruno Latour, *Reassembling the Social: An Introduction to Actor-Network-Theory* (Oxford: Oxford University Press, 2007).
34. The term more-than-human land community is an amalgamation of concepts by Aldo Leopold and David Abram. See Aldo Leopold, *A Sand County Almanac: With Other Essays on Conservation from Round River* (Oxford: Oxford University Press, 1966); David Abram, *The Spell of the Sensuous: Perception and Language in a More-Than-Human World* (New York: Vintage, 1997).

35. Following Endres and Senda-Cook, "place refers to particular locations (e.g., a city, a particular shopping mall, or a park) that are semi-bounded, a combination of material and symbolic qualities, and embodied." Places are always in articulation with broader spatial constructs—social space—that frame the way beings understand and interact with particular places. Danielle Endres and Samantha Senda-Cook, "Location Matters: The Rhetoric of Place in Protest," *Quarterly Journal of Speech* 97, no. 3 (2011): 286.
36. Abram, *The Spell of the Sensuous*; David Abram, *Becoming Animal: An Earthly Cosmology*, 1st ed. (New York: Pantheon, 2010); Leopold, *A Sand County Almanac.*
37. See note 33.
38. Cajete, Weaver, Deloria Jr., Medin, and Bang are careful to avoid essentialist notions that all Native Americans are ecologically superior to other racial, ethnic, or cultural groups, which would play into the Ecological Indian stereotype. Traditional indigenous ecological beliefs do not preclude exploitation, damage to the environment, or lack of alignment between spiritual ideals and actual practices within indigenous communities. While Native American cultures have changed over time and adapted, the roots of spiritual ecology as a cultural belief remain for many traditionalist Native Americans. Contemporary Native Americans often experience a dual existence between traditional spiritual ecological beliefs and the reality of living within a dominant society that does not recognize these beliefs. See Gregory Cajete, "'Look to the Mountain'"; Vine Deloria, *Red Earth, White Lies: Native Americans and the Myth of Scientific Fact* (Golden: Fulcrum Publishing, 1995); Vine Deloria, *God Is Red: A Native View of Religion* (Golden: Fulcrum Publishing, 2003); Douglas L. Medin and Megan Bang, *Who's Asking?: Native Science, Western Science, and Science Education* (Cambridge: MIT Press, 2014); Jace Weaver, "Introduction: Notes From a Miner's Canary," in *Defending Mother Earth: Native American Perspectives on Environmental Justice*, ed. Jace Weaver (Maryknoll, NY: Orbis Books, 1996).
39. Joe Sheridan and Roronhiakewen Longboat, "He Clears the Sky"; Dan Longboat, "The Haudenosaunee Imagination and the Ecology of the Sacred," *Space and Culture* 9, no. 4 (2006).
40. Abram's argument relies on a narrative of return to the traditional practices of indigenous cultures. Abram's concept is indebted to indigenous knowledge, even while his account romanticizes and historicizes indigenous cultures within what Kretch has called the Ecological Indian stereotype. See: Shepard Krech III, *The Ecological Indian: Myth and History* (New York: W. W. Norton & Company, 2000).

41. Cajete, "'Look to the Mountain,'" 4.
42. Ibid., 3.
43. Leopold, *A Sand County Almanac*, 204.
44. M. Nils Peterson, Markus J. Peterson, and Tarla Rai Peterson, "Environmental Communication: Why This Crisis Discipline Should Facilitate Environmental Democracy," *Environmental Communication: A Journal of Nature and Culture* 1, no. 1 (2007): 77.
45. Seegert, "Play of Sniffication," 164.
46. Richard A. Rogers, "Overcoming the Objectification of Nature in Constitutive Theories: Toward a Transhuman, Materialist Theory of Communication," *Western Journal of Communication* 62, no. 3 (1998): 244–272.
47. Ibid., 259.
48. Peterson et al., "Environmental Communication," 84.
49. For more on the tensions between environmentalism and environmental justice, see Phaedra C. Pezzullo and Ronald Sandler, "Introduction: Revisiting the Environmental Justice Challenge to Environmentalism," in *Environmental Justice and Environmentalism: The Social Justice Challenge to the Environmental Movement*, eds. Ronald Sandler and Phaedra C. Pezzullo (Cambridge: The MIT Press, 2007).
50. Ibid.
51. Jace Weaver, "Introduction: Notes From a Miner's Canary."
52. "Principles of Environmental Justice," Environmental Justice, http://www.ejnet.org/ej/principles.html.
53. Kyle Powys Whyte, "Indigenous Experience, Environmental Justice and Settler Colonialism," in *Nature and Experience: Phenomenology and the Environment*, ed. Bryan E. Bannon (New York: Rowman & Littlefield Publishers, 2016).
54. Tuck and Yang, "Decolonization Is Not a Metaphor."
55. Ibid., 4.
56. Barad, *Meeting the Universe Halfway*, 56.
57. See Rickert, *Ambient Rhetoric*; Rogers, "Overcoming the Objectification"; Samantha Senda-Cook, "Rugged Practices: Embodying Authenticity in Outdoor Recreation," *Quarterly Journal of Speech* 98, no. 2 (2012); Bridie McGreavy, "Resilience as Discourse," *Environmental Communication* 10, no. 1 (2016).
58. Foucault, *Archaeology*, 38.
59. Ibid., 68.
60. Barad, "Posthumanist Performativity," 820.
61. Barad, *Meeting the Universe Halfway*; Barad, "Posthumanist Performativity."
62. Barad, "Posthumanist Performativity," 818.

63. Alaimo, *Bodily Natures*, 146.
64. Fehner and Gosling, “Atmospheric Nuclear Weapons Testing,” 6.
65. Ibid., 5.
66. Kenneth A. Gould, “The Ecological Costs of Militarization,” *Peace Review* 19, no. 3 (2007): 331.
67. Brett Clark and Andrew K. Jorgenson, “The Treadmill of Destruction and the Environmental Impacts of Militaries,” *Sociology Compass* 6, no. 7 (July 1, 2012): 557.
68. Makhijani et al., *Nuclear Wastelands*.
69. Max Singleton Power, *America’s Nuclear Wastelands: Politics, Accountability, and Cleanup* (Pullman: Washington State University Press, 2008), 7.
70. Makhijani et al., *Nuclear Wastelands*.
71. For a fascinating photo-essay on nuclear testing, see Alan Taylor, “When We Tested Nuclear Bombs—The Atlantic,” *Atlantic*, May 6, 2011, https://www.theatlantic.com/.
72. Danielle Endres, “Sacred Land or National Sacrifice Zone: The Role of Values in the Yucca Mountain Participation Process,” *Environmental Communication: A Journal of Nature and Culture* 6, no. 3 (2012); Danielle Endres, “Animist Intersubjectivity as Argumentation: Western Shoshone and Southern Paiute Arguments Against a Nuclear Waste Site at Yucca Mountain,” *Argumentation* 27, no. 2 (May 1, 2013).
73. John McPhee, *Basin and Range* (New York: The Noonday Press, 1981).
74. Henry Nash Smith, *Virgin Land: The American West as Symbol and Myth* (Cambridge: Harvard University Press, 1950).
75. As cited in Pamela N. Limerick, *Desert Passages: Encounters with the American Deserts* (Albuquerque: University of New Mexico, 1985), 40.
76. As cited in Terrence R. Fehner and F.G. Gosling, *Origins of the Nevada Test Site* [DOE-MA 0518] (US Department of Energy, December 2000), 8.
77. Bruce Hevly and John M. Findlay, *The Atomic West* (University of Washington Press, 2000); Mark Fiege, *The Republic of Nature: An Environmental History of the United States* (University of Washington Press, 2012).
78. Ann Zwinger, “What Good is a Desert?” *Audubon*, March/April 1996, 42.
79. Fehner and Gosling, *Atmospheric Nuclear Weapons Testing*, 9.
80. Fehner and Gosling, *Origins of the Nevada Test Site*, 5, 1.
81. Ibid., 6.
82. Ibid., 20.
83. As cited in Gallagher, *American Ground Zero*, xxiv.
84. Endres, “The Rhetoric of Nuclear Colonialism; Endres, “From Wasteland to Waste Site.”

85. John Beck, "Without Form and Void: The American Desert as Trope and Terrain," *Nepantla: Views from the South*, 2, no. 1 (2001), 69.
86. Donald Worster, *Rivers of Empire: Water, Aridity, and the Growth of the American West* (New York: Pantheon Books, 1985), 69.
87. Bruce Stein, *States of the Union: Ranking America's Biodiversity, Nature Reserve* (Arlington, VA: NatureServe, 2002), http://www.natureserve.org/.
88. Nevada Natural Heritage Program, Biodiversity and At Risk Species, http://heritage.nv.gov/.
89. US Department of Energy, *Nevada National Security Site: Plant and Animal Life* [DOE/NV-894], (US Department of Energy, 2013), http://nnss.gov/docs/fact_sheets/DOENV_894.pdf.
90. Fehner and Gosling, *Atmospheric Nuclear Weapons Testing*, 9–10.
91. Frederick C.V. Worman, *Anatomy of the Nevada Test Site* (Berkeley: University of California Los Alamos Scientific Laboratory, 1965), 10.
92. Ibid., 8.
93. Endres, "The Rhetoric of Nuclear Colonialism"; William J. Kinsella, "One Hundred Years of Nuclear Discourse: Four Master Themes and Their Implications for Environmental Communication," in *The Environmental Communication Yearbook*, ed. Susan L. Senecah (Mahwah: Lawrence Erlbaum Associates, 2005).
94. William J. Kinsella, "Nuclear Boundaries: Material and Discursive Containment at the Hanford Nuclear Reservation," *Science as Culture* 10, no. 2 (2001).
95. A. Costandina Titus, *Bombs in the Backyard: Atomic Testing and American Politics* (Las Vegas: University of Nevada Press, 2001), 57.
96. Worman, *Anatomy of the Nevada Test Site*, 9.
97. Max Singleton Power, *America's Nuclear Wastelands: Politics, Accountability, and Cleanup* (Pullman: Washington State University Press, 2008), xiii.
98. About the Refuge, US Fish and Wildlife Service, https://www.fws.gov/refuge/Rocky_Flats/about.html.
99. Cathy Wills, ed., *Nevada National Security Site Environmental Report 2015* [DOE/NV/25946-2950] (Las Vegas: US Department of Energy Nevada Field Office, 2016), http://nnss.gov/docs/docs_LibraryPublications/DOENV_25946_2950_Final.pdf; Cathy Wills and W. Kent Ostler, eds., *Ecology of the Nevada Test Site: An Annotated Bibliography* [DOE/NV/11718-594] (Las Vegas: US Department of Energy Nevada Field Office, 2001), https://nnsa.energy.gov/sites/default/files/nv_sweis/appendixF/Wills%20and%20Ostler%202001.pdf.

100. Wills, *Nevada National Security Site Environmental Report 2015* [DOE/NV/25946-2950], 8–9.
101. Henry Brean, "Tortoises Find a Home at Nevada National Security Site," Las Vegas Review-Journal, September 23, 2012, http://www.reviewjournal.com/; National Security Technologies, LLC, Biological Assessment for the Effects of the National Nuclear Security Administration Activities on the Nevada Test Site on the Desert Tortoise (Las Vegas: National Security Technologies, LLC, 2008), https://nnsa.energy.gov/sites/default/files/nnsa/inlinefiles/nstec%20 2008.pdf.

CHAPTER 11

Toward a Geopoetical Rhetoric: The Transborder Immigrant Tool and Material Tactics

Anthony Stagliano

Introduction: The Situation, as It Stands Now

For those who enter the United States by foot, unauthorized, the journey has become even more dangerous than it already was. Since 1994's "Operation Gatekeeper," an Immigration and Naturalization Service (INS) program to stall immigrant foot traffic by increasing agent presence and constructing a massive fence along the San Diego area border, the flow of immigrant foot traffic has shifted from the urban areas along the Pacific coast to the unpopulated and inhospitable deserts and mountains to the east. As Rita Raley relates in her book *Tactical Media*, since the construction of "the fourteen-mile San Diego–Tijuana border fence, highway deaths are no longer as common as they once were—the scene of death has shifted eastward to the deserts and mountain." She explains that since the beginning of "Operation Gatekeeper" more than 3500 people have died attempting to enter the United States.[1] Anne Demo points to a shift toward border enforcement and away from apprehension

A. Stagliano (✉)
New Mexico State University, Las Cruces, USA
e-mail: staglian@nmsu.edu

B. McGreavy et al. (eds.), *Tracing Rhetoric and Material Life*,
Palgrave Studies in Media and Environmental Communication,
https://doi.org/10.1007/978-3-319-65711-0_11

strategies that started under the Carter administration, and then accelerated under the Clinton administration with programs like "Gatekeeper" at the San Diego area border and "Operation Blockade," the initial program at El Paso that was the model for Gatekeeper. Border enforcement, a material manifestation of discourses of national sovereignty, has not been equally applied to the southern and northern borders however. In 1995 there were only 330 Border Patrol agents assigned full-time to the northern border, while there were 4300 along the southwest border.[2]

This chapter examines the Transborder Immigrant Tool, a tactical media art project designed to provide safe travel through the desert around the US–Mexico border. The Transborder Tool is a hacked Motorola cell phone, with cellular functions turned off (so it cannot be tracked), and GPS left on (so it can provide guidance). The producers of the project also altered the visual interface, to make it easier to use as a guidance device for those unfamiliar with cell phones. In addition to its utility in providing pathfinding help through rough desert terrain, the developers of the project partnered with other aid organizations who place water stations in the desert for travelers, to include guidance to those water stations in the Transborder Tool's interface. The phone also has loaded on it recordings of poems welcoming the user to the United States. As a material, non-verbal form of rhetorical invention, responding not to discourses about the border bound up in policies like those above, but to the border apparatus itself, this project offers the chance to redeploy current developments in rhetorical theory that reimagine rhetorical situations in material ways. While there have been many creative, inventive, and tactical responses to the problem of the border, the Transborder Immigrant Tool offers a unique opportunity to view the border as a material rhetorical ecology which includes myriad apparatuses (not the least of which is GPS technology), and helps demonstrate that the rhetorical inventions and interventions that emerge in response to such an ecology emerge *from* it, with the border and its material-technological ecology playing a key role in the invention. Moreover, the developers of this tool describe a concept of *geopoetics*, which can be retooled for a tactical, material and ecological sense of rhetorical invention.

In what follows, I rethink the border apparatus as not *only* a resource for rhetorical symbolic invention, but as a material site with its own affordances that affect (and participate in) the inventions available. In examining the Transborder Tool as such a rhetorical invention, I also resituate the role of symbolicity within a chain of relations that links not just signs, subjects, and affects, but things, matters, places, and the like. I argue that its rhetorical and political effects are not to be found only in explicit discursive changes, but in the innovations of new structures of relations, new subjectivities, and new ways of being in the world among the many elements—including nonhuman entities—that are part of its rhetorical ecology. This is the geopetical potential of material rhetoric. I conclude by pursuing for this kind of rhetorical invention the implications of the concept geopoetics, coined by one of the Transborder Tool's developers.

This chapter unfolds in three movements. First, it charts the US–Mexico border as itself a material rhetorical ecology, materially effective, and embedded in ambient or ecological relations that make it irreducible to its symbolic or institutional elements. Out of this ecology, a rhetorical response emerges in the tactical media project called the Transborder Immigrant Tool. Made by an activist art collective, this project is at once explicitly *rhetorical*, in that it aims to do public political work, and explicitly *aesthetic*, in that its makers are invested in the practice of art-making as a means of invention, and thoroughly *ecological* in that its rhetorical, aesthetic, and political work emerges from and is made of the interactions of many different entities encountering each other in the Transborder Tool's movement within the ecology of the border.

The second movement of this chapter analyzes the Transborder Tool as an example of the material forms of rhetorical invention that emerge from rhetorical ecologies, afforded by the material circumstances of their emergence, but not wholly determined by them. In this section I extend the concepts of rhetorical ecology, developed by Jenny Edbauer, and rhetorical ambience, developed by Thomas Rickert, into such projects, to sharpen the conceptual utility of these ideas for explicitly political rhetorical inventions, like the Transborder Tool. Discussed in greater detail below, Edbauer retools the long-studied notion of "rhetorical situation" with that of "rhetorical ecology."[3] In this move, the elements typically *grounding* human action now actively participate, and rhetoric originates not *just* with a speaker or writer, to conclude with an audience, but instead manifests as a *viral* energy, infecting all the elements within its ecology. Likewise, Rickert, in *Ambient Rhetoric*, uses the concept of rhetorical ambience to argue a similar point: rhetoric's many things, objects, structures, and other materialities are not *mere* background or scene for human symbolic manipulation, but are themselves actively suasive.[4] Rickert, in emphasizing ambience, maintains the ecological sense found in Edbauer, arguing that persuasion is both material and relational at the same time, and is irreducible to simply present discrete elements.

Beyond the academic study of rhetoric, thinking the nature of change in these terms can have broad implications for tactically inventive responses to the world's problems. *If* political problems have the ecological character suggested by contemporary rhetorical theory, and are not reducible to the discourses, symbols, institutions, or intending actions of people *alone*, and *if* the inventive responses to such situations have that same emergent and material character, then rhetorical and political action is different than is commonly understood. Resistance, protest, micropolitics, on this view *may* take a shape that is not recognizable as such, but nevertheless effect real, if ephemeral, changes. And they may involve and implicate the actions of entities that are not human subjects. That is, the stakes of rhetorical and political action—of *tactical* work in the world—are altered if we take seriously what the materialist claims of contemporary rhetorical theory tell us about agency and materiality in action.

In other words, while recent materialist innovations in rhetorical theory are useful for developing new understandings of micropolitical or micro*rhetorical* tactics, that work remains to be done. That task is undertaken here in three articulations of how the emergence of the Transborder Tool affects and is affected by the border apparatus itself. In other words, if, as will become clear below, rhetoric is ecological in that it is emergent from an ecology of different kinds of things all attuned to each other, related and interacting, it is *also* the case that those relations can be *perverted*—altered, shifted, recast, *re-attuned*—available, that is, for the sort of geopoetical rhetorical invention that the Transborder Tool performs. To start to reveal this, I argue that the Tool intensifies the border, in an analysis alongside recent Deleuzian rhetorical theory. Next, I argue that it reassembles the border apparatus, connecting its intensification to developments of posthuman public rhetoric that move from ecologies to assemblages. Finally, I argue that it redistributes the affective powers of the border apparatus, connecting to Jacques Rancière's notion of the "distribution of the sensible" that governs the possibility for political change.[5] The argument here is synthetic. By reading these torsions in rhetorical theory together and in relation to micropolitical, inventive, and *tactical* responses, I am changing each, bringing into greater relief tendencies already there, pushing them in directions they had not yet gone, and altering these concepts in a fashion that mirrors what the Transborder Tool does to the border.

The final movement is a return to the zone of interference between art and rhetoric in which the Transborder Tool operates, via the concept of geopoetics suggested by one of its makers. For the makers of the Transborder Tool geopoetics as an inventive practice replaces geopolitics as a political technique. In this chapter, I suggest, the Transborder Tool appears as an instance of geopoetical rhetoric, which has three characteristics: a form of rhetorical invention *emergent* from a rhetorical ecology, rather than *responsive* to a situation; rhetorical work that is ephemeral and micropolitical while also material; and rhetorical work that implicates agents and actors who are not (all) human subjects.

Border Matters

While it seems clear that international borders generally are rhetorical, in that they organize through related discourses and material phenomena how we think, feel, and act about and around them, the specific characteristics of the US–Mexico border produce unique forms of rhetorical invention. Cutting through an inhospitable desert, this border is heavily policed at both the figurative and literal levels, being as it is one of the most scrutinized regions in American political discourse. On the south side of the border, drug cartel violence has marred already precarious rural and urban communities. On the north side politicians seek to gain rhetorical capital by exploiting the trope of invasion in describing the border as a bulwark against economic, social, and criminal ills. Considerable resources are dedicated to transforming the desert around the abstract line separating the two countries into a visibly controlled terrain, while "neoliberal market ideologies of liquid, free-flowing capital and open borders for commodities come up against new policing tactics to regulate the movements of people."[6] Rhetorical studies on the border have often attempted to make sense of the rhetorical situations that produce the border.

The border itself, in its current form, is not often theorized as a rhetorically active ecology inviting—*and participating in*—inventive responses. Thus, D. Robert DeChaine, in his editor's introduction to the volume *Border Rhetorics*, argues that the rhetorical power of borders "does not issue from borders per se, but rather from specific persons who call upon the figure of the border in specific ways in order to do specific things." The rhetorical power and symbolism of borders, DeChaine argues, is a potent "public *doxa*" that "informs cultural values, shapes public attitudes, and prescribes individual and collective actions." The work that volumes like *Border Rhetorics* does is important since, as DeChaine himself points out, much of current work on borders "assumes the existence of borders as static entities, as given objects to be examined for their effects on individuals and populations," and such work too tidily treats borders as *merely* "given, pre-symbolic entities" that are outside their status as "rhetorical *enactments*."[7] And yet, the framing of the border in that volume positions it either as a powerful symbolic resource for *human* inventive activity, or as brute material givenness. It is worth reframing rhetoric's relationship to the border in yet another way, to ask after the possibility of the border not being reduced *either* to symbolicity, dematerialized and entirely figural, *or* to mere facticity, a given material structure that *just is*, always effecting its force in the same way. Instead, I argue, any human rhetorical response that seeks to alter or redefine the apparatus of the border, its function in public discourse, or its effects on the communities who have to interact with it, must account for the complex interrelationships of its land, the material fact of the structures built and technologies deployed to police it, as well as the circulation of abstract concepts that coalesce to make the border *exist* as an apparatus and rhetorical ecology.

In other words, for whatever individuals or groups who see the border apparatus itself as a situation into which they need to intervene, that situation cannot be understood only as obtaining solely between autonomous *human* subjects. Rather, the most pressing situations of the day press not only on human subjects and populations, but also on nonhuman agents and actors. Any human intervention in social and political problems will have to reckon with nonhuman components and agencies in the assemblage(s) making up those problems, including the human-made informational networks, material structures, natural environments, energies, flows, and so forth that all trouble an expressly humanistic model of democratic change through deliberation. The border, taken as such a rhetorical situation, invites a complex, material form of non-verbal rhetorical response, which displaces human agency as the sole source of its movement. It presents itself as a rhetorical ecology as Edbauer theorized it, and one in which the rhetorical effects emerge in ambient relations among many entities, as Rickert has theorized such rhetorical relations.

The Transborder Immigrant Tool

A rhetorically inventive response to the current situation of the border would need to respond not only to national discourses of sovereignty, identity, and legality, but the very material realities of the border apparatus as it functions now. A rhetorical response to the *border*, then, would differ from a response to discourses about enacting the border and would thus produce different rhetorical effects. One such response is the Transborder Immigrant Tool, recently developed by the Electronic Disturbance Theater/b.a.n.g Lab, in a project led by University of California at San Diego art professors Ricardo Dominguez and Brett Stalbaum.[8] The Tool is a standard Motorola i455 cell phone, modified to aid those crossing the desert into the US. According to a promotional poster, "the project's main goal is to help reduce the number of deaths along the border by developing a common cell phone device into a navigation tool that will help migrants locate life saving resources in the desert such as water caches and safety beacons." The Tool's GPS compass guides its user through the desert to her destination, while providing the user with "tactile feedback," which "frees" her "from [the] phone display interface," permitting her to "concentrate on the surrounding environment."[9] Since many of the immigrants who might use this tool do not read either English or Spanish, and may be unfamiliar with mobile devices, the Tool's developers also simplified the cell phone menu. Dominguez notes, in an interview posted on the b.a.n.g. lab website, that "GPS itself does not require service and has free global coverage, courtesy of the United States government," meaning the devices need not be activated as phones.[10]

This project was, unsurprisingly, controversial when it first became public knowledge, inviting scrutiny from newspapers, legislators, the president of the University of California system, and even Glenn Beck.[11] What makes the Tool and its rhetorical situation baffling to Beck and others who responded to it with confusion and frustration is that its rhetorical and political effects are not to be found in discursive changes, but in the innovations of new structures of relations, new subjectivities, and new ways of being in the world among the many elements that are part of its rhetorical ecology. Critiques of the Tool, for the most part, miss the Tool's target, seeing it *only* as a subversion of the (naturalized) border and national sovereignty, omitting its performative intervention into and perversion of a larger ecology. Placing a critical emphasis only on discursive or embodied threats (the flow of unauthorized people) refuses an opportunity to encounter the border's rhetorically ecological character. While the controversy surrounding the Tool, and whether it subverts the law, is not central to the argument here, it works in the background, detailing the Tool as a different sort of response to a different sort of rhetorical situation. It is that difference, I maintain, and not whether distributing the Tool breaks the law, which drove the short-lived controversy that the Tool inspired.

As Dominguez describes it, though, the Tool does more than merely guide people safely to water and through the desert. Citing Donna Haraway's notion of the cyborg, he suggests that the intervention into the border staged by the Transborder Tool is considerably more complex. Part of the project "is to call into question the northern cone's imaginary about who has priority and control of who can become a cyborg or 'trans' human—and immigrants are always presented as less-than-human and certainly not part of a community which is establishing and inventing new forms of life." Further claiming that the Tool queers communication technology, Dominguez states:

> This gesture dislocates the techno-political effect with aesthetic affects that become something other than code: a performative matrix that fractalizes and reverses the disorder of things with excessive transbodies acting from the inside-out of those enforced borderless borders. These affects assemble new empirico-tran(s)cendental forms of multi-presence(s) incommensurable with the capitalist socius of the so called "immaterial" Empire. As the Zapatistas say, "we do not move at the speed of technology, but at the speed of dreams"—the heart of the trans-border-borg.[12]

Though Dominguez claims aesthetic affects take the place of political effects, we will see throughout this analysis that the political is *effected* through those very affects and their rhetorical work, through the qualitative changes in the user's experience with the Tool, the Tool's relationship to the apparatus of the border, the role GPS and communication technologies play in the transnational capitalism which, with NAFTA, has necessitated the border in its current form. The Tool, then, does not rhetorically intervene only by altering the course of discursivity summoning, surrounding, enabling, and enacting the border, but instead emerges from the relationships among elements in the rhetorical ecology itself. These elements, and that situation, including the material border apparatus, the technologies embodied by the Tool (the Motorola i455), GPS, the modifications made to the device by EDT/b.a.n.g lab, the immigrant's own body, and nativist, racist rhetorics such as Glenn Beck's, all come together to form a rhetorical ecology that is perverted by the Transborder Tool's geopoetical rhetorical invention.

Geopoetics and Artivism

To help produce what Dominguez calls its "aesthetic affects," the Transborder Tool is loaded with pre-recorded poetry that the user may listen to on her journey. "Immigrants should not only be able to move safely, find water, and hear poetry," Dominguez is quoted in a *Chronicle of Higher Education* article, "but they should also be able encounter the landscape in a way that American painters have approached the landscape: as a sublime object." In the same article, Amy Sara Carroll, the project's poet, claims the aim of her poems is "interrupting the aesthetic of hate and fear that dominates the debate about immigration."[13] What is interesting here is not necessarily the utility (or even the wisdom) of listening to recorded poetry during a clandestine, illegal journey through hostile desert, but the slippage in Dominguez and Carroll's statements between artistic creation and rhetorical invention. For many, the project of changing a general mood and climate of hate and fear is a rhetorical project, and furthermore, many would call that climate a *rhetoric* of hate and fear. It is to this overlapping relationship between artistic and rhetorical invention that I return at the very end of this chapter.[14]

Dominguez, a former member of the activist art collective Critical Art Ensemble, calls his practice "artivism," and thus, the motives for rhetorical invention and artistic invention are, in practice, intermixed in his work. On a video posted on the b.a.n.g. Lab website, demonstrating the Transborder Tool in use (going *into* Mexico), an off-camera voice, which sounds like Dominguez, calls their practice a "geopoetic system."[15] In this way, the Transborder Tool employs an unusual mode of rhetorical engagement, the provocative and suggestive geopoetics, using inventional strategies from art practices toward rhetorical ends, and doing so in an explicitly (geo-) material and ecologically oriented way. But if the Tool replaces what Dominguez calls political *effects* with something he distinguishes and separates as aesthetic *affects*, it helps us question what consequences and changes count as effective rhetorical (or political) activity, given the nature of the situation intervened into here, and what results may remain imperceptible if we hold to traditional understandings of what activities and situations are rhetorical (or political), while also blurring the distinctions that hold apart the work of politics, rhetoric, and aesthetics.

In the next section, I begin to answer this question by turning to the twinned question of rhetorical ecologies and the emergent, ephemeral, and tactical forms of invention that come with accepting affect, materiality, and relationality as effective loci of rhetoric's movements. Starting with a look at Edbauer's theory of rhetorical ecologies, and turning to Christine Harold's analysis of *affective* and material forms of rhetorical protest, I will show that the Tool does indeed *do* something rhetorical, and that what it does is not just what its makers or its putative user do, but it, the Tool, in emergence from the ecology of the border, enters into its own rhetorical relations with the elements of that ecology.

Three Transformations of the Border

In rhetorical studies the concept of the rhetorical situation has been an important tool for understanding how rhetorical action responds to and affects the world, and what analyzable conditions are present where rhetoric exists. The concept, meanwhile, has undergone many productive transformations, beginning with Lloyd Bitzer's articulation in the inaugural issue of *Philosophy & Rhetoric.* Bitzer argued that rhetoric is always "situational," that is, as suasive discourse it is brought into being "in order to effect change" in situations. The rhetorical situation, according to Bitzer, is a "complex of persons, events, objects, and relations presenting an actual or potential exigence." Prior to the invention of the rhetorical discourse, Bitzer writes, there are three characteristics of the situation: *exigence*, the need; the *audience* who needs to be persuaded to change the situation; and *constraints* on their ability to be persuaded or make change.[16] Bitzer's concept enabled rhetorical studies to analyze a wider range of elements in the moment of rhetorical action, while also opening the conceptual door for expanding theories of rhetoric beyond the action of intending human subjects, which in turn widens the possibilities of thinking *political* tactics that do not originate in intending human subjects.[17]

In her 2005 article, "Unframing Models of Public Distribution: from Rhetorical Situations to Rhetorical Ecologies," Edbauer articulates an alternate version of the rhetorical situation to complicate Bitzer's, one that imagines its elements not as discrete, autonomous entities, but as emergent and interconnected in "rhetorical ecologies." Edbauer claims part of her intention is to "add the dimensions of history and movement (back) into our visions/versions of rhetoric's public situations, reclaiming rhetoric from artificially *elementary* frameworks." In this ecological view of rhetorical activity "place becomes a space of contacts, which are always changing and never discrete," and "the contact between two people on a busy street is never simply a matter of these two bodies: rather these two bodies carry with them the traces of effects from whole fields of culture and social histories."[18] Accounting fully for its ecological and material nature, as Edbauer encourages, such a contact would also have to include in those fields of culture and social histories the street, the traffic, the buildings, the conceptual apparatus of the city, and so forth.

Borrowing a concept from Gilles Deleuze and Felix Guattari's *A Thousand Plateaus*, Edbauer argues that the rhetorical activities she means "operate within a viral economy." The sense of the viral and the contagion, as Deleuze and Guattari describe it, is a "transversal communication between different lines," which "scramble[s] the genealogical trees."[19] The virus transfers genetic material between different, seemingly unrelated species—in their image, baboons and domestic cats—creating an assemblage that is irreducible to the orderly hierarchy in "genealogies." In other words, there are relationships among elements in an ecology (or an assemblage to use the Deleuzian term), that cannot be accounted for in the genealogical model. According to Edbauer, rhetoric, from this perspective, "is not *contained* by the elements that comprise its rhetorical situation (exigence, rhetor, audience constraints). Rather, a rhetoric emerges already infected by the viral intensities that are swirling in the social field." In such a rhetoric "what is shared among" its elements "is *not* the situation, but certain contagions and energies."[20] This conception of a contagious rhetoric of affective intensity emerging within an ecology, an assemblage, is clearly not reducible to instrumental, suasive speech alone, but admits of the agency and activities of nonhuman actors, which help spread the intensity.

Viewing the Transborder Tool as part of such an emergent contagious rhetoric, we see its user, traversing the desert in a clandestine movement, and holding this device that may be her only lifeline to safe arrival, being *infected* by its rhetoric, becoming more intensely attuned to the border and contemporary technology, relating to them differently. Provocatively, one of the claims made in EDT's promotional poster for the Tool, was that it "provides the user with greater situational awareness," which can have both a simple, instrumental meaning and the ecological, material connotation I am teasing out here.[21] The Tool *gives* its user a new relationship to her situation and greater awareness of that relationship. Meanwhile, in this interaction, the Motorola phone is changed completely into the Tool, and as its old functions are pushed away by its use in its new set of relations with its user, it too is infected by the project's intensive rhetoric. The transversal communication that Deleuze and Guattari describe can be seen here not as happening only between natural species, but also between technological device, the border apparatus, and the human user, all of whom have their relations to each other rhetorically reshuffled, as the relations that kept them discretely separate are transevered by the Tool's rhetorical lines of flight. In the sections that follow I elaborate a micropolitics of material, ecological rhetoric by tracing how this rhetorical contagion, the Transborder Tool, transforms the border in three specific ways.

Intensifying the Border, or Surfing the Desert

Importantly, one thing the Transborder Tool does not do is openly *oppose* the border. That is, it is not an attempt at destruction or a literal attack on the border. The Tool does not, say, try to interfere with or jam Border Patrol activities or communications. Dominguez and the Electronic Disturbance Theater developed the tactic of denial-of-service attacks on government and corporate websites as a mode of activism, and Dominguez himself has used the same tactic to protest the privatization of public higher education in California.[22] Thus, it is well within the technical and creative power of the Tool's developers to use direct confrontation as one of their rhetorical strategies. Instead, the Transborder Tool operates as geopoetical rhetoric, *making* a rhetorical viral "contagion" out of the materials it has embedded within its ecology.

There are theoretical criticisms, though, of such technologically mediated invention. Political theorist Jodi Dean, for instance, is quite critical of the politically disempowering effects of what she terms "communicative capitalism" and the fetishization of technology standing in for political action.[23] Dean argues that with the expansion of networked communication technologies—like those used by the Transborder Tool—"ideals of access, inclusion, discussion, and participation come to be realized in and through expansions, intensifications, and interconnections of global telecommunications." But instead of making the world more democratic and equitable, she argues, these "rhetorics of access, participation, and democracy work ideologically to secure the technological infrastructure of neoliberalism, an invidious and predatory politico-economic project that concentrates assets and power in the hands of the very, very rich, devastating the planet and destroying the lives of billions of people."[24] While Dean's concern is with a general tendency to mistake expressing opinions on social media sites, blogs, and so forth for political action, it is worthwhile grappling with her concerns about the use of communication technologies produced in contemporary capitalism for subversive politics, and wondering if the Transborder Tool is *doing* political work. Certainly, we can imagine that its particular users—who undoubtedly come from the billions of people whose lives are being destroyed by neoliberal capitalism—benefit from its use. This perspective, though, leaves the impression that the Transborder Tool simply *moves through* the border region, leaving that ecology unaffected. Instead, I am arguing, as it travels (actually and virtually), it geopoetically transforms the border.

For one, it *intensifies* the border. In her book *Ourspace*, Christine Harold identifies three broad strategies employed in resistance practices: sabotage, appropriation, and intensification.[25] As Harold argues, both *sabotage*—favored by the Situationists and their stylistic heirs, such as *Adbusters*—and *appropriation*—preferred by hackers—are caught by their own logic, which attributes to capitalism a monolithic unity it does not have, while preserving discrete borders between it and the public sphere, which is idealized as a utopian space uninfected by the market. Attempting to escape the logic of capitalism by supposing a monolithic capitalism "out there" to oppose inscribes a dialectic that maintains capitalism's existence. *Intensification*, however, works inside the operations of the market, turning its forces and energies into useful tactics for resistance.[26] Harold draws on Deleuze, developing his claim that in control societies "*surfing* has taken over from all the other *sports*" into a full-fledged metaphor for the kind of rhetorical movement happening with the Transborder Tool. The surfer, Harold tells us:

> does not *conquer* a wave the way a football player conquers territory. For the surfer, the ground beneath her feet is no ground at all. It is unstable, it is fluctuating, it is permeable. The surfer succeeds neither by *overcoming* the ocean as if it were an obstacle, nor by appropriating it from another

> surfer. She succeeds by cultivating a way of being in the world that affords her the capacity to *respond* kairotically to what comes next. A good surfer must know something about the wind, tides, ebbs and flows and how they work together to create waves.[27]

Thus, the user of the Transborder Tool does not simply find herself in a situation, but rather in a material world where activities are *intensive*—she comports herself to a different way of being in the world, one in which her responses do not oppose the border but intensively engages it differently. In this process, she, the Tool (and all the complex flows of capital that concentrate in it), and the border apparatus (and all the complex flows of capital that insist on its reality), *all* change in their relations to each other. In other words, the Tool hooks its user into the border assemblage and makes it (the border) do something else. Recall that, as Dominguez noted above, GPS, the crucial technology in the Tool, is a development of the US government, and in its general use, from smartphone applications to Google-powered maps, it participates in the global flow of capital that also necessitates the border apparatus in its current form. The Tool and its user do not protest the border assemblage by trying to conquer it, or destroy it, but resist it by *deterritorializing* its systems, such as GPS, and *intensifying* it by putting it to different use.

Dean's criticism comes from a different perspective than Harold's, or Deleuze's, for that matter, in that she finds objectionable efforts that do *not* unravel or oppose the logic of late capitalism, and are content with intensification. Harold, via Deleuze, refutes that view, and claims that Deleuze "describes the political subject who is best able to respond to the new economy" because, as Harold argues, "political strategies that rely on a dialectical *repudiation* of capitalism hinder from the start their own ability to *respond* to the speed and elasticity of the marketplace."[28] As both these perspectives suggest, the problems of responding rhetorically to the marketplace and of responding to national politics are mutually imbricated, and the border presents itself as a rhetorical ecology in which capital flows and national boundaries are inseparable elements. In this way, the Transborder Tool's rhetorical response to the political/economic problem of the border is to "surf" both the border itself and the capitalist communications technologies that enable both border policing and the Transborder Tool at the same time. The Transborder Tool invites us to build on Harold's work by bringing a material dimension into her account of ecological rhetoric/intensification, by attending to material agency and potency.

While both Harold and Dean are concerned with the effects of technology in late capitalism, subtly, however, in both perspectives, the role of technological devices is assumed merely either to aid or impede the formation of an effective social and political sphere for human actors to resolve pressing political and economic matters. For Harold and Dean, contemporary technology is a tool for human use and is not activated with agency in the rhetorical ecologies they describe. However, the Transborder Tool's rhetorical activity is not easily reducible to a human agent, as the border itself expresses agency. That is, the border and all the elements in its ecology, when we look at the Tool, remind us of what Latour admonished, that we remember in composing an "object-oriented democracy," that "each object—each issue—generates a different pattern of emotions and disruptions, of agreements and disagreements" and, thus, "each object gathers around itself a different assembly of relevant parties."[29] It also recalls another caution from Latour, that the term "social" designates only "*a type of connection* between things that are not themselves social," and that to replace the familiar sense of the word, we should imagine it only as "a very peculiar movement of re-association and reassembling."[30] The movement that traces new social connections between things that are themselves "not social"—i.e., the objects theorized in an object-oriented politics—is the same intensive, geopoetical and ecologically emergent movement of the nocturnal and secret path that the Transborder Tool and its user potentially cut across the southern California desert. On the one hand, the border expresses agency by presenting itself as a *real* object of concern for those who come up against its corrugated metal walls, razor-wire fences, networked surveillance, and heavily policed desert trails. On the other hand, the emergence of the Transborder Tool, from within the technologies, infrastructures, and sets of relations that enable and make real the border, expresses a different rhetorical agency in its very movement, and changes the way those things relate to each other—*are* a common ecology—by intensifying their interactions.

Re-Attuning the Border, or the Transborder Tool's Scordatura

The place of the border, in this sense, is a *generative*, rhetorically inventive ecology, partly natural (the desert, the wind, the heat), partly built (the fences, the walls, the water stations for travelers), and partly symbolic (the discourses of sovereignty, of security, of territory). None of these elements can be separated out as discrete, nor can the border be taken only as static. Nor can the practice of rhetoric be understood as taking place in language and symbolicity separate from these other elements. Thus, a deeply rhetorical materiality exists within the things that come together, with or without the active origination of human subjects. In *intensifying* the elements that compose the border, the Transborder Tool can be seen as altering what, say, GPS *is* and *does*, not just what it means. That change has material effects, meanwhile, on the user, whose own being is altered, and who moves differently through the place that the border is and has become, and is disposed to it differently.

Rickert calls this disposition "attunement." On his reading, attunement is the "dynamic unfolding by an originary worldly rhetoricity, an affectability inherent in how the world comes to be."[31] Attunement, which Rickert adapts from Heidegger, is "one's disposition in the world, how one finds oneself embedded in a situation," which, however, is not *subjective* or *personal*, but instead "results from the co-responsive and inclusive interaction that brings out both immersion (being with) and specificity (the way of our being there)."[32] This means that, for Rickert, all things have a material rhetoricity, an affectability that is itself rhetorical. Things can affect *how* one another *are* in the world, and *how* things are in the world emerges from interaction with other things: the "attunement" of a thing (even a person) is irreducible to its choices alone. But as Rickert, through Heidegger, argues, "we are always already attuned, there are only changes in attunement," meaning that the *orientation* or *disposition* that attunement is, is the only site of change.[33]

As the Transborder Tool demonstrates, while the border apparatus is a rhetorical ecology and a place that is emergent from its attunements—the ways it relates to all manner of other human and nonhuman entities—it is also *pervertible* in real ways. For his part Rickert argues that the only way forward to a world ecology that has any chance of lasting is a global change in "human dwelling," that form of attunement of human being in the world that orients our species more often than not toward the world as resource or useful. In that way, he argues Latour's *Dingpolitik* "sneaks humanism back in" when it is supposed to evacuate it, because it is still caught in a representational model of democracy.[34] On Rickert's view, the re-attunement of human dwelling with nonhumans is not about "electing" nonhumans to a parliament, but in a radical shift in our species' orientation (its attunement) in (and to) the world.

This leaves the question open of what kind of (at)tuning may be performed by the movement of the Transborder Tool. While it is clear it does not re-attune our whole species-being, it can be seen as effecting an alteration in the being of the elements it involves, tuning them in unexpected ways. In music, this is called scordatura, and is an intentional, inventive *bad tuning*.[35]

More *tactical* forms of altering our attunement are available—scordatura-like forms—and these may indeed be part of the overall strategy for changing how we, as a species, relate to each other and the other entities we live with on this planet. This, I argue, is one of the lessons we can learn from the Transborder Tool, that its relationship with the border is a form of invention that *perverts*, misaligns, or re-attunes itself and the border, and, in that way, the orientation of the user's body not only to the desert, but to the symbolic possibility of the United States itself.

To find a sense of this re-attuning, it is worth returning to the Deleuzian notion of assemblage. Byron Hawk connects the assemblage to materialist, posthuman public rhetoric, arguing that Deleuze and Guattari's assemblage is a "connected collection of animate and inanimate bodies, actions and passions, and enunciations and statements in constant motion."[36] To understand the public rhetorical implications of this, Hawk turns us back to Latour's *Dingpolitik* to rethink the materiality of publicness, and argues that the "public rhetoric" that emerges there, in which things become public, "opens the assemblage back onto the outside world in order to make new connections with it possible, continually enacting the process through the invention of new rhetorics."[37] In Latour's material sense of the public, which reorients agency and displaces speaking human subjects from their position as sole agents, *things* are active, agential forces, and through the assembly that forms in debating matters, their public itself emerges. On this reading of *Dingpolitik*, it is not sneaking as much humanism in as Rickert reads it, but opens up a view of *material* publics, in which the human–nonhuman assemblages emerge as sites of public rhetoric.[38] Hawk argues that Latour does not provide a new *theory* for public rhetoric, but only the tools needed for its emergence; each public rhetoric would emerge from its particular material circumstances.[39] In this way, by recentering the *process* of assembling, Hawk helps make more visible the counter-movements of nonhuman rhetorical agency within rhetorical ecologies.

Taken together these insights suggest that on its path through the desert, in constant communication with the US Government's GPS satellites, the Transborder Tool is *qualitatively* reproducing the desert itself and, importantly, the GPS system. The border, an ecology involving all of these elements and more (including discourses of sovereignty), is *reassembled*, and *re-attuned*, by this. Viewing the Tool's movement in this way, we can see the apparatus of GPS enrolled into service in helping people navigate the border, in surfing the border. This is a perversion, of course, of GPS' typical function. GPS, as a tool of US governmentality, is deeply entwined in neoliberal capital and functions as part of a surveillance and policing apparatus when it comes to the border. For consumers in the global north GPS has commercial properties, aiding them with geolocation apps in, say, smartphones, and helping them conduct their daily lives. The "billions of people" Dean noted who are crushed under by neoliberal capital do not typically get to *use* GPS in such ways, and are more often targets of GPS-supported surveillance and policing enterprises. The Tool, as has become clear here, *reinvents* GPS, if qualitatively and tentatively, for a different use. This is both a rhetorical and an aesthetic operation. It is also, as will become clear in the next section, the essence of politics.

Redistributing the Border

The Transborder Tool reassembles the border as a process of the political itself. That is, it is not only, as the makers of the Tool suggest, replacing political work with aesthetic work, but also conducting actual political work, through its geopoetical rhetorical movement—re-attuning, intensifying, emergent. The Tool, in its existence either in putative movement through the desert or in its virtual movement as a vibrant possibility, *redistributes* the positions that make possible what can be done, seen, said.

Jacques Rancière, in articulating his redefinitions of the terms police and politics, argues that it is police, and *not* politics, that is "the set of procedures whereby the aggregation and consent of collectivities is achieved, the organization of powers, the distribution of places and roles, and the systems for legitimizing this distribution." Police is, thus, "first an order of bodies that defines the allocation of ways of doing, ways of being, and ways of saying, and sees that those bodies are assigned by name to a particular place and task."[40] *Politics*, however, is a break with the police order that is "manifest in a series of actions that reconfigure the space where parties, parts, or lack of parts have been defined." In short, "Political activity is whatever shifts a body from the place assigned to it or changes a place's destination."[41] The political is a rend in an existing police order, a particular "distribution of the sensible," which shakes up what is visible, sayable, and possible. The distribution of roles, and "parts," that constitutes the "distribution of the sensible" always constitutes some sort of miscount, and there are those who "have no part" in the society—i.e., they are not politically or socially agents in the existing distribution of the sensible, what is actually *perceptible* within the society is determined. Rancière argues:

> Politics exists because those who have no right to be counted as speaking beings make themselves of some account, setting up a community by the fact of placing in common a wrong that is nothing more than this very confrontation, the contradiction of two worlds in a single world; the world where they are and the world where they are not, the world where there is something "between" them and those who do not acknowledge them as speaking beings and the world where there is nothing.[42]

Thus, those who have no part are not visible at all. A redistribution of roles and arrangements must take place for them to be seen and heard, for their speech to count as political speech and not merely noise.

This redistribution, the actual *movement* of the change in and reassembly of what is sensible, is conceptually akin to what I have been arguing here. Democracy, or the political, is the part of those who have no part in the existing police order. In Rancière's use, *police* does not have the same valence it generally does, but resonates with the Deleuzian suspicion of "territorializing" movements, the "state order" or "molar formations."[43] Taken together with what the Transborder Tool has shown us, this answers more cogently Dean's concern that such actions are not genuinely political. From Rancière's perspective, existing subjectivities massing together to protest particular arrangements within a given distribution of the sensible is not in itself the political, and wholly within the existing police order. Mobilizing Rancière's concepts of the political and the movement that redistributes "the sensible" along with the concepts in circulation above allows us to recognize the movement of the Tool as itself democratic action.

If we connect the Tool's movement with the *ecological* insights I am arguing for here, then this movement is not *just* political action, but is expressly *rhetorical* action, and a material, affective form of rhetorical invention. As Hawk has it, the value of Latour's *Dingpolitik* for retheorizing rhetoric is that it is "grounded in the gathering and movements of assemblages and enacted through the manifold assignments of the objects gathered and the human, interpretive involvement in this disclosing of a world."[44] Rhetorical invention in this key is not wholly *directed* by, nor does it wholly *originate* within, the human subject involved. Despite DeChaine's insights that the border is a symbolic *resource* for human social invention, for "sense-making" and constituting a "public *doxa*," what we see here, in analyzing the Transborder Tool, is the emergence of different worlds, different assignments, destinations, and directions for the things involved.[45] Their re-assembling in the presence of the Tool is what constitutes political action for Rancière and emerges as the very form of rhetorical invention. This version of public rhetoric does not reduce the many things, places, technologies, codes, affects, and the like to *resources* for intentional symbolic manipulation. Neither does it, though, take these things, as DeChaine worried, as *merely* "given, pre-symbolic entities" that are outside their status as "rhetorical *enactments*."[46] Where this conceptualization parts with DeChaine is in imagining that the *action* of rhetoric that enacts the border is not wholly human. In this sense, the border apparatus itself and the Transborder Tool emerge as important rhetorical agents in a productive *re*assembly of the border, which rhetorically invents new worlds, new possibilities for *all* its elements, not the least of which is the Transborder Tool's user in her furtive travels.

Meanwhile, although it is not clear that Rancière would argue for the active agency of the Transborder Tool or the border itself, it is safe to argue that the redistribution of associations, the *reassembly* involved with the Tool, adds its users and participants in the project to the rolls of those who *actively* have a part in that rend in the police order that he calls politics. A fair response to this is that while we can concede the qualitative change in relations among the elements within what we mapped as the Tool's rhetorical ecology, it does not necessarily follow that its movement effects the new "distribution of the sensible" Rancière describes above, wherein the speech of undocumented immigrants is taken as political *speech*. But what is described there is not the wholesale acceptance of the new subjectivity as legitimate, but as in Rancière's example of the plebeian revolt in ancient Rome, the irrefutability of the *existence* of this subjectivity. The Tool's existence, as a project at least, can be seen as activating, in some small part, such a difference. These actions constitute Rancière's political "rearrangement." Or, in Latour's terms, "we have to reshuffle our conceptions of what was associated together because the previous definition has been made somewhat irrelevant. We are no longer sure about what 'we' means; we seem to be bound by 'ties' that don't look like regular social ties."[47] What a consideration of the Transborder Tool helps us do, then, is to develop a robust ecological sense of political and rhetorical activity that draws on Rancière, Latour, and Deleuze, and accounts for the rhetorical and political force of qualitative changes in attunement and rearrangements in the relations between elements in whatever assemblage is under discussion.

If the Transborder Tool's only political and rhetorical effect were on its user, altering her relationship to the border assemblage and the networked communication technologies of late capitalism, these effects would be limited, to say the least. Likewise, there is no concrete evidence that the Tool has ever even been *used* in the way opened by Electronic Disturbance Theater.[48] The Tool does, however, have the capacity to generate other effects, which are more wide-reaching than individualistic changes in each user (should those users even exist). Like many recent projects called, variously, tactical media, interventionist art, critical art, the Tool experimentally opens up new rhetorical tactics and strategies for resistance within late capitalism that are not founded on the flawed oppositional structures that Harold (through Deleuze) describes.

Conclusion: Affirming Geopoetical Rhetoric

We can now look more closely at the notion of a geopoetical rhetoric, as sketched several times above. For the makers of the Transborder Tool, geopoetics as an inventive practice replaces geopolitics as a political technique. A crucial feature of a geopoetical rhetoric, to my mind, is the *inventiveness* available to all to address matters of deep concern. In mapping the three ways the Transborder Tool transforms the border, I argued that it, the Tool, as well as the border (composed of many different materialities and elements) were inventively altered by the Tool's movement. In intensifying, reassembling, and redistributing the border, the Tool performs its geopoetical work. This alteration of the border, meanwhile, is not reducible to symbolic transformations alone, in what the border or a Motorola cell phone *mean*, but rather, ontologically and affectively, the Tool and the border *are* different. Such a geopoetical rhetoric, then, is a set of tactics of invention, instead of the application, like geopolitics, of pre-existing schemata to emergent situations and subjectivities—*poesis* standing in for the familiar use of *politics*. On Rancière's view, the standard form of politics, discussion, consensus, and opinion—that is, the deliberative model of democratic change—is a form of the police. What is needed instead of "choosing from among responses proposed" within the existing distribution of the sensible is "the invention of a question that no one was asking."[49] Political invention "decomposes and recomposes the relationships between the ways of *doing*, of *being*, and of *saying* that define … the relationships between the places where one does one thing and where one does something else."[50] From this perspective, geopoetical rhetorical tactics, then, are modes of rhetorical invention that experimentally explore different modes of political and rhetorical activity. They are materially and ecologically embedded. The Transborder Tool's actual and virtual encounter with the border performs its geopoetical rhetorical work. Emergent from its ecology, blurring distinctions between aesthetic and rhetorical invention, making its work (*poesis* in the classical sense of the word) out of the elements in its ecology, the Tool's rhetorical and political work that reassembles and redistributes sensible positions of action, in all of this, enacts a *geopoetical rhetoric*.

Deleuze and Guattari provide another articulation of such an experimental/inventional political tactic:

> This is how it should be done: Lodge yourself on a stratum, experiment with the opportunities it offers, find an advantageous place on it, find potential movements of deterritorialization, possible lines of flight, experience them, produce flow conjunctions here and there, try out continuums of intensities segment by segment, have a small plot of new land at all times.[51]

These modes may not be immediately recognizable as rhetoric. They may collapse audience, rhetor, and situation into an ecological assemblage that makes common forms of rhetorical analysis incomplete; they may use electronic media and objects not as prostheses for speech, but as active participating agents in their rhetorical activity; they may produce content that is not expressly political but nevertheless effects alterations in the ways one can be in the world. But these modes of activity nevertheless increase the supply of tactics available to those who feel the need to respond and do not see existing frameworks of political activity as adequate. Geopoetical rhetorical tactics, in short, are more experimental than revolutionary, even when the possibility of revolutionary effects remains open.

In this way, the Transborder Tool allows us to view the politically inventive force of material rhetoric, that rhetoric that emerges from ecologies or assemblages of human and nonhuman, as being very much *geopoetical.* Geopoetical rhetoric is a form of rhetorical invention *emergent* from a rhetorical ecology, rather than *responsive* to a situation; rhetorical work that is ephemeral and micropolitical, while also material; and rhetorical work that implicates agents and actors who are not (all) human subjects. The implications of rearticulating material rhetoric in terms at once explicitly *political* and explicitly *aesthetic* are to energize a sense of material rhetorical projects, favoring the experimental, deterritorializing thrust of rhetorical work like the Transborder Tool's.

Notes

1. Rita Raley, *Tactical Media* (Minneapolis: University of Minnesota Press, 2009), 32–33.
2. Anne Demo, "Sovereignty Discourse and Contemporary Immigration Politics," *Quarterly Journal of Speech* 91, no. 3 (2005): 296–297.
3. Jenny Edbauer, "Unframing Models of Public Distribution: From Rhetorical Situation to Rhetorical Ecologies," *Rhetoric Society Quarterly* 35, no. 4 (2005): 5–24.
4. Thomas Rickert, *Ambient Rhetoric: The Attunements of Rhetorical Being* (Pittsburgh: University of Pittsburgh Press, 2013).
5. Jacques Rancière, *Disagreement*, trans. Julie Rose (Minneapolis: University of Minnesota Press, 1999).
6. Raley, *Tactical Media*, 32.
7. D. Robert DeChaine, ed., *Border Rhetorics: Citizenship and Identity on the US–Mexico Frontier* (Tuscaloosa: University of Alabama Press, 2012), 2–3.
8. "Bits, Atoms, Neurons, Genes." The lab is housed at the University of California San Diego, directed by Dominguez, accessed June 1, 2017, http://bang.calit2.net/.
9. See http://www.tacticalmediafiles.net/mmbase/attachments/4950/xborder.pdf, accessed June 1, 2017 .
10. Lawrence Bird, "Global Positioning: An Interview with Ricardo Dominguez." b.a.n.g. lab, last modified October 15, 2011, accessed December 10, 2011, http://www.furtherfield.org/features/global-positioning-interview-ricardo-dominguez.
11. For an overview of the controversy, see Evan Goldstein, "Digitally Incorrect," *Chronicle of Higher Education*, 8 (2010); Richard Marosi, "UC San Diego Professor who Studies Disobedience Gains Followers—and Investigators," *LA Times.com*, May 7, 2010, accessed June 1, 2017, http://articles.latimes.com/.

12. Bird, "Global Positioning."
13. Goldstein, "Digitally Incorrect."
14. While it is beyond the scope of this chapter, there is much work on the relationship between rhetoric and aesthetics. For two different perspectives, see John Poulakos, "From the Depths of Rhetoric: The Emergence of Aesthetics as a Discipline," *Philosophy and Rhetoric* 40, no. 4 (2007): 335–352; Janet Atwill "Bodies and Art," *Rhetoric Society Quarterly* 36, no. 2 (2006): 165–170. What is needed, as I see it, is to follow the conclusions drawn in the present chapter to a deeper analysis of rhetoric as a public art *like* tactical media or other social practice art making, in which rhetoric's traditional function as a public *art* of democracy blends with the emerging art-making practices of intervention and world-building that use tools different from speech and writing. If rhetoric, that is, *is* a public art irreducible to language, it becomes worthwhile to analyze its relationship to current trends in public art that have become rhetorical. For studies of social practice art, see Shannon Jackson, *Social Works: Performing Art, Supporting Publics* (New York: Routledge, 2011); Grant Kester, *Conversation Pieces: Community and Communication in Modern Art* (Berkeley: University of California Press, 2004).
15. "Transborder Tool Crosses into Mexico," b.a.n.g. lab, last modified June 6, 2011, accessed November 30, 2011.
16. Lloyd Bitzer, "The Rhetorical Situation," *Philosophy and Rhetoric* 1, no. 1. (1967): 4–6.
17. Of course, Bitzer's conceptualization of the rhetorical situation opened a *host* of other theoretical doors too, most immediately Richard Vatz's rejoinder to Bitzer and most compellingly Barbara Biesecker's extension. See Richard Vatz, "The Myth of the Rhetorical Situation," *Philosophy and Rhetoric* 6, no. 3 (1973); Barbara Biesecker, "Rethinking the Rhetorical Situation from within the Thematic of 'Différance'," *Philosophy and Rhetoric* 6, no. 3 (1973).
18. Edbauer, "Unframing Models," 9–10.
19. Quoted in Edbauer, "Unframing Models" 14.
20. Ibid., emphasis in original.
21. Available here http://www.tacticalmediafiles.net/mmbase/attachments/4950/xborder.pdf, accessed June 1, 2017.
22. See Dominguez's faculty bio on the University of California at San Diego website, for information on EDT and "virtual sit-ins," http://visarts.ucsd.edu/~gd2/faculty/ricardo-dominguez, and for the "sit-in" staged on the UC President's website, see Marosi.
23. Jodi Dean, *Democracy and Other Neoliberal Fantasies: Communicative Capitalism and Left Politics* (Durham: Duke University Press, 2009).
24. Ibid., 23.

25. Christine Harold, *Ourspace: Resisting the Corporate Control of Culture* (Minneapolis: University of Minnesota Press, 2009), xxviii.
26. Ibid., 160–161.
27. Ibid., 161.
28. Ibid.
29. Bruno Latour, "From *Realpolitik* to *Dingpolitik*," in *Making Things Public: Atmospheres of Democracy*, ed. Bruno Latour and Peter Weibel (Cambridge: MIT Press, 2005), 5.
30. Bruno Latour, *Reassembling the Social: An Introduction to Actor-Network-Theory* (Oxford: Oxford University Press, 2005), 5–7.
31. Rickert, *Ambient Rhetoric*, 8–9.
32. Ibid., 9.
33. Ibid.
34. Ibid., 239.
35. In avant-garde music like that of Glenn Branca or his followers in the band Sonic Youth, scordatura makes for an often unrepeatable sound, as instruments are tuned in all manner of haphazard or intentionally unusual ways.
36. Byron Hawk, "Reassembling Postprocess," in *Beyond Postprocess*, eds. Sidney Dobrin, J.A. Rice, and Michael Vastola (Logan: Utah State University Press, 2011), 78.
37. Ibid., 91.
38. Public rhetoric scholarship has begun taking up this call, if not yet a turn toward micropolitical rhythms. See Nathaniel Rivers, "Tracing the Missing Masses," *Enculturation* 17; Scot Barnett and Casey Boyle, eds., *Rhetoric Through Everyday Things* (Tuscaloosa: University of Alabama Press, 2016).
39. Hawk, "Reassembling," 91.
40. Rancière, *Disagreement*, 28–29.
41. Ibid., 30.
42. Ibid., 27.
43. Gilles Deleuze and Felix Guattari, *A Thousand Plateaus: Capitalism and Schizophrenia*, trans. Brian Massumi (Minneapolis: University of Minnesota Press, 1987), 303.
44. Hawk, "Reassembling," 89.
45. DeChaine, *Border Politics*, 3.
46. Ibid.
47. Latour, *Reassembling*, 6.
48. The *San Diego Union-Tribune*'s editors, arguing for Dominguez's academic freedom during public (and university) scrutiny of the project, doubt the Tool was ever intended for use. *San Diego Union-Tribune*

April 10, 2010. Likewise, the video putatively showing the Tool going *into* Mexico, in reality *shows* no such thing. Whether it did happen, such movement is not visible in the footage as assembled ("Transborder Tool Crosses…").

49. Rancière, *Disagreement*, 33.
50. Ibid., 40.
51. Deleuze and Guattari, *A Thousand Plateaus*, 161.

CHAPTER 12

Stabilizing Energies: Intersections Between Energy Promotion Texts and Rhetorical Theory

Brian Cozen

A common refrain from discourse that promotes sustaining and expanding fossil fuel use argues that these resources are vital and should be harnessed for the purposes of human potential. Fossil fuels, in these terms, permit the energetic spirit of human existence to flourish. For instance, the campaign Fueling U.S. Forward, an industry-related offensive effort to publicly brand the cultural narrative of fossil fuels, argues that energy enables the activities of daily existence: "It brews our coffee and paves our roads. It turns on the nightlight and gets us to soccer practice. It lifts people out of poverty and offers new beginnings. It fuels innovation and saves lives."[1] In this empowerment framework, humans awaken the vitality in the natural world and bend it toward serving human potential.

This chapter analyzes similar fossil fuel promotional discourse, in particular ExxonMobil's strategic campaign, Energy Lives Here, to

B. Cozen (✉)
California State University, Fresno, CA, USA
e-mail: bcozen@csufresno.edu

B. McGreavy et al. (eds.), *Tracing Rhetoric and Material Life*,
Palgrave Studies in Media and Environmental Communication,
https://doi.org/10.1007/978-3-319-65711-0_12

interrogate how modern fossil fuel powered systems are rhetorically named and captured as the source of humans' energetic potential. In this rhetoric of stability, discourse channels the feel of the energetic and domesticates, or stabilizes, its source. Such fossil fuel promotional texts circulate amid broader applications of the energy trope, in which energy stands in as a vital force that generates lively activity, including rhetorical action. As such, energy systems—networks, infrastructures, and practices related to the extraction, circulation, and consumption of energy resources such as oil and natural gas—prepare the field of human potential. That is, the promotion of energy systems also promotes the idea that human potential is enabled or made possible via these systems. In this formulation, energy, channeled through the processes of extracting fossil fuel resources and maintaining modern infrastructures, names the harnessed vitality of the world that ignites the vitality inherent in humans. Energy systems prepare the field; they enable the unleashed kinetic energies of human potential.

This logic maintains a desire for fossil fuels as the naturalized foundation from which the modern world builds. That is, fossil fuels denote the current, empowering energy system. Fossil fuel promotion proffers a sense of energy-as-vital, which serves to legitimate maintaining the system that is in place and appears stable. If fossil fuels make possible the field of human potential, the logic follows, they are the prepared *grounds* or foundations from which human vitality, creativity, and activity spring.[2] Playing off Brian Ott's discussion of stabilizing structures and destabilizing energies, this chapter argues that energy promotion texts are forwarding a conception of stabilizing energies.[3] First, this framing situates energy systems as the stabilizing structures of modern infrastructures and a general grid logic of order and stability. Second, in that the energy system is said to enable human vitality by harnessing and channeling the vitality of the world, it stabilizes the terrain of affective energies. In sum, this textual worldview states that energy enables the field or environment for human potential; nature's vitality lies dormant until prepared for human activity. In all, energy promotional texts attempt to rhetorically circulate an understanding of fossil fuel practices as an entrenched network readily justified as necessary for life to function as is. And, as Håvard Haarstad and Tarje I. Warnik argue, to conceive of energy's role in society as "systemic" is in part to reinforce this sense of stability.[4]

This chapter juxtaposes an analysis of fossil fuel promotional texts with contemporary rhetorical theories that draw upon the language of energy to highlight how energy promotion rhetorics, specifically fossil fuel texts, categorize energy and theorize its social role in the world. Via this energy trope, energy promotion texts name how the energetic needs its systems in place. In some ways, these texts give an ecological sense to agency, but one that foregrounds the energy company as the preparer of a promoted ecology bent toward serving human agency. Energy systems—socialized moments of the energetic channeled into particular power structures—are figured as domestications of nature's vital qualities. A rhetorical analysis of these texts can inform how such systems are invariably linked to human vitality, the natural world, and the energy company mediator.

Further, an engagement with how the energy trope circulates can inform how public discourses "may be shaping and underwriting certain theoretical trends" like defining rhetoric in terms of energy.[5] To further interrogate the potential political implications of our theoretical trends, this chapter invites other rhetorical theorists to attend to energy resource and environmental politics when building conceptual frameworks around chance, change, force, intensities, and emergence—concepts often referred to as energies. Scholars conceptualizing rhetoric ecologically, highlighting rhetoric as a dynamic or *energetic* process of emergence and transformation, can better scrutinize the implications of our own conceptual categories by examining how these categories are deployed in other settings. Theorizing how rhetorical energies become stabilized around energy resource systems, or rhetorically invoked as naturally emanating through these systems, can further engage scholars of rhetoric and materiality in the ecological questions of nature, culture, energy, mediation, capacities, and the emergence of socially and environmentally just practices.

This chapter offers a rhetorical perspective on how energy systems are promoted as necessary grounds for unleashing human vitality. In what follows, I first sketch the rhetoric-as-energy trope as found in rhetorical theory. Next I explain the logical framework through which fossil fuel promotional texts deploy the energy trope, focusing in particular on how the energy company mediator converts the vitality of the dormant natural world towards human vitality. I illustrate this logic through a couple of textual examples before focusing analysis on ExxonMobil's Energy

Lives Here campaign. The conclusion revisits the relationship between rhetoric-as-energy and energy systems promotional rhetoric, calls for further work to critically examine the public circulation of theoretical concepts, and suggests that paths for rhetorical energies—toward acting and doing work in the world—can emerge outside the fossil fuel system.

The Energetic in Rhetorical Theory: Rhetoric's Energy Trope

This section tracks various uses of the term energy in rhetorical theory. Across rhetorical theories, energy often denotes, in broad terms, a force between encounters, or effects that result from individual effort to more complex, multidirectional interactions. All these theorists are trying to conceptualize how rhetorical action occurs. The energy trope helps them conceive of how such activity is assembled, or to answer the question: How do elements come together to spark action? Categorizing what energy names in these theories—specifically, as the exertion of individual or distributed agency, as circulating affect in a given social field, and as transformational versus harnessed potentiality—can inform the persuasive logics of promotional fossil fuel rhetorics, which claim that energy systems actualize modern life. In the chapter's conclusion, I posit that such promotional, circulating rhetorics may, in part, be shaping these theoretical trends.

From Individual to Distributed Agency

Agency, as an ability to act, has long been considered a distinguishing feature of rhetoric.[6] In theorizing rhetorical agency, the energy trope often names rhetoric as an actualizing art. This section traces the energy trope in evolving understandings of agency from an individual faculty of a rhetor to a distributed capacity across human and nonhuman actors.

Energy and cognate terms are often used to describe a faculty for symbolic action. Some of these associations relate back to rhetoric's canonical origins, where the classical (and male) orator persuades or moves an audience; as Girke and Meyer put it, "*rhetoric* calls to mind the successful, energetic orator who bends the audience to his will."[7] In classical terms such as *energeia*—as bringing before the eyes—and *dunamis*—as faculty, capacity, power, and force[8]—we see rhetoric emphasized as an actualizing art. Debra Hawhee draws a parallel between *pouvoir*, the French term for power that also incorporates a sense of can-do-ness, and *dunamis* to highlight the performative, relational elements in language or *logos*.[9] Rhetoric is symbolic work, in action; rhetors act in the world, through language.[10] *Energeia* puts this faculty for persuasion into action, transferring a vision or structure to an argument. In more recent theory, George Kennedy defines rhetoric *as* energy, in that words have a material basis in which people expel energy to enunciate utterances for particular persuasive purposes. Such an approach to rhetorical energy highlights how symbolic expression is a physical act one performs to engage another. Energy, for Kennedy, defines inducement as a material action.[11]

Katherine Miller and Ralph Cintrón draw upon this traditional lineage, and Kennedy's updated framework, to emphasize how individuals exert energy upon an audience dialogically engaged in the process of invention; energy, for these theorists, names rhetorical activity.[12] First, Miller discusses the distinction between the faculty for and the enactment of effective discourse as the difference between potential energy and kinetic energy. *Ergon*, Miller notes, is the capacity for work or potential energy, while *energeia* illustrates how composition can be understood as kinetic energy, in which rhetorical invention is enacted in speech or writing, its effects actualized. *Energeia* is "action itself."[13] Kinetic energy, here, is the result of interactions between rhetor and audience, or the interactive doing that compels (symbolic) action.[14] The work falls on the writer to be cognizant of this interactive context, exercising her or his capacity "to produce kinetic energy in a textual performance" by engaging the agency an audience brings to its production.[15] Second, Cintrón explains topoi, or rhetorical lines of argument, in terms of energy. In theorizing democracy, Cintrón argues that commonplaces are "storehouses of social energy."[16] These "templates of argument," in rhetorical theory, have a "generative dimension" in that they function as the key way a rhetor compiles a public argument.[17] These linguistic reserves of social energy are the building blocks upon which a rhetor brings before the eyes inventional resources to actualize an issue and move people to act. In sum, and similar to Miller, the enactment of effective discourse requires unleashing stored energy.

Kennedy also influenced ecological understandings of rhetorical agency, in that he conceived this faculty for symbolic action as not reserved for humans but as a generalized bodily capacity of any animate being engaged with its environments. The energy trope, especially the energetic body, highlights Kennedy's corporeal conception of rhetoric as a bodily capacity. Girke and Meyer discuss the "energetic effort and involvement" of rhetoric as a bodily activity[18]; rhetoric "is an energetic and bodily, at least as much as a semiotic and mental, phenomenon… our intentions are constantly transformed in such energetic dialogues."[19] For Hawhee, whose work often focuses on the corporeal dimensions of thought, this generative potential of encounters between mobile bodies within an environment is often explained as energetic, or as transformational action. For instance, Hawhee describes the role of rhythm in Kenneth Burke's thought as "an enlivening force—sheer energy—with a unique capacity to mingle with and transform bodily energies and rhythms already churning, humming, and moving."[20] Bodily energies are ignited by the play of movements and rhythms that transform them. Scholars drawing on Kennedy's corporeal rhetoric, therefore, often use energy to connote the capacities and dynamism of a more distributive rhetorical agency, focusing on symbolic action in terms of ecological capacities, and how organisms adapt to and alter their environments. This focus has encouraged theorists to consider the capacity for action beyond the individuated human or, in the least, emphasize how the individual can act within a larger ecology of forces.

Capacities for symbolic action are therefore not simply a rhetor marshaling linguistic resources but involve nesting levels of active agents. That is, rhetorical capacity demarcates levels of complexity deeper than individual agency.[21] For instance, Byron Hawk asks how vitalism in terms of complexity can inform a rhetorical heuristic that treats the writing process as situated in a vibrant context of emergence. Hawk details how this view of vitalism conceives life as situated within ecological relations, or that life springs from "the relationships among…a complex combination of material, biological, historical, social, linguistic, and ultimately technological" bodies and their forces or processes of emergence.[22] The fundamental generative element of life, emerging between these multifaceted combinations, is described as "force, energy, will, power, or desire."[23] For Hawk, then, energy is a fundamentally generative force central to the constitution of life. This force relates to humans and nonhumans. In a similar line of thinking, Jane Bennett highlights the agency in human-nonhuman assemblages, using the term "thing-power" to underscore the "energetic pulse" between various vital elements and the "inexplicable vitality or energy" within ostensibly inanimate things.[24] In all, the energy trope spans the action of individual exertion to a generative force in the encounters between various elements.

Fields of Potential and Affective Energies

The energy trope helps theorize rhetoric ecologically, in which capacities for action distribute across a variety of channels. Further theoretical inquiries, especially those interested in circulation and emergence, turn to the energy trope to name the dynamic (and rhetorical) element in a field of encounter. For instance, in Jenny Edbauer's conception of rhetorical ecologies, rhetorics emerge within a structure of feeling or a social field of potential. Energies are the circulating affective qualities that permeate and intensify this social field. The social field's dynamism is characterized by "viral intensities that are circulating in the social field… What is shared between them is not the [rhetorical] situation, but certain contagions and energy."[25] Rhetorics emerge as effects of such circulating energies. Edbauer's emphasis on intensification and contagions situates rhetorical energies as forces, positioning rhetorical agency as decentered and emergent from the capacities available in the ecology of relations circulating and actualized. Rhetorical action emerges from the environment or plane of possibility; energy defines the force of possible action; taken together, structures of feeling, or the social field of potential, are the effects of energy, the latter defined as the dynamic activity of rhetorical processes.

Other scholars will refer to rhetoric's generative energies in terms of the overdetermined forces in any field of encounters. Catherine Chaput argues that one cannot know how the complexity within a social field of forces will be actualized.[26] For Chaput, energy names rhetoric's "connective source," aligning rhetoric and energy as both effects of circulation.[27] Persuasion, here, is a capacity of the social field from which actualized effects emerge out of the forces already circulating. For Chaput, an emphasis on circulation suggests how "some element moves throughout material and discursive spaces," and this element that moves is the energy of Chaput's primary concept, "rhetorical energy."[28] Energy names the mediating, connective, "persuasive power affectively sustaining the overdetermined ecology of our life worlds."[29]

These foci on rhetorical ecologies and capacities for action within a social field or ecology of relations relate to broader affect theories. Affect refers to capacities for action, distributed amid human and nonhuman actors engaged in complex encounters.[30] In affect theory that emphasizes encounters among humans, such as Teresa Brennan's work, affect functions as mediation that moves and draws bodies to (or away from) one another in the transmission of energies between them.[31] Affect overall, though, shifts the source of rhetorical action from a specific entity to capacities within an ecology or field. Jenny Rice and Catherine Chaput have referred to this shift as "affective energies."[32] Rice points out how "affect describes an energetics" not delimited to signification.[33] This affective energy suggests that the process of rhetorical circulation energizes and creates meaning making by tapping into overdetermined forces; in specific situations, affective energy takes flight in various directions. Energy, here, is a generative, emergent force. The next section explores how such emergence has been conceived in terms of transformative social change and in terms of controlling affective energies; it is the latter—in terms of fossil fuel promotional campaigns naming the energy company's fashioned ecology as enabling individuals' vital emergence—that defines these discourses' rhetorical materialism.[34]

Affective Energy as a Field of Transformation or Harnessed Potential

In that affective energy denotes possible actions that may or may not arise in a given social field, the energy trope has also named the possibility of transformation, both as a destabilizing force and as a new form of power. An emphasis on *overdetermination*—also referred to as a wrinkle, swerve, and other terms that suggest spillover from the expected, stabilized structures of meaning and activity—underscores the first set of considerations, in which energy denotes the transformational potential of rhetoric. For many rhetorical scholars reviewed here, energy often names a destabilizing force that allows for unexpected—but always potentially present—actions, responses, emotional resonances, and so forth. Rhetorical capacity comes, in part, from what circulates in the social field, or the energies that move in this space. Such energetic forces animate the social field, opening the possibility for redirection and change. For instance, Donovan Conley theorizes the swerve, the dialectic to gridded networks and logics, as an animating force or "raw energy of social transformation."[35] Christine Harold calls for the rhetorical prankster to jam with the social field in order to redirect or channel energies: "resistance can also enable and direct energy flows rather than merely thwart them."[36] Each of these treatments of energy-as-transformation reflect Edbauer's ecological conception of rhetorical energy, in which animate bodies engage in their environments and, within given environments, produce anew wrinkles in the social environment. The energy trope here represents a contrast with seemingly stabilizing structures and discourses, as in Brian Ott's reference to "the stabilizing structures and destabilizing energies that animate art."[37]

To turn Ott's phrase, what of stabilizing energies? Various stabilizing [power] structures also seek to harness energy in its dual sense—that is, as resources redistributed and channeled through energy systems, and as affect. For instance, Clough, Goldberg, Schiff, Weeks, and Willse refer to the explicit relationship between shifts in energy science and shifts in capitalist control. They argue that, while Marx wrote in a context in which the laws of thermodynamics informed efforts to minimize expended energy from the individual laborer, now capital seeks ways to learn from information-as-energy and preemptively control the field of what the authors label as "affect-itself."[38] Here, instead of Harold's conceptual framework in which energy flows are rhetorically channeled and redirected, energies themselves can be rhetorically captured and work for purposes of capitalist control.[39] It is my contention that explicitly turning to case studies related to such purposes can denaturalize a potential assumption that the energy trope is inherently destabilizing and transformational. Specifically, fossil fuel promotional texts are capturing energies, and naming their practices as necessary for individual agency and for enlivening a field of potential as itself a generative life force. Such rhetorical practices, in other words, preemptively control definitions of social activity as only understood through energy systems and practices. We can call these rhetorics of preemptive control stabilizing energies.

In all, how we talk about rhetoric mediates how we understand rhetorical action, agency, and capacities; in general, the scholarship engaged here employs the energy trope to explain ecological understandings of rhetorical capacity. Further, how we talk about energy resource systems also mediates rhetorical flows, informing how people think about bodies in social environments powered by such energy systems. The materialist texts of energy system promotional campaigns, therefore, can also be understood in terms of their rhetorical energies. John Sloop's rhetorical materialism illustrates this relationship through the energy trope, defining rhetoric as "the energy or flow mediating between bodies, body prosthetics, and 'semiotic conditions'."[40] Energy names the active elements of rhetorical mediation. While the rhetorical flows of this mediation will be overdetermined, it is still useful to examine how textual articulations promoting fossil fuels frame energy systems and circulate positive affective energies related to these systems. In their persistent circulation, such texts at least hold the potential to influence how people conceive of their everyday electrified and mobile practices. Such textual circulations, in other words, empower particular energy practices by naming these systems as the grounds of bodily capacities for acting in the world. In the next section, I discuss how this naming logic functions through another rhetorical conceptualization of potential and actualized energies.[41] This logic rhetorically captures the energy trope, naturalizing energy systems as generative and stabilizing enablers of affective energies.

The Energetic in Fossil Fuel Promotional Texts

Examining how texts promote energy systems illuminates some of the circulating rhetorical energies that infuse modern fossil fuel use with a sense of constancy and naturalness. The fossil fuel company does not simply have the capacity to circulate energy resources; it also circulates rhetorical energies that promote and legitimize ways of thinking about energy-related practices. In other words, the fossil fuel company does not simply supply the power source circulating within material infrastructures; it also supplies circulating discourses that name that which both enables action *and* constitutes vitality. These textual articulations turn on the trope of energy, linking various associations between energy systems and energies, work, action, and agency from an ecological-yet-mediated perspective. In fossil fuel promotional logics, "energy" names the generative force between an individual and her or his environment—that which affords life's vitalities—with that environment prepared by the company. Bodily agency, this logic suggests, emanates from this grounding.

This section is divided into two phases. First, I illustrate the oil company promotional logic of stabilizing energies, or how energy systems are said to offer generative infrastructures, legitimized as something worth maintaining and expanding. Second, Eric Jenkins' discussion of visual rhetoric, specifically the conditions of possibility from which discourse circulates, informs this pro-fossil fuel logic, which defines human potential through energy regimes.

Illustrating Energy Regimes as Stabilizing Energies

Rhetorical theory, as shown in the previous section, often foregrounds the energy trope as representative of destabilizing forces that open rhetorical spaces for change or redirection. Energies as intensities and instabilities, however, are also channeled into the stabilizing structures of purportedly generative activity. The framework of this turn, when promoting fossil fuels, moves from destabilizing to stabilizing energies. To turn Brian Ott's mention of "stabilizing structures and destabilizing energies" in art, the stable structure of energy regimes, the promotional argument follows, enables those energies that animate the lifeworld.[42]

A recent Chevron campaign illustrates this logic of solidity and facilitation. Typing in "doers.com" may take you to a longer URL: https://www.chevron.com/Doers/USA/. Overlaid on an aerial shot of highway arterial flows, the top of the lead page states, "It takes doers and a whole lot of energy to keep America doin'." There is an accompanying video that explains various articulated links between America, energy sources, and action.[43] Chevron's fast-paced video is characterized by a series of cuts and narrative digressions that reflect a sense of kinetic energy that defines American ingenuity. The narrative begins, "This is a story about doers," and proceeds to bounce around numerous elements that characterize American achievements and how "doing got this nation done," highlighting canals, artificial hearts, electric guitars, rockets to the moon, assembly lines, paper clips, dams, railroads, bridges, skyscrapers, hula hoops, blue jeans, "and that little thing we call the interstate highway system." This "land of the doers" is characterized by hard and innovative work, as "doers built this country." This frenetic narrative sets the stage of human activity, specifically through an American ideological framework that emphasizes individualized human agency and innovation.

This set-up then turns its attention to the capacities that enable such agency: "And all this doing: It takes energy." "All this doing" stands in for all of human activity, insight, and innovation—that is, the generative forces that constitute human life. "It takes energy" suggests strategic ambiguity or a dual referent. On the one hand, the phrase refers to an individualized sense of agency or the energetic spirit that is required in an individual's life to accomplish greatness. On the other hand, the phrase highlights the resources and energy systems required as grounds for these actions. "Doing" things, achieving, and activating one's dynamic potential are all possible in this world of action. At the same time, there are caretakers of this capacity. Chevron offers the necessary mediation between energetic doers—the human world—and energy resources—the vibrancy lying dormant in the natural world. The ad continues: "Energy got us here, and it's our job to make sure there's enough energy to keep doers doin' the stuff doers do to keep us all doing what we do." "Energy got us here" again functions in two ways. First, the energetic spirit characterizes human activity, and this spirit brought to America the fruits of such actions. Work "got us here" in the modern moment. Next, resources enabled this moment. That is, energy does not merely refer to a spirit of innovation—or at least not solely—but also to the continuous practice of fossil fuel extraction, harnessed and channeled into systems, and used. Energetic workers have put this vitality of the world to work. Therefore, it is "our job," or the work of Chevron's Doers, to maintain these grounds that support energetic activity. Energy rhetorics emphasize this dual function of the term: energy both as an energetic human spirit of hard work and as the energy company workers who prepare a world that enlivens that spirit. The energetic spirit, in other words, requires the stability of energy work.

Fossil-Fueled Conditions, Human-Centered Potentials

Energy, in this rhetoric, naturalizes the conditions for any potential human activity by emphasizing the work put into preparing and maintaining energy resource systems. In particular, the current system powered by fossil fuel resources is said to empower individuals of the modern world. Across promotional texts we see a common refrain: a series of people engaged in some form of social activity that, at least in part, relies on sources of energy. Oil Company Valero, in the commercial "What We Use," specifies and concretizes this source as the gasoline station.[44] The identifying "we" functions as an American identity grounded in that which gasoline has supplied, imaged by a series of supposedly everyday Americans standing (or sitting) in front of their vehicles: an individual male, a family, an ice cream truck driver, paramedics, a motorcyclist, and a military worker approaching a jet. Valero's "clean, quality fuels," the ad then proclaims, "is what we use to explore the open road…to keep our nation strong…[and for] powering the American dream." This literal power resource empowers individual goal attainment, whatever those goals may be. These everyday activities are depicted as the source of happiness; the energy system is then the source of these activities, and the energy trope the pivot on which energy systems become understood as the grounds for energetic activity.

To further explain this fossil fuel promotional logic—of the energy company and its energetic workers preparing the ground for people to actualize their energetic potential—I turn to Eric Jenkins' discussion of the difference between conditions of possibility and fields of potential.[45] Jenkins borrows this distinction from Brian Massumi, who uses the example of lightning.[46] The flash is the culmination, while the atmosphere that produces the flash is revealed as a field of potential that always held the conditions for a lightning strike—in addition to the conditions for other possibilities. That is, atmospheric conditions hold out multiple possibilities that may or may not reveal themselves. Conditions of possibility are therefore not actual manifestations but in the realm of the virtual, or the conditions that can make certain outcomes possible. As certain elements interact with others, some of these possibilities reveal themselves. In their enactment—as when lightning strikes, through specific atmospheric encounters—one sees one manifestation of potential outcomes. Jenkins turns to questions of the virtual and potentialities in seeking to understand how cultural phenomena are actualized and circulated. Jenkins' example is the Internet meme, only possible amid technological conditions that enable its circulation. The field of potential, on the other hand, suggests that how memes develop intersects with energies that were *already* circulating in the social field.[47]

Fossil fuel companies claim both to lay the conditions of possibility and to discursively ascribe the energy resource/system as merely enabling the consumer to unlock the fields of potential that were already capable of manifesting. The energy company, in its articulated logic, is the technological possibility, or the mode that makes modern life possible and that awakens the dormant vitality of the natural world. You, the consumer of this logic, are constituted as the individualized, autonomous user going about your everyday, dynamic, mobile life. Your everyday encounters with energy systems unleash vibrant opportunities. The world mobilized around energy systems, like atmospheric conditions, is revealed as a field of potential that always held the conditions for one to enact the generative energies that animate a person's life. Using gas from Valero to "power the American dream" reveals an encounter between the individual's desires and, similar to Hawk's discussion, their relations of complexity[48]; here, Valero sets the conditions in place for however that encounter might manifest. Put another way, in the first stage of Valero's (and others') logic, energy resources function as conditions of possibility. The car does not run without its fuel supply, a supply that does not arrive without a series of economic and political global networks. The second stage of this logic states that fossil fuel resources and their infrastructural networks ignite the actualized capacities already circulating in the social field; the resource is discursively ascribed as the source of culminating, dynamic, worldly practices, or the source of agentic potential. "Powering the American Dream" is described as an act of human agency, individually enacted, but only once the grounds—of gas stations and other infrastructural capacities—have laid their groundwork. Energy systems, here, combine elements and spark action.

In sum, energy promotional texts delimit the conditions of possibility—the vital world as prepared by the energy company—while suggesting that affective energies are abundant and life is infinitely generative *once these conditions are met*, or once the modern world has been prepared, maintained, and advanced via the fossil fuel company. Similar to themes in rhetorical theory, these associations conceive of energy in terms of human agency, a field of potential and ecological capacities, and transformations. However, in this rhetorical capturing of "energy-itself" the potential for transformative energies is only possible in a fossil fuel powered world, effacing the possibilities of a world outside these conditions.[49]

Energy, Actualized: ExxonMobil's *Energy Lives Here* Campaign

ExxonMobil's Energy Lives Here is part of this larger trend in energy promotional discourse, elaborated here because of the campaign's synthesis of the various elements at play as well as ExxonMobil's size and reach. The largest publicly traded multinational oil and gas corporation, ExxonMobil is a consolidation of the previously dissolved Standard Oil, recombining Standard Oil of New Jersey (Exxon) and Standard Oil of New York (Mobil). In late 2009 ExxonMobil announced its purchase of XTO, which effectively made the company's natural gas reserves nearly on par with its oil reserves.[50] Due to these recently expanded ventures, ExxonMobil began to heavily promote natural gas as the fuel of the future, able to power transport, digital devices, and a combination of the two.[51] The Energy Lives Here advertising campaign, lauded by the industry[52], extended ExxonMobil's promotional efforts by combining the company's historical focus on oil with its recent investments in natural gas, endorsing their work that supplies transport fuel and the expanding digital world's electricity.[53]

Below I highlight selected representative texts from the campaign, with particular emphasis on the introductory commercial that began its primary thematic, associating the energy company with that which constitutes life, and energy's ubiquity in the environment and in people's everyday practices. In addition to this commercial, I also analyzed the initial campaign web pages, including their embedded videos. Further, I examined a few additional examples as the campaign continued to expand on its themes: specifically, the commercials "Lights Across America," "Enabling Everyday Progress: Egg," "To Do List," and "Gordon is Energy." I retrieved the opening commercials and webpages through video and screen captures, and subsequent commercials were chosen in relation to how often they were broadcast.

Setting the Stage: Capturing an Ambient World of Energy-Itself

The campaign frequently employs an inventive visual scheme, specifically a visual light display, to thread together a relationship between energy sources, unseen at consumption point, and the mobile lives they claim to (em)power. One critic describes the campaign as a series of imagistic arguments "that helped connect quality of life in America with the energy that makes it all possible."[54] The campaign's particularly innovative component occurs in how it visualizes this connection between the daily devices that stand in for this "quality of life" and the energy resources that power them. The campaign employs a visual scheme that is difficult to describe verbally, but that might best be understood as an aurora. The Northern Lights (Aurora Borealis) are one well-known aurora phenomenon, a natural light display resulting from combined atmospheric conditions from the Earth and the Sun. From the particular vantage point beyond the Earth's magnetic poles, color schemes of light become visible. ExxonMobil deploys this scheme as its main motif to visualize energy resources as a sort of aurora. This production design saturates the whole campaign with a cool, tranquil lighting scheme that surrounds, encompasses, and diffuses the campaign's depicted networks of digital objects, transportation, and other everyday objects requiring energy resources.

For instance, the commercial that introduces the campaign opens with an up-close shot of a young female's finger pressing her iPad play button. This action causes an oval-shaped streak of light to quickly stretch across the iPad screen, resembling a wavelength force with undulating ripples, like a pebble's force stretched across the surface of a pond. Energy is activated, and the device awakens. Out of the central location where the finger initiated this charge, the lighting, like the flames of a fire arising in a brilliant hue, slowly ascends outward from the iPad, begetting further ripple effects spreading across the iPad screen. At the epicenter, where the finger activated the iPad, a white light shines. Panning back, the camera now shows the youth who initiated the action smiling at the mystical vitality occurring in front of her eyes. The fire-like light, flowing with yellow streaking lines, reddish waves, and an outline of blues and purples that adds a softer cooler shape to the lighting, enchants the user of the once dormant and now lively device. The commercial then cuts to an overhead shot of a cityscape. Colorful lights streak across the gridded roads. Purple and green vertical lines—perhaps reflective of emission lines—shoot up from the ground and vibrate. At the top of the screen, a purple and blue aurora extends out on the horizon. White lights spring forth from the ground and dissipate into the atmosphere above. The camera pans in, and the narrator begins: "If we could see energy, what would we see?"[55]

The focus turns to the uses of energy in a mobile, digital world, yet its added emphasis visualizes the energy presence that makes these uses, and the consumer's activation of the dynamic presence of these objects, possible. The rest of the commercial depicts various devices and cityscapes as they would look if one could see the consumed energy that flows through it. As the narrator puts it, "We'd see all the electricity, flowing, through the devices that connect us, and teach us," as well as "the billions of gallons of fuel that get us to work," the manufacturing of medical plastics, and the industry that supports it all (as well as "American jobs"). The "devices that connect us" are illustrated through iPads surrounded by a force field of light as well as lights that shoot out from the screens, as museum-going children read from these screens. When the narrator proclaims, "life takes energy," a concert scene visualizes sonic waves pouring from the stage. Concertgoers raise their phones to capture video of the show, and radio waves—currently imaged in the popular imaginary through the Wi-Fi icon—emanate from these phones. Later, a couple sits in a park sharing a pair of ear buds. Energy radiates out of the ear buds, and a streak of white light extends from this object as a colored, gridded plane, resembling mathematical visualizations, emerges out of this white light. Each visual offers a warm, pulsating glow to the objects from which they derive.[56]

This aurora trope follows the campaign throughout, from further cityscapes pulsating with these vibrant hues to the colored lights emanating from an air conditioning system as a little girl feels its cool air on her outstretched hands and face. Other daily scenes depicted include workers taking the subway and bus to work, people using smartphones to capture videos and look up information, and running a marathon while equipped with a wearable fitness tracker. In a subsequent commercial, "Natural Gas," aired early on in the campaign, the narrator proclaims that the fossil fuel source "can help provide the power for all this." "All this" remains open-ended but refers to whatever the consumer seeks to activate in the dormant object world of transport and the digital.[57]

Such visualization attributes an ambient quality to the singular, grounding source of energy systems. Thomas Rickert situates his theory of ambient rhetoric in relation to ancient theorizing on "the power of our surroundings, both material and spiritual."[58] ExxonMobil offers the aurora light display as both a means to visualize the power of the mobile world's material objects, and to suggest how energy resources act as a circulating power. This visualization calls forth a general mood to the environment of mobile communication and transportation systems, proffering a sense of ambience, yet one mediated by ExxonMobil and enacted by the individual. Energy physically carries the rhythms and sensations of the landscape. It emanates out of and animates the environment, through a colorful, flowing light starting with and surrounding various objects. These images of animating light function as conduits, with circuits of movement visible within the force fields and flows saturating the modern landscape. ExxonMobil visualizes this vibrancy or "power of our surroundings" toward two purposes: one, to suggest that such power is only possible through the continued practices of the energy company; and two, this power is merely the plane, prepared and maintained, on which the individual moves through the world. ExxonMobil applies technology, and the individual activates energy. The source merely needs maintenance so that activation can spread across the visual plane, animated by whatever uses humans make of this energetic field.

In offering the etymological origins of his treatise on ambient rhetoric, Rickert first considers how ambience has been defined as support, or "the arrangement of accessories to support the primary effect of a work."[59] Through its aurora imagery, ExxonMobil's campaign suggests that energy supports the primary work of infrastructural networks of transport and communication and the individual's use of these objects. "But," Rickert continues, the concept of ambience "does significantly more than that. It begins to convey more elusive qualities about a work, practice, or place. Often these are keyed to mood or some other form of affect."[60] The campaign calls forth these affective qualities to then channel them through the prepared grounds of energy systems. Its mood lighting embraces objects to represent their otherwise elusive qualities, such as their energetic qualities of movement, dynamism, and vitality. Renzi Stone, chairperson of a rival ad agency, attempts to explain the visual scheme as a combination of X-ray vision and astral plane, seeing beyond the eye's visible plane to an almost deified depiction of the urban scene.[61] The X-ray descriptor suggests how light, here an aurora, always surrounds us, whether or not we can see it. Potential, emergent capacities are brought before the eyes (*energeia*), but only because of the a priori work prepared by ExxonMobil.

The campaign highlights mobile daily habits to visualize the link between, on the one hand, an American way of life defined through daily practices, and, on the other hand, capacities enabled by the energy systems in place.[62] That is, the capacities in the encounters between individual and technological objects here are rendered possible by the mediating presence of energy systems and energy companies. The utility of these objects adds something to everyday life but this vitality, the argument follows, is awakened through the energy that flows through them. If we could see that, as the campaign depicts, we would see that the objects that brighten our world are enlivened through energy infrastructures. Another example from the beginning of the campaign highlights how this perspective on energy visualizes not the oil derrick or hydraulic fracturing of production, but the rhythms, movements, and sensations of consumption. On ExxonMobil's website, a textual synopsis explains the campaign through presence and sight: "Energy lives in everything we do. If you could see energy, where would you find it? The answer: everywhere."[63] Whether one sees it or not, the lumen of energy is visible in daily social practices, ever-present in the habits and spaces of consumption.[64] Amid this rarefied glow of energy—its illumination—the city and the people are abuzz. "Energy lives" in the animating presence of pulsations and vibrancies. ExxonMobil visualizes this presence of energy systems to promote corporate actions as the stable, immersive ground. ExxonMobil's presence is "everywhere."[65] The company argues that life requires ExxonMobil's supply, and indeed, one just needs to see that it has always been all around us, and hence will continue to be so.

Circulating Energies: Continuing the Energy Lives Here Thematic

The campaign has continued, adding new texts that circulate the company's themes of energy constituting life, and of ExxonMobil's ubiquitous presence stabilizing the field of potential activation. "Lights Across America" is a sixty-second commercial that relates the energy company to the plane on which the (American) consumer goes about day-to-day functions.[66] The advertisement opens with a moving panoramic image approaching a lighthouse. As the camera is about to pass by the lighthouse, a light turns on. Then a series of hands flip on various switches at work, home, a coffee shop, and a softball field. As two runners are passing down a dark road, their movements triggering sensors and activating the lampposts above, the narrator states, "You might not even think about the energy that lights up your world. But we do." The nighttime runners—even less so than the person who consciously flips the light switch—may not consider the source of light as they go about an activity that, unto itself, does not initiate electrification but does depend upon such a system being in place. ExxonMobil, the argument follows, continues to think about that which "you" may not. "We," specified as ExxonMobil, are said to secure the infrastructure that maintains "the energy that lights up your world," visually depicted as a three-dimensional modeling image of an extraction site, to an actual extraction site, a generating station, an electric grid, and city lights.[67] They "do" this work "because turning on the lights isn't as simple as just flipping on a switch," a claim made as a mother turns on her toddler's nightlight.

"The energy that lights up your world" classifies energy as the ground on which you, the consumer/viewer, activate a field of possible actions. Here, that ground is promoted, attended to, and given a source. Consider the mother who turns on her child's nightlight. She is focused on the light her child needs. She represents you and your agency to activate that ability—to, say, offer security to a child—because of the series of conditions that ExxonMobil supplies, from conception to gridded infrastructure. This field of activation or potential is conceived as the dynamism, or the diversity, of life and its experiences, whether in the home (the child), or work (office space), play (softball field), or both (coffee shop). Energy constitutes the field on which one lives a modernized life and through which possibilities occur. Here we have a first level of the energetic found in fossil fuel promotional logics. Energy is the vibrant world of possibility. The second level occurs when ExxonMobil claims that one's consumerist world of emergent potential starts with the company's presence and activity. ExxonMobil argues that it itself is the ground, or that individual agency is possible through ExxonMobil. The argument is not that ExxonMobil gives energies so much as that it supplies the infrastructure on which one is enabled to act. These stabilizing energies are depicted to empower the *possibility* of human energies, or individual agency actualized.

The company and energy system, on these terms, is the connective tissue between the energetic environment and human activity. The advertisement "Enabling Everyday Progress: Egg" juxtaposes a woman boiling an egg with a montage of the corporation finding and exporting the energy required to transport the carton of eggs, to heat the stove, and so on.[68] Here, ExxonMobil constructs its persona less around supplying oil and other fossil fuels and more around supplying the infrastructural capacity to convert that oil into the heated water that boils one's egg. ExxonMobil supplies the grounds on which individuals are able to act—for instance, turn the knob and boil the egg—and hence offer the claim that the company is "enabling everyday progress." The commercial "To Do List" further illustrates this point, as the ad depicts ExxonMobil workers who engage in activities ranging from mapping oceans to fueling the global economy, concluding, "And you thought we just made the gas." Instead, their work centers on "one purpose: power the world/ responsibly."[69]

The totality of the campaign goes further, arguing that this empowering ground is something the audience does not think about and, because ExxonMobil is hard at work on preparing this ground, does not have to consider beyond a surface level. That is, the campaign invites its audience to ascend to the narrative argument that the vitality of one's everyday life is rendered possible by modern energy systems. How a person creates one's life within that framework is up to the individual; that is, individual agency is exercised against the backdrop of that which ExxonMobil supplies. Some commercials were structured around an individual's daily habits and concluded by defining that person in terms of energy. Each of these commercials is titled, "[Name] is Energy." For instance, "Gordon is Energy" opens with a moving car, and its visualized energy, against a cityscape saturated in a gray, muted filter. "This," the narrator begins, "is Gordon," shuttling downtown on work calls. We then see Gordon in his car, looking off in the distance. The outside lining of his car is dotted with what looks like miniature flashbulbs of light along with swirls of purples and greens and a blue tint. On the windowpane, streaks of bowed purple lines seem to catch Gordon's attention. Then the shot cuts to his car making a quick right turn, with a multi-covered streak of light functioning as a sort of jet stream, engulfing his vehicle and tailing off as the car moves past the spaces it left behind. The car and its visualized streaks of energy are offered in quick succession as Gordon goes to "meeting, after meeting, after meeting." All this travel, the narrator claims, is why he chooses ExxonMobil and its efficient fuels. At this point, the commercial switches from work to play. Filling his tank, Gordon loosens his tie, opens his trunk, and drops his work clothes next to his drum set. Gordon relies on ExxonMobil to also do what he loves; ExxonMobil supplies reliable fuels "so he can make it to the meeting that matters most," that is, music practice. With the car back on the road, blue streaks to its side, the narrator proclaims, "Gordon is Energy." The energy systems animate energetic activities, whether at work or whatever one hopes to do in life. Again, the commercial ends, "energy lives here," with energy denoting the energetic individual *and* the presence that makes it all possible.[70]

In all, the campaign's circulating discourse energizes a way to read ExxonMobil as the naturalized support for autonomous individual action. This rhetoric functions to stabilize energies, literally around energy systems. That is, the campaign names human vitality as funneled through the fossil fuel company, the supplier and igniter of individual agentic potential.

Conclusion

Taking an ecological approach to rhetorical criticism, this chapter follows Sloop's rhetorical materialism heuristic by examining how circulating fossil fuel discourse encourages audiences to think about themselves as made vital by their energetic environments. As seen with Sloop defining rhetoric as the "energy" between such discourses, bodies, and bodily prosthetics, this chapter further focuses on the ubiquity of this energy trope to elaborate on how energies are named in fossil fuel promotional rhetorics.[71] In particular, contemporary fossil fuel promotion names oil and gas companies as energy companies. This tropic turn highlights a naming practice that associates energetic human potential with the conditions made possible by the energy company. This turn profits by conjoining vitality (energies) with specified grounds for such vitality (energy systems). Rhetorically invoking the generative emergence of affective energies emanating only once the system—maintained through oil and gas companies—is in place, the energy company, as such, prepares the ecology of potential activity. These campaigns circulate positive energies, amid bodily practices that continuously consume energy, and define the contours through which human agency emerges. This argument assembles dual conceptions of energy seen in rhetorical theory: as individual agency actualized by engaging in one's environment; and as the dynamic presence available in an environment of relations and encounters. The argument combines these conceptions in a way that naturalizes energy systems as the environment from which relations and agency—life's activity—emerge.

This chapter calls on theorists to consider how our concepts circulate in public discourse. Such theorizing is not removed from energy practices but is part of the rhetorical climate that encourages people to think about the vital quality of both life and of fossil fuels, in that each is categorized as energetic.[72] This chapter does not necessarily argue that rhetorical theory ignores the stabilizing function of rhetorical energies; instead, it supports further theorizing how this stabilizing function intersects with rhetorics *about* energy. As Roger Stahl concludes his analysis of the information bomb metaphor, "[Rhetorical theorists] should be mindful of the ways public discourse…may be shaping and underwriting certain theoretical trends…the task is to denaturalize the metaphor, call it out, and turn it into an object of scrutiny."[73] In general terms, Stahl reminds theorists that critical inquiry should strive for perpetual reflexivity. More specific to this chapter, I argue that the task here is to draw attention to the theoretical and public circulations of energy, both as a trope and as a sociopolitical practice. What circulates in public discourse is the energy trope that turns the oil and gas company into the source of affective energies. I call on rhetorical theorists to further recognize and denaturalize how the energetic circulates materially and discursively.

The enchanted, sanitized version of Energy Lives Here and other promotional campaigns, of course, effaces the full reach of fossil fuel ubiquity. If we could see energy, we indeed would find it everywhere: in the mining operations and pervasive sites of extraction; in the pollution spikes and fine particulates adjacent to refineries; in the greenhouse gases saturating the atmosphere; in the numerous oil spills small enough to legally avoid public acknowledgement; in the fracturing of earth to unleash natural gas below ground, threatening water contamination. Concurrently, if we could see rhetorical energies, we too would see everywhere transformative potential for social change. Conditions for such transformative energies are not predetermined; activity need not pass through modern energy systems. However, material practices, including the materiality of circulating promotional discourses, help naturalize current energy systems as conditions of possibility.

Promotional discourses capture a vision not merely of the fossil fuel system but also of what is meant by *energy* itself. If we can conceive of our episteme as a fossil fuel era, should we be surprised that theoretical questions of what constitutes life, vitality, and energy would fold back on to and through fossil fuel resource systems and their rhetorical circulations?[74] That said, the question of what constitutes life, in its complexity and generative elements, is not automatically tied to this particular moment's political, social, and economic makeup, particularly in those terms—of individualized agency predicated on automobility[75]—championed in the textual examples above. Haarstad and Warnik's argument, previewed earlier, refers to how social science scholarship theorizes energy in terms of systems, a default focus that implicitly naturalizes the stability of our current energy practices; instead, Haarstad and Warnik offer, energy practices have the potential for sudden transformations.[76] Attending to our theoretical tendencies and to the rhetorical energies of energy practice—both how they are stabilized in circulating discourses, as analyzed here, and how they may be transformed—can denaturalize fossil fuel systems as well as reveal transformations outside of these systems.

Notes

1. "About," Fueling U.S. Forward. Accessed November 3, 2016. http://www.fuelingusforward.com/about/.
2. Grounds also can refer to electricity and conduction with the earth. Like the stabilizing forces that redistribute electrical charges, this chapter examines how rhetorical discourse redistributes this specified relationship between fossil fuels, companies, and human potential. Alternative ways of distributing agency, in the context of energy discourse, must consider these argumentative grounds.

3. Brian L. Ott, "The Visceral Politics of *V for Vendetta*: On Political Affect in Cinema," *Critical Studies in Media Communication* 27, no. 1 (2010), 41.
4. Håvard Haarstad and Tarje I. Warnik, "Carbonscapes and Beyond: Conceptualizing the Instability of Oil Landscapes," *Progress in Human Geography* 0, no. 0 (2016), 1–19.
5. Roger Stahl, "Weaponizing Speech," *Quarterly Journal of Speech* 102, no. 4 (2016), 390.
6. Cheryl Geisler, "How Ought We to Understand the Concept of Rhetorical Agency? Report from the ARS," *Rhetoric Society Quarterly* 34, no. 3 (2004), 9–17.
7. Felix Girke and Christian Meyer, "Introduction," in *The Rhetorical Emergence of Culture*, ed., Christian Meyer and Felix Girke (New York: Berghahn Books, 2011), 7.
8. Debra Hawhee, "Kairotic Encounters," in *Perspectives on Rhetorical Invention*, eds., Janet M. Atwell and Janice M. Lauer (Knoxville: The University of Tennessee Press, 2002), 16–35.
9. Ibid.; also see Barbara Biesecker, "Michel Foucault and the Question of Rhetoric," *Philosophy and Rhetoric* 25, no. 4, (1992): 351–364.
10. Similar to how numerous scholars have emphasized definitions of rhetoric around what rhetoric does (as opposed to what it is), this emphasis implies that the consequences invited by discourse make rhetoric empirically observable; see, John Louis Lucaites and Celeste M. Condit, eds., *Contemporary Rhetorical Theory: A Reader* (New York: Guilford Press, 1999).
11. George A. Kennedy, "A Hoot in the Dark: The Evolution of General Rhetoric," *Philosophy and Rhetoric* 25, no. 1 (1992), 1–21.
12. Each state their use of the energy trope "parallels" or is "reminiscent" of Kennedy's definition; Ralph Cintrón, "Democracy and its Limitations," in *The Public Work of Rhetoric: Citizen-Scholars and Civic Engagement*, eds., John M. Ackerman and David J. Coogan (Columbia: University of South Carolina Press, 2010), 101; Carolyn R. Miller, "What Can Automation Tell Us About Agency?" *Rhetoric Society Quarterly* 37, no. 2 (2007): 154.
13. Miller, "What Can Automation," 147.
14. This heuristic sets up how automated grading systems create new understandings of these agentic interactions; Miller, "What Can Automation." In explicit contrast to Miller, Marilyn Cooper argues "that deeds are always done by someone, and replacing the doer of the action, the agent, with an amorphous force like kinetic energy leaves us with no basis for assigning responsibility for actions." Marilyn M. Cooper, "Rhetorical Agency as Emergent and Enacted," *College Composition and*

Communication, 62, no. 3 (2011), 438. Here I argue how, not simply an amorphous force, energy is often named.

15. Miller, "What Can Automation," 149.
16. Cintrón, "Democracy and its Limitations," 100.
17. Ibid., 101.
18. Felix Girke and Christian Meyer, "Introduction," 6.
19. Ibid., 14.
20. Debra Hawhee, *Moving Bodies: Kenneth Burke at the Edges of Language* (Columbia: The University of South Carolina Press, 2009), 28; also see page 29.
21. On capacity versus agency, see: Nathan Stormer and Bridie McGreavy, "Thinking Ecologically about Rhetoric's Ontology: Capacity, Vulnerability, and Resilience," *Philosophy & Rhetoric*, 50, no. 1 (2017), 1–25.
22. Byron Hawk, *A Counter-History of Composition: Toward Methodologies of Complexity* (Pittsburgh: University of Pittsburgh Press, 2007), 5.
23. Ibid, 5.
24. Jane Bennett, *Vibrant Matter: A Political Ecology of Things* (Durham, NC: Duke University Press, 2010), 24; 18.
25. Jenny Edbauer, "Unframing Models of Public Distribution: From Rhetorical Situation to Rhetorical Ecologies," *Rhetoric Society Quarterly* 35, no. 4 (2005), 14.
26. Catherin Chaput, "Rhetorical Circulation in Late Capitalism: Neoliberalism and the Overdetermination of Affective Energy," *Philosophy and Rhetoric* 43, no. 1 (2010), 1–25.
27. Ibid., 13.
28. Ibid.
29. Ibid.
30. Seigworth and Gregg define affect as "the capacities to act and be acted upon" as bodies encounter one another; Gregory J. Seigworth and Melissa Gregg, "An Inventory of Shimmers," in *The Affect Theory Reader*, eds., Melissa Gregg and Gregory J. Seigworth (Durham, NC: Duke University Press, 2010), 2.
31. Teresa Brennan, *The Transmission of Affect* (Ithaca, NY: Cornell University Press, 2004). The area of spiritual communication includes such understandings of the energetic. For instance, Jason Del Gandio theorizes the human vibe as an "energetic" force, or a quality of shared vibrations "of bodily *force*, *energy*, and/or *presence* that is emitted by one particular body and felt by another particular body" in given situations (emphasis in original); Jason Del Gandio, "From Affectivity to Bodily Emanation: An Introduction to the Human Vibe," *PhaenEx* 7, no. 2 (2012), 49.

32. Chaput describes rhetoric as a mobile encounter amid complexity—the circulation and "social connectivity of affective energy" in which various audiences attach added value to the flow of discourse; Chaput, "Rhetorical Circulation," 8.
33. Jenny Edbauer Rice, "The New 'New': Making a Case for Critical Affect Studies," *Quarterly Journal of Speech* 94, no. 2 (2008), 201.
34. John M. Sloop, "People Shopping," in *Rhetoric, Materiality, and Politics*, eds., Barbara A. Biesecker and John Louis Lucaites, 67–98 (New York: Peter Lang, 2009).
35. Donovan Conley, "Grid and Swerve," *Critical Studies in Media Communication* 27, no. 1 (2010), 25.
36. Christine Harold, "Pranking Rhetoric: 'Culture Jamming' as Media Activism," *Critical Studies in Media Communication* 21, no. 3 (2004), 197.
37. Ott, "Visceral Politics," 41.
38. Patricia Ticineto Clough, Gregg Goldberg, Rachel Schiff, Aaron Weeks, and Craig Willse, "Notes Towards a Theory of Affect-Itself," *Ephemera* 7, no. 1 (2007), 60–77.
39. Rice further states that types of work function "as affective energy that drives new forms of capitalism," with capitalist workers producing "the circulation of consumptive energies"; Rice, "The New 'New'," 7.
40. Sloop, "People Shopping," 70.
41. Eric S. Jenkins, "The Modes of Visual Rhetoric: Circulating Memes as Expressions," *Quarterly Journal of Speech* 100, no. 4 (2014), 442–466.
42. Ott, "Visceral Politics," 41.
43. "America, Land of the DOERS (30s)," Chevron. Accessed April 12, 2017. https://www.chevron.com/Doers/USA.
44. "What We Use," Valero. Accessed April 12, 2017. https://www.ispot.tv/ad/ASSj/valero-what-we-use.
45. Jenkins, "Modes of Visual Rhetoric."
46. Brian Massumi, "Introduction: Like a Thought," in *A Shock to Thought: Expression After Deleuze and Guattari*, ed., Brian Massumi (London: Routledge, 2002), xiii–xxxix.
47. Jenkins, "Modes of Visual Rhetoric."
48. Hawk, *A Counter-History*.
49. Borrowing from "affect-itself," Clough et al., "Notes Towards."
50. Steve Coll, *Private Empire: ExxonMobil and American Power* (New York: Penguin Press, 2012). For the long history of "associated gas" or gas from an oil field, from a perceived waste product to becoming cost-competitive, see: Morgan Downey, *Oil 101* (New York: Wooden Table Press, 2009); and Martin V. Melosi, *Coping with Abundance: Energy and Environment in Industrial America* (Philadelphia: Temple University Press, 1985).

51. “ExxonMobil: Unlocking a Century’s Supply of Natural Gas,” ExxonMobil. Accessed November 5, 2016. https://www.youtube.com/watch?v=XDfVycbnaBc.
52. Matt Shaw, “Campaigns We Loved: A Look Back at 2013,” Council of Public Relations Firms. Accessed September 16, 2016. http://prcouncil.net/voice/2013/campaigns-we-loved-a-look-back-at-2013.
53. Their commercials end with the logos for Exxon, Mobil, and XTO.
54. Shaw, “Campaigns We Loved,” para. 6.
55. “Energy Lives Here™ Anthem,” ExxonMobil. Accessed April 12, 2017. https://www.youtube.com/watch?v=FZ3S2EOBbwE. ExxonMobil now refers to this video as “Energy in Our Lives”; see ExxonMobil. Accessed April 12, 2017. http://corporate.exxonmobil.com/en/company/multimedia/energy-lives-here/video-energy-in-our-lives. See author for screen and video captures of the campaign circa 2014.
56. Ibid.
57. “Natural Gas,” ExxonMobil. Accessed April 12, 2017. https://www.youtube.com/watch?v=_s80Hbac2b8.
58. Thomas J. Rickert, *Ambient Rhetoric: The Attunements of Rhetorical Being* (Pittsburgh, PA: University of Pittsburgh Press, 2013), 6.
59. Ibid., p. 6.
60. Ibid.
61. Shaw, “Campaigns We Loved.”
62. As interpreted by Shaw, “Campaigns We Loved.”
63. The version of the site in November 2016 includes the line, “The need for energy is universal.” “Energy Lives Here: An Introduction,” ExxonMobil. Accessed November 3, 2016. http://corporate.exxonmobil.com/en/company/multimedia/energy-lives-here.
64. See Martin Jay, *Downcast Eyes: The Denigration of Vision in Twentieth-Century French Thought* (Berkeley: University of California Press, 1994), 29.
65. “Energy Lives Here: An Introduction,” ExxonMobil.
66. “Lights Across America,” ExxonMobil. Accessed April 10, 2017. https://www.ispot.tv/ad/AIOJ/exxon-mobil-lights-across-america.
67. While overall this chapter argues that such depictions are sanitized, they are not completely absent, just secondary to the visualization of consumption.
68. “Enabling Everyday Progress: Egg,” ExxonMobil. Accessed October 30, 2016. https://www.youtube.com/watch?v=WrH19H-J4pU.
69. “To Do List,” ExxonMobil. Accessed November 3, 2016. http://corporate.exxonmobil.com/en/company/multimedia/energy-lives-here/video-the-many-jobs-of-exxonmobil.

70. "Quality Fuels: Gordon is Energy," ExxonMobil. Accessed April 12, 2017. https://www.youtube.com/watch?v=0iAORxk5Ybk.
71. Sloop, "People Shopping," 70.
72. See in particular, Hawk, *A Counter-History.*
73. Stahl, "Weaponizing Speech," 390.
74. Thomas Princen, Jack P. Manno, and Pamela L. Martin, eds., *Ending the Fossil Fuel Era* (Cambridge: MIT Press, 2015).
75. For an elaboration on cultural politics linking individualism, oil, the meaning of life, and the reproductive realm of capitalism, see Matthew T. Huber, *Lifeblood: Oil, Freedom, and the Forces of Capital* (Minneapolis: University of Minnesota Press, 2013).
76. Haarstad and Warnik, "Carbonscapes and Beyond."

CHAPTER 13

Afterword: Working in an Ecotone

Nathan Stormer

Even within the circle of the special science we may find diversities of functioning not to be explained in terms of that science. But these diversities can be explained when we consider the variety of wider relationships of the pattern in question.

Alfred North Whitehead, *Modes of Thought*

Not everything within the circle of ecology is ecological and, similarly, not everything within the circle of rhetoric is rhetorical. As discussion of ecology within rhetoric blossoms, this volume demonstrates that each functions within the other so we ought to "consider the variety of wider relationships of the pattern in question," as Whitehead advises. One can and should view this collection as a series of steps in a rapidly moving dance between two fields, but I prefer to consider the way this volume epitomizes a distinctively fluid, transitional space that brings together scholars who would not normally collaborate. Doing so, I discuss the commingling of rhetoric and ecology as a fluctuating margin that forms its own environment.

N. Stormer (✉)
University of Maine, Orono, ME, USA
e-mail: nathan@maine.edu

B. McGreavy et al. (eds.), *Tracing Rhetoric and Material Life*,
Palgrave Studies in Media and Environmental Communication,
https://doi.org/10.1007/978-3-319-65711-0_13

Identified circa 1900, an ecotone is a transitional environment that bounds two adjacent communities of terrestrial species, such as two kinds of forest, although its usage is wonderfully plastic and has been molded to fit ecosystems like riverbanks and estuaries, as well as varying scales of environment. The suffix, tonus, names a zone of strongly varying stresses. In more rhetorical terms the passage between vital communities generates different styles of being marked by a particular tonal dynamism, which expresses those communities but is nonetheless distinct. It is ecology in the middle voice.

I drafted part of this text during a short residency near Schoodic Point, a rocky, intertidal ecotone in Maine, the type of environment that Bridie McGreavy discusses and which Diane Keeling and Jennifer Prairie mention. Orange-pink granite streaked with black basalt lies in a massive, broken jumble stretching over a kilometer, stoically engaging in a daily give and take with the Atlantic. Of this kind of "tension zone," van der Maarel writes:

> The ecotone connects the always wet lower system with the always dry upper zone, itself being alternatively wet, dry, or intermediate. Only few species are adapted to such strong fluctuations and thus the ecotone is poor in species and, as far as plant species are concerned, their patterns are coarse-grained...The species assemblage may contain elements from either system as well as [its] own species.[1]

The tides flow in and out, yet the relations between marine and terrestrial zones are highly changeable (wave patterns, the leading edge of wet-dry phases, are regular yet mesmerizingly original, for example). The coastline undergoes constant fluctuation within a relentless tidal cycle.

Ecotones are "anomalous" in Deleuze and Guattari's terms and epitomize a "phenomenon of bordering":

> a multiplicity is defined not by the elements that compose it in extension, not by the characteristics that compose it in comprehension, but by the lines and dimensions it encompasses in "intension." If you change dimensions, if you add or subtract one, you change multiplicity. Thus there is a borderline for each multiplicity; it is in no way a center but rather the enveloping line or farthest dimension. . .[2]

As the enveloping line between multiplicities, the tensions ecotones display exemplify *becoming-intense*, meaning that the mingling of communities changes both. In the case of Schoodic Point, we have land becoming-ocean and ocean becoming-land depending on the hour of the day, the day of the month, the month of the year. Deleuze and Guattari write, "[w]aves are vibrations, shifting borderlines" that deterritorialize the adjacent zones even as they bound land and sea.[3]

Heeding Keeling and Prairie's call to view rhetoric and ecology symbiotically, it is profitable to think not only of specific unions between them but also of the dynamics that characterize the variety of their union, vibrating between *rhetorical ecology* and *ecological rhetoric*. In her watershed article, Jennifer Edbauer (Rice) defines rhetorical ecology as "co-ordinating processes moving across the same social field and within shared structures of feeling" that cannot be "properly segmented into audience, text, or rhetorician." She describes affect ecologically, such that feeling constitutes environments of "distribution, concatenation, and encounter" that "comprise and extend rhetorics."[4] Rhetoric's intension, the galvanizing dynamism of its multiplicity, is changed by recognizing an ecological dimension to the rhetorical situation, as Edbauer (Rice) argues. A distinction between human social environments and nonhuman ones remains active for her, if muted. This is ecology becoming-rhetorical where rhetoric is closer to high tide and ecology to a transitional minimum.

Ecological rhetoric names an intensification of rhetoric becoming-ecological; where rhetoric approaches low tide and ecology its transitional climax. Kenneth Burke's invocation of ecology, which McGreavy develops, is a useful example. Influenced by ecological explanations for the Dust Bowl of the 1930s, Burke incorporated a sense of ecosystem balance into his critique of capitalism, including his reading of neo-Malthusianism's political economy (wherein an otherwise healthy population produces collapse if it is imbalanced with its environment). Moreover, according to Marika Seigel, Burke's comic frame is fundamentally ecological in that it aims at creating balance (sustainability and resilience in contemporary terms).[5] Indeed, the aesthetics of balance/equilibrium that permeate ecology, continually contrasted with imbalance/disequilibrium, is an unexplored frontier of ecological rhetoric.[6] Tonally, the tension between society and nature is slack in ecological rhetoric, if it exists at all. In this way, rhetoric is an aspect of living among and between communities of existents in the world. Did Burke not say that all living things are critics, noting that the trout learns to avoid the hook?[7] Conversely, Seigel observes that early ecologists viewed living systems sociologically, using community and society as concepts, which evokes thoughts today of assemblage or actor-network theory.[8] Rhetorical theory increasingly entertains more ecological conceptions of rhetoric, notably Rickert's *Ambient Rhetoric*, but we have not yet seen low tide, I think.[9]

The tensions on display in an ecotone are not headed toward a final synthesis because an ecotone is an environment, not a dialectic. High tide is not the antithesis to low tide's thesis so one does not resolve an ecotone into a terminal abstraction; one inhabits it. In that spirit, rather than paint a tidy picture of an amalgamated rhetoric and ecology, the authors of this volume enact tonal variations of two fields phasing into one another.

First and foremost, all authors insist (in their own ways) on mutual immanence.[10] Rhetoric does not overtake ecology or vice versa; each is embedded in the other, existing together, not apart. Deleuze and Guattari expose the fact that multiplicities like rhetoric and ecology change one another through their mutual grasp, establishing a farthest dimension for each that affects their shape and character. Danielle Endres provides a clear example of mutual immanence in her analysis of the Nevada Test Site (NTS) and nuclear colonialism. She describes nuclear colonialism in terms of a material-discursive rhetoric that entwines humans, the more-than-human world, land communities, and technology. Beyond traditional forms of discursivity lies the material, animate rhetoric of violence and injustice from 928 nuclear detonations. The existence of the NTS as a wasteland, for instance, depends on the material-discursive entanglement of bombs, creatures, and land, just as the resilience of life within the NTS enacts a material-discursive resistance to nuclear colonialism.

Brian Cozen makes a parallel move in his discussion of Exxon Mobil's Energy Lives Here campaign when he argues for "rhetoric-as-energy and energy-as-rhetorical," which conserves but strategically confuses the difference between energy and rhetoric (as Endres does with materiality and discourse). Cozen argues that rhetoricity channels affective resources into and through discourse. Rhetoric is a connective force that conducts energy circulating at large, making rhetoricity at once a resource to manage and the practice of managing that resource. Energy, in the end, is that which affords movement where rhetoric is the shaping, discriminating circulation of energy. We have two forces coexisting, yet discriminated, that catalyze and direct affect within and across vital communities.

Instead of materiality or energy per se, Joshua Ewalt accentuates dynamism, particularly the movement of different materialities as they arrange and re-arrange networks of intra-active relationships. Such fluctuating arrangements afford an ecological, regional view of rhetorical invention because, Ewalt explains, invention is incited by changing spatial configurations. He uses the example of a guided prairie walk to describe how a shifting assemblage of actants constitutes contingent performances of rhetoric, all of which emerge in response to (and elicit responses through) their tangled intra-activity. What separates one performance from another as rhetoric is a functional arrangement of affective intensity that changes as things move together, modifying time and place.

Endres, Cozen, and Ewalt make the premise of mutual immanence their focal point: Rhetoric emerges from material environments, and environments, in their materiality, are rhetorical. It is worth noting that in their works the expression of rhetoric is experienced as some sort of extra-textual discursivity, whereas the expression of ecology is experienced as milieu. Endres stresses the material connectivity between rhetoric and ecology; Cozen stresses the vitality, the causative potential that binds the two; and Ewalt stresses shifting arrangements that cultivate ever-changing rhetorical moments and that affect the "ecology of encounter" that sustains those very arrangements.

Rhetoric and ecology are coeval dimensions of processes by which the world changes, grows, and perishes in response to itself, so accounting for the conjunctures of these two dimensions as well as their distinctions is the challenge. It follows that if rhetoric and ecology are mutually immanent, they must occupy ontological commonplaces, with bombs, fuel, and prairie identified as such in the previous examples. Urban neighborhoods, dammed rivers, jugs, footprints, mimic octopi, bicycles, GPS, and mudflats also make their appearance in these essays, but these do not explain the ontological commonality between rhetoric and ecology. The commonplaces I speak of are not specific to physical locales or things, but to convergent qualities: appetition, relationality, affectability. These qualities are expressed stochastically, showing up differently contingent on the actual occasion one is concerned with.

Conversely, these fields must continuously engage in ontological separation in that some distinction between rhetoric and ecology arises persistently, despite the character of the distinction also being subject to stochastic variation. The separation occurs not in this or that exact circumstance but through ontological qualities that diverge, continuously and differently: addressivity and interdependency. One can say that being addressed is conditioned by a milieu, but the sense of importance is lost if we say that an act of address is an environment. Here I would include a vole convinced by fire to seek other lodging as being addressed, just as I would a voter convinced by a demagogic nominee to find another political party. The fire/demagoguery proposes action, but it's proposition is not environmental unless it is situated in a field of interdependencies. The ontological accent on relations, affect, and appetite enables a rhetorical emphasis on addressivity to mingle with but not disappear into an ecological emphasis on webs of connectivity, and vice versa. Qualitatively, the intertidal is always powerfully similar in its difference; a rhetoric-ecology ecotone is so as well.

Where Endres, Cozen, and Ewalt are most concerned with appetition amid affective relations (animacy, resource husbandry, and invention being ways of describing ontological drive), Samantha Senda-Cook and George McHendry, Jr. highlight the affective immersion of relationality with their focus on habitus. They explain that activists in Omaha, who organized a bicycling and local food event, embody a tactically resistant rhetoric to urbanization and to environmentally unfriendly development. The event of the activists' movements empowers addressivity across groups of residents (producing ideological identification and cultural capital) at the same time that they rely on and feed back into the interdependency of urban Omaha. Habitus allows one to see how activist rhetoric and urban ecology converge, ontologically dependent on the same affective relations that make up life in Omaha.

A similar concern, with a different tonality, characterizes Anthony Stagliano's analysis of the Transborder Tool created to assist immigrants trying to cross the US–Mexico border. He locates rhetorical potential in the ways that the affective intensities circulating in an environment are re-molded into new forms of association, such that changing relations between entities of all sorts aesthetically realizes rhetorical transformation. The desert, the phone, the border apparatus, and the human user all become different through new conjunctures (or, more precisely, become-intense differently), "infecting" one another and so creating a new attunement to border crossing, a unique form of rhetoric. Further, the ecology is itself rhetorically generated by the composition of affective relations. The chora affords new rhetorics to emerge and those rhetorics afford changes in the chora.

An even stronger emphasis on interdependency in relation to address marks Caroline Gottshalk Druschke's and Candice Rai's meditation on humans, dams, and herring acting on one another. Narrative, fish swimming upstream, policy argument, a dam blocking the herring's passage, calls for help among humans, tossing fish over the dam: All of these elements are part of a complex ecology crystallized in an unusually vivid display of cross-species obstruction and co-operation. Rhetoric here is endemic to the forces coursing through these encounters in that actants of many sorts are moved by each other, not just by being attuned to the forces involved but by being engaged with and by those forces. If rhetoric is "equipment for living," then rhetoric here is experienced as ecology for ecology. Or, instead of conceiving rhetoric as epistemic, it is feeling rhetoric as ecosystemic.

Thomas Rickert stresses ecological attunement in relation to rhetoric through the cosmologies of Empedocles and Heidegger, which resonates with Gottshalk Druschke and Rai, although he emphasizes that addressivity cannot be limited to human polemical forms and must also be understood as a felt dimension coincident with dwelling (in a Heideggerian sense). Address is co-extensive with the ways that being gathers together and discloses conflicting tensions among four forces that comprise being, poetically rendered as earth, sky, mortals, and gods. These forces can be rendered in widely different ways, such as in Braidotti's posthuman ethics, but the significance is that tensions between the forces expressed through a thing's existence function as "attractors" that incite new forms of attunement, new ways of being. Rickert uses his ecological view of rhetoric-as-attunement to argue for an ancient attunement, "ecosophy," that may act as a lure for more a positive future.

The tone shifts across these examples from more of a rhetorical ecology in Senda-Cook and McHendry, Jr. to an ecological rhetoric in Gottshalk Druschke and Rai as well as Rickert, with Stagliano somewhere between them. A caution: Pinpointing them on a continuum so as to affix a firm label is beside the point and utterly useless in the end. Ecotonal variety does not lie on a strict continuum; it is a contingent field of contrasts whose shades change the more contrasts there are.

There is a discernible mutualism of address and interdependence amid the wildly diverse, affective relations that sub-tend existents doggedly pursuing that very existence. Mutuality inevitably gives way to contrast because addressivity and interdependence are nonetheless distinct qualities that can and will be accentuated relative to the other.

Ontological separation, then, is not a process of purification so much as "enjoyment" in a Whiteheadian sense, meaning it is the completion of a tendency within dynamic, entangled relationships.[11] Finding clear distinctions between rhetoric and ecology does not mean that one is purged of the other. McGreavy describes the "enjoyment" of ontological separation through rhythmic commonplaces of embodied memory. The bodily metaphors used by clammers she interviewed as part of her ongoing ethnography demonstrate the conjunction of living rhythms (among humans, clams, ocean acidification, mudflats, and so on) that engender metaphor. The rhythmic memory of harvesting shellfish is performed together by all the actants; metaphors separate from these rhythms as expressions of that collaborative memory just as the clams are expressions of it. As such, metaphor is reimagined as a kind of rhythmic orientation to the world, opening and closing relationships between entities. The same rhythms embody the remembering/forgetting of togetherness/separation, such that we may feel both ecological interdependency and rhetorical addressivity in varying degrees of intensity.

Nathaniel Rivers provides another example of rhetoric and ecology enjoying a certain distinction through their mutuality. He presents an elliptical trajectory for the "footprint" synecdoche, tracing its rise as a measurement construct for ecological and economic modeling in the 1990s, only to deflect it back toward motivating action in an immersive, interdependent world. As an image of statistical thought, the trope of footprints separated out from dynamic, affective relations is a way of addressing that interdependence. Through Ingold's concept of wayfaring, Rivers folds the trope back into a way of navigating our environments such that a figurative contrast between individuated action and milieu performatively shapes what McGreavy would call new rhythms.

As McGreavy's and Rivers' essays suggest, addressivity and interdependence, because they are immanent to relations, affects, and appetites human and otherwise, only transcend one another in contrast, not in essence. Keeling and Prairie make this particularly clear through their etymology of "tropes" within rhetoric and ecology. Understood broadly as relational "turns" or patterned ways of being, tropes are complexly distributed throughout the living world, and rhetoric and ecology have both drawn on tropology to explain action within communities of living beings. The authors offer tropes as a handle for grasping ontological commonplaces that gather the living world together. Importantly, they note that various tropisms are not always perceptible or affective across communities simply because they enact homologous modes of turning. Tropes may be ontologically commonplace rather than rarefied, but distinct tropisms are also polymorphic and so transcend that commonality, separating into distinct rhetorical opportunities that are not reducible to the ecology from which they arise or to other such rhetorical opportunities.

Keeling and Prairie reinforce McGreavy's and Rivers' comments about tropes but add something crucial that connects every work in the volume regarding the irony that rhetoric and ecology separate ontologically via their mutual immanence. They argue, based on the etymology, that tropes are trophic or nutritive. Tropes nourish future affective relations, feeding into them as different elements in an ecology fold into one another. Given that "nourishment" at its broadest means "promoting growth," one could say that trophic dynamics help explain the transduction of potential. Rivers makes note of ductus, a rhetorical concept of "leading" that is the root of seduction, abduction, transduction, and so forth. Brian Massumi, and Erin Manning after him, have developed the idea of transduction as a way of describing how forces experienced by a body are transformed in movement.[12] In regard to trophic, or nutritive, qualities of tropes, it would not be that rhetoric "leads us along," but that tropes promote our growth. This is not far from Spinoza's oft-remarked understanding of appetition as the striving to persist (conatus).[13] He argued that all things transform the forces that affect them into new potential, growing their ability to affect and be affected. Of course, appetition has many names (orexis in Aristotle, will to power in Nietzsche, creative advance in Whitehead, élan vital in Bergson, motive in Burke, etc.), but in each variation, a trophic dimension is required. The affective relations between entities must contribute to the growth of those entities. The potential actualized tropologically must nourish the potential to trope again. Without this, there is no novelty, no dynamism, no movement either in interdependence or in address.

The moment when rhetoric and ecology, existing together in an ontological commonwealth, separate from each other is hard to imagine let alone pinpoint. John Ackerman writes at length about the phase-shift between addressivity and interdependence, the former considered through "configurations of textuality" and the latter through "moments of presence" experienced as a body navigates its environment. Working from ethnographic fieldwork, he uses walking in an urban milieu as his case, where the walker "oscillates" between a textural, hermeneutic orientation and the vague immediacy of a larger world's presence as s/he moves along. This is a "double-movement" meaning that textuality is not a derivative of presence or the other way around. Instead, affective potential is separated out, transformed in both directions as inhabitation and as significance, which affectively vitalize (nourish) each other. The moment when presence transcends textuality and textuality presence he calls "the jump," and it names a tonal shift within an ontological commonplace that produces a kind of trophic dynamic. The oscillation between presence and meaning is generative, creating potential rather than spending it down. Also important is that we make "the jump" continually as part of everyday living. Oscillation is not the property of specialists.

Considered as ecotonal, rhetoric and ecology express aspects of co-existence that continually separate from one another yet which share ontological commonplaces (affect, relationality, appetite). Of course, we are not discussing an intertidal ecotone so the difference between an academic transition zone and a place like Schoodic Point needs to be tidied up in parting. The authors offer two related but distinct ways to understand the ecotonality of rhetoric and ecology. The first is as communities of scholars doing research (as disciplines that study co-existence) and the second is as phenomena occurring regardless of the presence of scholars (as spectra of co-existence).

As to the first, rhetoric and ecology border each other wherever specialists work together in collaboration or at a distance by incorporating one another's ways of working. Rob Nixon describes scholarly ecotones "as interpenetrating fields [that] proliferate at the borders between once separate disciplines, at times creating new dynamic combinations while also, depending on one's perspective, inflicting casualties through habitat fragmentation."[14] Keeling and Prairie literally embody this in their essay, as do McGreavy and Gottshalk-Druschke in their fieldwork. The other authors do also to the extent that "ecological thought" infuses their research at varying intensities.

As to the second, rhetoric and ecology express distinct tendencies of togetherness regardless of scale or location. The rhetoric and ecology of being are not adjacent in the way the ocean and the rocks are. They are proximate as qualities of existence that occupy the same world. Building from Ackerman, it is quite possible, even expected, to be attuned to one or the other rather exclusively, to addressivity or to interdependence. Yet there are moments in the course of events when these two qualities commingle, becoming the other. These moments are not locations, though they are located; they are anomalous zones of opportunity where the qualities of address and interdependence are felt as one, in differing degrees of confluence, tending from rhetorical ecology (rhetoric at high tide) to ecological rhetoric (rhetoric at low tide). No fixed ontological core provides the raw material from which rhetoric and ecology form together in shared commonplaces, nor does an intrinsic telos determine how, emanating from the same "substance," they become distinct.

I want to underscore that within such transition zones, the ontological collocation and allocation of rhetoric and ecology should be accounted for stochastically, not essentially. An ecotone is characterized by variable randomness, or creative diversity. In those transition zones where the mutual immanence of rhetoric and ecology is expressed as such the form the mutuality takes will be emergent, either between communities of scholars or at moments of oscillation between dimensions of co-existence within vital communities. In that regard, the essays of this volume are not landmarks on a map (and that extends to others working the transition zone between rhetoric and ecology). They are distinct adaptations that thrive in transition as waves break around them with all their predictability and uniqueness. As Nixon remarks, an ecotone supports a certain "variety of life-forms, but at the expense of less-adaptable, specialist species."[15] One can anticipate new adaptations of affect, relationality, and appetite that stress addressivity and interdependence conjointly, as opposed to those which emphasize one to the exclusion of the other. These adaptations are the enveloping borderland between rhetoric and ecology, introducing change into both.

Notes

1. Eddy van der Maarel, "Ecotones and Ecolines are Different," *Journal of Vegetation Science* 1, no. 1 (1990): 135.
2. Gilles Deleuze and Felix Guattari, *A Thousand Plateaus: Capitalism and Schitzophrenia*, trans. Brian Massumi (Minneapolis: University of Minnesota Press, 1987), 245.

3. Ibid., 252.
4. Jennifer Edbauer, "Unframing Models of Public Distribution: From Rhetorical Situation to Rhetorical Ecologies," *Rhetoric Society Quarterly* 35, no. 4 (2005): 20, 19.
5. Marika Seigel, "'One little fellow named Ecology': Ecological Rhetoric in Kenneth Burke's Attitudes toward History," *Rhetoric Review* 23, no. 4 (2004):390–401.
6. Claire Colebrook, *Death of the Posthuman: Essays on Extinction: Volume 1* (Ann Arbor: Open Humanities, 2014), 17–18, 55–56, 106–108, 125–127.
7. Kenneth Burke, *Permanence and Change: An Anatomy of Purpose*, 3rd ed. (Berkeley: University of California Press, 1984), 5.
8. Seigel, "'One Little Fellow,'" 389–390.
9. Thomas Rickert, *Ambient Rhetoric: The Attunements of Rhetorical Being* (Pittsburgh: University of Pittsburgh Press, 2013).
10. "The world is included in the occasion in one sense, and the occasion includes the world in another sense" such that "the togetherness of things involves some doctrine of mutual immanence." See Alfred North Whitehead, *Modes of Thought* (New York: Free Press, 1968), 163, 164.
11. Ibid., 26–29.
12. Brian Massumi, *Parables for the Virtual: Movement, Affect, Sensation* (Durham: Duke University Press, 2002), 42–43; Erin Manning, *Relationscapes: Movement, Art, Philosophy* (Cambridge: MIT Press, 2012), 71–72.
13. Gilles Deleuze, *Expressionism in Philosophy: Spinoza*, trans. Martin Joughin (New York: Zone Books, 1992), 230.
14. Rob Nixon, *Slow Violence and the Environmentalism of the Poor* (Cambridge: Harvard University Press, 2013), 30.
15. Ibid.

Erratum to: Tracing Rhetoric and Material Life

Bridie McGreavy, Justine Wells, George F. McHendry, Jr. and Samantha Senda-Cook

Erratum to:
B. McGreavy et al. (eds.), *Tracing Rhetoric and Material Life*, Palgrave Studies in Media and Environmental Communication, https://doi.org/10.1007/978-3-319-65711-0

In the original version of the book, in the affiliation of "B. McGreavy", university name "Creighton University" has been changed as "University of Maine" in Chapters 1 and 4.

The updated online version for this book can be found at
https://doi.org/10.1007/978-3-319-65711-0_1
https://doi.org/10.1007/978-3-319-65711-0_4
https://doi.org/10.1007/978-3-319-65711-0

B. McGreavy et al. (eds.), *Tracing Rhetoric and Material Life*,
Palgrave Studies in Media and Environmental Communication,
https://doi.org/10.1007/978-3-319-65711-0_14

Index

B. McGreavy et al. (eds.), *Tracing Rhetoric and Material Life*,
Palgrave Studies in Media and Environmental Communication,
https://doi.org/10.1007/978-3-319-65711-0

F

FSC
www.fsc.org
MIX
Papier | Fördert gute Waldnutzung
FSC® C083411

Kolibri 360 GmbH
Ferdinand-Jühlke-Straße 7
99095 Erfurt, Deutschland
produktsicherheit@kolibri360.de